# Applications in Electronics Pervading Industry, Environment and Society

# Applications in Electronics Pervading Industry, Environment and Society — Sensing Systems and Pervasive Intelligence

Editors

**Sergio Saponara**
**Alessandro De Gloria**
**Francesco Bellotti**

MDPI • Basel • Beijing • Wuhan • Barcelona • Belgrade • Manchester • Tokyo • Cluj • Tianjin

*Editors*
Sergio Saponara
Università di Pisa
Italy

Alessandro De Gloria
University of Genoa
Italy

Francesco Bellotti
University of Genoa
Italy

*Editorial Office*
MDPI
St. Alban-Anlage 66
4052 Basel, Switzerland

This is a reprint of articles from the Special Issue published online in the open access journal *Sensors* (ISSN 1424-8220) (available at: https://www.mdpi.com/journal/sensors/special_issues/applepies).

For citation purposes, cite each article independently as indicated on the article page online and as indicated below:

LastName, A.A.; LastName, B.B.; LastName, C.C. Article Title. *Journal Name* **Year**, *Volume Number*, Page Range.

**ISBN 978-3-0365-0478-0 (Hbk)**
**ISBN 978-3-0365-0479-7 (PDF)**

# Contents

# About the Editors

**Sergio Saponara** is an Italian scientist, engineer, and entrepreneur. He is Full Professor of Electronics at the University of Pisa, Italy, where he is also President of the BSc and MSc degree programs in Electronic Engineering. He got his Master and PhD degrees cum laude in Electronic Engineering in 1999 and 2003, respectively. He started his research activities in the fields of digital electronics in 1999, first with a research grant from the National Research Council and then a grant from STMicroelectronics. In 2002, he was a Marie Curie Research Fellow at IMEC, Leuven, Belgium. Since 2005, he has been responsible for the teaching of courses in Automotive Electronics and Electronic Systems for Robotics and Automation at the University of Pisa, where he contributed to the foundation of the MSc degree in Cybersecurity, being responsible for HW and Embedded Security. Since 1999, he has also taught Electronics at the Italian Naval Academy. As an entrepreneur, he co-founded the company IngeniArs in 2014, winner of several innovation prizes such as the H2020 SME Instrument. Since 2016, he has been the founder and director of the Summer School in Enabling Technologies for Industrial IoT, awarded in 2017 and 2018 by the IEEE CAS society. Since 2020, he has been director of the Automotive Electronics and Powertrain Electrifications specialization course. He has co-organized more than 150 conferences and he is Associate Editor of the journals Energies, Sensors, Designs, IEE Electronic Letters, Springer Nature Journal of Real Time Image Processing, IEEE Canadian Journal of Electrical and Computer Engineering, IEEE Consumer Electronics Magazine, and IEEE Vehicular Technology Magazine. Currently, he coordinates the University of Pisa in the H2020 European Processor Initiative, where he is WP leader for security HW implementation and member-elected of the steering committee. He also has ongoing projects with MIT on power converters and e-drives for electric/hybrid vehicles. He has co-authored around 300 scientific articles, 5 books, 10 journal special issues, and 20 patents. In his technology transfer activity, he has collaborated with Marelli, Maserati, STMicroelectronics, Renesas, Intel, Huawei, Infineon, Ericsson, P2P, Solari, Trenitalia, AMS, RiCo, INTECS, IDS, Sitael, and Leonardo. He has worked as an evaluator for R&D projects for the governments of Italy, Romania, Kazakhstan, and the European Commission. He is an expert of the European Institute for Science Media and Democracy and has been included in Stanford University's list of the world's top 2% scientists.

**Alessandro De Gloria** is Full Professor in Electronic Engineering at the University of Genoa, where he leads the ELIOS Lab research group at the Dept. of Naval, Electrical, Electronic and Telecommunications Engineering (DITEN). He graduated cum laude in Electronic Engineering and received his PhD from the University of Genoa. His main research interests are in the field of modeling and simulation, virtual reality, game technologies, computer architectures, and human–machine interaction. He is Editor in Chief and founder of the International Journal of Serious Games, and the founder of the Serious Games Society. He is part of the "Società Italiana di Elettronica" (SIE), the Italian association of Electronic Engineering university professors. He has been responsible for the Applications and Systems sector at a national level, and for the overall Genoa local unit. He was one of the founders of the Liguria District of Technologies (SIIT). He is a member of the Directive Council of the Genoa University Advanced Simulation and Education Service Center (SIMAV). He has been responsible for more than 30 research projects in the fields of automotive, game technologies, and technology-enhanced learning. He has authored more than 250 papers in international scientific journals and conferences.

**Francesco Bellotti** is Associate Professor at DITEN, University of Genoa, teaching Cyberphysical systems and Edge computing on the MSc course in Electronic Engineering. He received his MSc in Electronic Engineering (cum laude) and PhD in Electrical Engineering from the University of Genoa. He is on the didactic board of the Interactive Cognitive Environments PhD program. His main research interests concern intelligent transportation systems; edge computing; Internet of Things; cyberphysical systems; human–computer interaction. He has been responsible for the WPs of several industrial research projects particularly in intelligent transportation systems. He is on the editorial board of MDPI's *Energies and Sensors*, Hindawi's aHCI, and the *Int J. Serious Games*.

*Editorial*

# Recent Trends on Applications of Electronics Pervading the Industry, Environment and Society

**Sergio Saponara [1,*], Alessandro De Gloria [2] and Francesco Bellotti [2]**

[1] DII (Dipartimento di Ingegneria della Informazione), University of Pisa, via G. Caruso 16, 56122 Pisa, Italy
[2] DITEN (Electrical, Electronics and Telecommunication Engineering and Naval Architecture Department), University of Genoa, via Opera Pia 11/a, 16145 Genoa, Italy; adg@elios.unige.it (A.D.G.); franz@elios.unige.it (F.B.)
* Correspondence: sergio.saponara@unipi.it

Received: 10 December 2020; Accepted: 13 December 2020; Published: 18 December 2020

---

This Editorial analyzes the manuscripts accepted, after a careful peer-reviewed process, for the Special Issue "Applications in Electronics Pervading Industry, Environment and Society—Sensing Systems and Pervasive Intelligence" of the Sensors MDPI journal. The Special Issue was co-organized by the University of Pisa (Professor Sergio Saponara at the Department of Information Engineering) and University of Genoa (Professors Alessandro de Gloria and Francesco Bellotti at the Department of Electrical, Electronics and Telecommunication Engineering and Naval Architecture) in Italy. Most of the papers were selected as the best papers of the 2019 edition of the "Applications in Electronics Pervading Industry, Environment and Society" (Applepies) Conference that was held in Pisa in September 2019. All these papers were significantly enhanced with novel experimental results, as we show in the following.

The selected papers give an overview of the trends in research and development activities about the pervasive application of electronics to the industry, environment and society. The focus of the papers is on cyber physical systems (CPS) with research proposals for new sensor acquisition and ADC (analog-to-digital converter) methods, high-speed communication systems, cybersecurity and data processing, including emerging machine-learning techniques.

For each paper, the physical implementation aspects are always discussed, as well as the trade-off to be found between functional performances and hardware costs is exhaustively analyzed.

The Special Issue is characterized by 13 original research papers [1–13] that we briefly introduce in the following.

The first paper [1] is entitled "A Portable Support Attitude Sensing System for Accurate Attitude Estimation of Hydraulic Support Based on Unscented Kalman Filter", written by Xuliang Lu, Zhongbin Wang, Chao Tan, Haifeng Yan, Lei Si and Dong Wei from the School of Mechatronic Engineering, China University of Mining and Technology, Daxue Road, Xuzhou 221116, China.

The paper proposes the design of a support attitude sensing system composed of an inertial measurement unit (IMU) with MEMS (microelectromechanical system) sensors. In the classis attitude, a control system's yaw angle estimation with magnetometers is disturbed by the perturbed magnetic field generated by a coal rock structure and by high-power equipment. On the other hand, roll and pitch angles are often estimated using a MEMS gyroscope and accelerometer, and the accuracy is not reliable with time, usually due to long-term bias instability problems. In order to eliminate the measurement error of the sensors and to obtain an accurate attitude estimation, the paper proposed the use of an unscented Kalman filter based on quaternion, according to the characteristics of complementation of the magnetometer, accelerometer and gyroscope. Then, the gradient descent algorithm is used to optimize the key parameter of the unscented Kalman filter—namely, processing the noise covariance—to improve the accuracy of the attitude calculation. An industrial application shows that the average measurement error of the yaw angle is less than 2° and that of the pitch angle and roll angle are less than 1°, which proves the efficiency and feasibility of the proposed cyber physical system.

The second paper [2] is entitled "Analysis and Comparison of Rad-Hard Ring and LC-Tank Controlled Oscillators in 65 nm for SpaceFibre Applications", written by D. Monda et al. from the University of Pisa and by INFN (the Italian National Institute for Nuclear Physics).

This work presents a comparison between two voltage-controlled oscillators (VCOs) designed in a commercial 1.2-V 65-nm CMOS (complementary metal oxide semiconductor) technology to address the needs of SpaceFibre, a recent standard for high-speed communications in space applications. The first architecture based on a ring oscillator (RO) was designed using three current mode logic (CML) stages connected in a loop, while the second one was based on an LC-tank resonator. This analysis aimed to choose a VCO architecture able to be integrated into a rad-hard phase-locked loop to meet the specifications of the SpaceFibre protocol, supporting frequencies up to 6.25 GHz. The paper presents the full custom schematic and the layout designs. The single-event effect simulation results, performed according to an IMEC (Interuniversity MicroElectronics Center, Belgium) model with a double-exponential current pulses generator, are also discussed. The performance of the RO-VCO are quite attractive in terms of technology scaling and reduced area occupation. However, the RO-VCO solutions suffer from larger frequency spreading due to process variations and due to operations in harsh space conditions. On the contrary, the LC-VCO solution is characterized by a lower sensitivity to PVT (process–voltage–temperature) variations. Hence, the LC-VCO architecture is the one selected to fulfill the specifications of the new SpaceFibre aerospace standard.

The third paper [3] is entitled "Analysis and Design of Integrated Blocks for a 6.25 GHz SpaceFibre PLL" and is written by M. Mestice at al., with authors from the University of Pisa and by INFN (the Italian National Institute for Nuclear Physics).

Additionally, this paper refers to the SpaceFibre standard for high-speed communication in space applications and presents the design of the key blocks for a phase-locked loop (PLL) to generate the clock reference up to 6.25 Gbps: triple-modular redundancy phase/frequency detector, charge pump and a passive loop filter. Modeling and simulation activities were carried out in the ADS (Advanced Design System) RadioFrequency environment and in the Cadence Virtuoso environment. The results achieved proved that the PLL can be fully integrated on-chip in a commercial 1.2-V 65-nm CMOS technology with an area size dominated by the passive loop filter. Both system-level and layout-level rad-hard techniques were proposed. The results achieved showed that a compact (0.09 mm$^2$) and low-power (about 10 mW) dead zone-free and rad-hard PLL can be obtained with a phase noise below −80 dBc/Hz @ 1 MHz and targeting the 6.25-Gbps maximum data rate of the SpaceFibre standard.

The fourth paper [4] is entitled "Machine Learning on Mainstream Microcontrollers" and is authored by F. Sakr at al. from University of Genoa.

This work addresses the emerging problem of implementing machine-learning (ML) techniques in edge devices. More in detail, the paper introduces the edge-learning machine (ELM), a machine-learning framework for edge devices. The goal is managing the ML training phase on a desktop computer, while the inference phase is implemented on a STM32 (STMicroelectronics 32-bit) microcontroller. By using a platform-independent C language, the paper deals with several supervised ML algorithms (a support vector machine with a linear kernel, k-nearest neighbors and decision tree) and exploits the capability of the STM X-Cube AI to implement artificial neural networks (ANNs) on STM32 Nucleo boards. Multiple datasets are considered for classifications and regression. The results of the research work prove that the edge platforms reach the same performance score of a desktop computing platform with a similar time latency. To support the community of developers and makers, the ELM framework is released as an open source.

The fifth paper [5] deals with cybersecurity. It is entitled "Cryptographically Secure Pseudo-Random Number Generator IP-Core Based on SHA2 Algorithm" by L. Baldanzi et al., a group of authors from the University of Pisa.

The adoption of advanced sensors and systems for autonomous driving, combined with an increased connectivity of vehicles, robots and drones, is increasing the importance of embedding security features in computing and communication platforms. To this aim, RNG (random number generation) has a crucial role in ensuring the robustness of the security chain. RNG is a key technology to generate the encryption keys to be used for ciphers. Hence, any weakness in the key generation process will lead to leaks of information and will increase the probability to breach even the strongest cipher. More in detail, the paper shows the architecture of a CSPRNG (cryptographically secure pseudo-random number generator) macrocell that was validated by using the official Statistical Test Suite of the NIST (National Institute for Standard and Technology) to assess the degree of randomness of the numbers generated. The proposed CSPRNG macrocell was characterized by both FPGA (field-programmable gate array) and ASIC (application-specific integrated circuit) standard-cell technologies.

The sixth paper [6] is entitled "Data Processing and Information Classification—An In-Memory Approach" and is authored by M. Andrighetti et al. from Politecnico di Torino.

In the Internet of Things (IoT) era, an enormous amount of data, generated by billions of electronic devices full of sensors that constantly acquire data, must be processed and classified. A classic approach is transferring these data to servers that elaborate them remotely in the cloud. This approach is energy-inefficient (there is a huge battery drain due to the high amount of information to be transferred) and is affected by latency problems in safety-critical time-sensitive applications. Data may be processed locally in edge devices, near the sensor itself, but this solution requires a high-performance computation (HPC) and memory capability that is often missing in mobile microprocessors and microcontrollers. To address these issues, the paper presents a PIM (processing-in-memory) approach where new memories are designed to elaborate the data inside them, overcoming the well-known "memory wall" issue. More in detail, the work, with reference to a bitmap indexing case study, presents a hardware accelerator designed in CMOS technology around the PIM approach. The hardware accelerator is capable of implementing the bitmap indexing algorithm and can also be reconfigured to implement other tasks. The achieved results show that the PIM approach allows to process and classify huge amounts of data locally, with a very low power consumption.

The seventh paper [7] is entitled "Digital Circuit for Seamless Resampling ADC Output Streams" and is authored by M. D'Arco et al. from the Federico II University of Naples.

This work first presents DSP (digital signal processing) techniques to change the sampling rate of digital storage oscilloscopes (DSOs) by means of digital resampling approaches. Then, it proposes a new digital circuit to be included in the acquisition channel of a DSO, between the internal analog-to-digital converter (ADC) and the acquisition memory, that allows the user to select any sampling rate lower than the maximum one with fine resolution. The new circuit exploits both digital-filtering techniques with dynamically generated coefficients and ad-hoc memory management strategies. For the circuit, both FPGA and ASIC implementations are evaluated.

The eighth paper [8] is entitled "Embedded Bio-Mimetic System for Functional Electrical Stimulation Controlled by Event-Driven sEMG" and is authored by F. Rossi et al., a group of authors from Politecnico di Torino.

This paper deals with an assistive technology application and, particularly, with the surface electromyographic (sEMG) signal for controlling the functional electrical stimulation (FES) therapy, a technique that is being widely accepted as an active rehabilitation method for the restoration of neuromuscular disorders. To this aim, the paper proposes an embedded implementation of the average threshold crossing (ATC)-FES control system. The system was characterized and validated by analyzing the computing core and memory usage in different operating conditions, as well as measuring the system latency. Experimental results on a testing population of 11 subjects was also carried out.

The ninth paper [9] is entitled "Fast Approximations of Activation Functions in Deep Neural Networks when using Posit Arithmetic" and is authored by M. Cococcioni et al., a group from the University of Pisa and the company MMI srl.

The paper addresses the problem of an arithmetic format to achieve real-time computing for deep neural networks (DNNs). Overcoming the limits of a classic IEEE 754 standard floating-point representation, the paper presents a new numerical format called Posits. While waiting for the widespread availability of hardware-native Posit Processing Units, the paper shows that it is possible to exploit the Posit representation and the currently available arithmetic-logic unit (ALU) in every microprocessor to speed up DNNs by manipulating the low-level bit string representations of Posits. To this aim, the paper presents a new class of Posit operators called L1 operators, which consists of fast and approximated versions of existing arithmetic operations or functions, such as the hyperbolic tangent (TANH) and the extended linear unit (ELU), while being adopted as activation functions in DNNs. The achieved results show that using either 64-bit ARM processors or x86 Intel 10-bit Posits can represent an exact replacement for 32-bit floats, while 8-bit Posits could be an interesting alternative to 32-bit floats, since their performances are a bit lower, but their high-speed and-low storage properties are very appealing. This size reduction will lead to a lower bandwidth demand and more cache-friendly codes. Moreover, with only 8- or 10-bit Posit operations, they can be tabulated in a very efficient way.

The tenth paper [10] is entitled "Steerable-Discrete-Cosine-Transform (SDCT): Hardware Implementation and Performance Analysis" and is from R. Peloso et al., a group of authors from Politecnico di Torino.

This paper deals with the hardware acceleration of new, efficient video compression methods and, particularly, the steerable discrete cosine transform (SDCT) that were proposed to exploit directional DCT using the basis of having different orientation angles. With respect to classic solutions, the SDCT leads to a sparser representation and, hence, to an improved compression efficiency. The hardware accelerator for SDCT processing proposed in this paper is able to work at about 200 MHz with a throughput of 3G sample/s and can support an 8k UHD (ultra-high definition) format at 60 frames per second.

The eleventh paper [11] is entitled "Distillation of an End-to-End Oracle for Face Verification and Recognition Sensors" and is authored by F. Guzzi et al., a group of authors from the University of Trieste, the Elettra Sincrotrone Trieste S.C.p.A and the Abdus Salam International Centre for Theoretical Physics.

This paper deals with face recognition functions that are important for many applications, including security systems, inclusion devices and others. In this work, a distillation technique is applied to a complex model to enable fast recognition on low-complex hardware face recognition sensors. The proposed biometric systems are examined for the two problems of face verification and face recognition in an open set by using training/testing methodologies and datasets.

The twelfth paper [12] is entitled "A Model-Based Design Floating-Point Accumulator. Case of Study: FPGA Implementation of a Support Vector Machine Kernel Function" and is authored by M. Bassoli et al., a group of authors from the University of Parma.

This paper presents a novel model-based floating-point accumulation circuit (relying on the state-of-the-art delayed buffering algorithm) to accelerate ML functions, such as the kernel function of support vector machines. The proposed model was implemented in a Simulink environment and then implemented, by means of an HDL (hardware description language) design, in FPGA technology. The simulation results showed that it has a better performance in terms of speed and occupied area when compared to other solutions. To better evaluate its figure, a practical case of a polynomial kernel function was considered.

Finally, paper thirteen [13], "Managing Big Data for Addressing Research Questions in a Collaborative Project on Automated Driving Impact Assessment", is authored by Bellotti et al., an international group of researchers collaborating in the L3Pilot project, which is the piloting of Society of Automotive Engineers (SAE) level 3 automated vehicle functions.

The paper presents the development of a set of tools for a big data management process involving several project actors (vehicle manufacturers, research institutions, suppliers and developers), with different perspectives and requirements. In order to implement a reference methodology, the authors highlight the importance of (i) a common data format to process all the source data coming from proprietary sources, (ii) a measurement-oriented application programming interface (API) for storing and retrieving data and of a tool to synthetize meaningful data from the original, proprietary vehicular time series.

**Author Contributions:** All guest editors contributed equally to this editorial. All authors have read and agreed to the published version of the manuscript.

**Conflicts of Interest:** The authors declare no conflict of interest.

## References


1. Lu, X.; Wang, Z.; Tan, C.; Yan, H.; Si, L.; Wei, D. A Portable Support Attitude Sensing System for Accurate Attitude Estimation of Hydraulic Support Based on Unscented Kalman Filter. *Sensors* **2020**, *20*, 5459. [CrossRef] [PubMed]
2. Monda, D.; Ciarpi, G.; Saponara, S. Analysis and Comparison of Rad-Hard Ring and LC-Tank Controlled Oscillators in 65 nm for SpaceFibre Applications. *Sensors* **2020**, *20*, 4612. [CrossRef]
3. Mestice, M.; Neri, B.; Ciarpi, G.; Saponara, S. Analysis and Design of Integrated Blocks for a 6.25 GHz Spacefibre PLL. *Sensors* **2020**, *20*, 4013. [CrossRef]
4. Sakr, F.; Bellotti, F.; Berta, R.; De Gloria, A. Machine Learning on Mainstream Microcontrollers. *Sensors* **2020**, *20*, 2638. [CrossRef]
5. Baldanzi, L.; Crocetti, L.; Falaschi, F.; Bertolucci, M.; Belli, J.; Fanucci, L.; Saponara, S. Cryptographically Secure Pseudo-Random Number Generator IP-Core Based on SHA2 Algorithm. *Sensors* **2020**, *20*, 1869. [CrossRef]
6. Andrighetti, M.; Turvani, G.; Santoro, G.; Vacca, M.; Marchesin, A.; Ottati, F.; Roch, M.R.; Graziano, M.; Zamboni, M. Data Processing and Information Classification—An In-Memory Approach. *Sensors* **2020**, *20*, 1681. [CrossRef] [PubMed]
7. D'Arco, M.; Napoli, E.; Zacharelos, E. Digital Circuit for Seamless Resampling ADC Output Streams. *Sensors* **2020**, *20*, 1619. [CrossRef] [PubMed]
8. Rossi, F.; Ros, P.M.; Rosales, R.M.; Demarchi, D. Embedded Bio-Mimetic System for Functional Electrical Stimulation Controlled by Event-Driven sEMG. *Sensors* **2020**, *20*, 1535. [CrossRef] [PubMed]
9. Cococcioni, M.; Rossi, F.; Ruffaldi, E.; Saponara, S. Fast Approximations of Activation Functions in Deep Neural Networks when using Posit Arithmetic. *Sensors* **2020**, *20*, 1515. [CrossRef] [PubMed]
10. Peloso, R.; Capra, M.; Sole, L.; Roch, M.R.; Masera, G.; Martina, M. Steerable-Discrete-Cosine-Transform (SDCT): Hardware Implementation and Performance Analysis. *Sensors* **2020**, *20*, 1405. [CrossRef] [PubMed]
11. Guzzi, F.; De Bortoli, L.; Molina, R.S.; Marsi, S.; Carrato, S.; Ramponi, G. Distillation of an End-to-End Oracle for Face Verification and Recognition Sensors. *Sensors* **2020**, *20*, 1369. [CrossRef] [PubMed]
12. Bassoli, M.; Bianchi, V.; De Munari, I. A Model-Based Design Floating-Point Accumulator. Case of Study: FPGA Implementation of a Support Vector Machine Kernel FunctionH. *Sensors* **2020**, *20*, 1362. [CrossRef] [PubMed]
13. Bellotti, F.; Osman, N.; Arnold, E.H.; Mozaffari, S.; Innamaa, S.; Louw, T.; Torrao, G.; Weber, H.; Hiller, J.; De Gloria, A.; et al. Managing Big Data for Addressing Research Questions in a Collaborative Project on Automated Driving Impact Assessment. *Sensors* **2020**, *20*, 6773. [CrossRef] [PubMed]


**Publisher's Note:** MDPI stays neutral with regard to jurisdictional claims in published maps and institutional affiliations.

*Article*

# Managing Big Data for Addressing Research Questions in a Collaborative Project on Automated Driving Impact Assessment

Francesco Bellotti [1,*], Nisrine Osman [1], Eduardo H. Arnold [2], Sajjad Mozaffari [2], Satu Innamaa [3], Tyron Louw [4], Guilhermina Torrao [3], Hendrik Weber [5], Johannes Hiller [5], Alessandro De Gloria [1], Mehrdad Dianati [2] and Riccardo Berta [1]

[1] Department of Electrical, Electronics and Telecommunication Engineering and Naval Architecture (DITEN), University of Genova, 16145 Genova, Italy; Nisrine.Osman@elios.unige.it (N.O.); alessandro.degloria@unige.it (A.D.G.); riccardo.berta@unige.it (R.B.)

[2] Warwick Manufacturing Group (WMG), University of Warwick, Coventry CV4 7AL, UK; E.Arnold@warwick.ac.uk (E.H.A.); Sajjad.Mozaffari@warwick.ac.uk (S.M.); M.Dianati@warwick.ac.uk (M.D.)

[3] VTT Technical Research Centre of Finland Ltd., P.O. Box 1000, FI-02044 VTT Espoo, Finland; Satu.Innamaa@vtt.fi (S.I.); Guilhermina.Torrao@vtt.fi (G.T.)

[4] Institute for Transport Studies, University Road, University of Leeds, Leeds LS2 9JT, UK; T.L.Louw@leeds.ac.uk

[5] Institute for Automotive Engineering, RWTH Aachen University, Steinbachstr 7, 52074 Aachen, Germany; hendrik.weber@ika.rwth-aachen.de (H.W.); johannes.hiller@ika.rwth-aachen.de (J.H.)

* Correspondence: franz@elios.unige.it

Received: 28 September 2020; Accepted: 19 November 2020; Published: 27 November 2020

**Abstract:** While extracting meaningful information from big data is getting relevance, literature lacks information on how to handle sensitive data by different project partners in order to collectively answer research questions (RQs), especially on impact assessment of new automated driving technologies. This paper presents the application of an established reference piloting methodology and the consequent development of a coherent, robust workflow. Key challenges include ensuring methodological soundness and data validity while protecting partners' intellectual property. The authors draw on their experiences in a 34-partner project aimed at assessing the impact of advanced automated driving functions, across 10 European countries. In the first step of the workflow, we captured the quantitative requirements of each RQ in terms of the relevant data needed from the tests. Most of the data come from vehicular sensors, but subjective data from questionnaires are processed as well. Next, we set up a data management process involving several partners (vehicle manufacturers, research institutions, suppliers and developers), with different perspectives and requirements. Finally, we deployed the system so that it is fully integrated within the project big data toolchain and usable by all the partners. Based on our experience, we highlight the importance of the reference methodology to theoretically inform and coherently manage all the steps of the project and the need for effective and efficient tools, in order to support the everyday work of all the involved research teams, from vehicle manufacturers to data analysts.

**Keywords:** research data collection and sharing; connected and automated driving; deployment and field testing; vehicular sensors; impact assessment; knowledge management; collaborative project methodology

---

## 1. Introduction

Solving grand challenges, such as automated connected driving, often requires collaboration across multiple domains and technical areas. A key factor in the success of collaborative research projects is the methodology and the relevant tools used to address the challenges. Some collaborative tools are well established (e.g., for managing the project and its risks, sharing documentation and source code, etc.), while others are less general and more related to application-specific tasks. The literature is rich in guidelines, techniques and tools for general project challenges (e.g., [1,2]), but there is a lack of specific information and tools for different partners to deal with sensitive data in order to answer a set of research questions (RQs) at project level. This task is gaining relevance in the current industrial research context, where there is a growing focus on big data, especially from ever more pervasive and sophisticated sensors and on extracting meaningful information from them. However, while there is significant work published on application-oriented data analysis (e.g., [3]), we found a lack in data management for assessing the impact of the new technologies' adoption.

We thus intend to investigate how to organize a robust workflow for quantitatively addressing RQs in a collaborative project sharing sensitive data among various partners, while ensuring methodological soundness and data validity and protecting partners' intellectual property (IP).

We think that the automated driving sector represents a highly significant investigation domain given the huge amount of research that is being carried out in the field (e.g., [4–6]). As an example use case, we thus discuss our experience in a 34-partner EU-funded project, L3Pilot, which is assessing the impact of Society of Automotive Engineers (SAE) Level 3 (L3) and Level 4 (L4) automated driving functions (ADFs). Tests are being conducted in pilots in 10 European countries, with vehicles provided by 13 vehicle owners (original equipment manufacturer (OEM), suppliers or research facilities).

The L3Pilot RQs cover different leading-edge ICT adoption impact assessment areas, including (i) technical performance of the tested L3 ADFs, (ii) user acceptance and behaviour, (iii) impact on traffic and mobility and (iv) societal impacts (see L3Pilot Deliverable D3.1 [7]). This paper presents how we implemented a reference methodology for large scale pilots and field operational automotive tests—namely Field opErational teSt supporT Action (FESTA) [8]—in order to get the quantitative information needed to answer the project's research questions (RQs). A key novelty in this process is the use of the Consolidated Database (CDB), which allows data from all the pilot sites to be shared anonymously and securely amongst project partners to facilitate data analysis aimed at answering the project's RQs.

This paper presents the challenges we have faced in implementing the methodology and consequently developing a coherent, robust workflow. First, we needed to quantitatively capture each RQ's requirements in terms of raw data to be collected during the tests, so as to allow a proper investigation. Then, we set up the needed tools in an iterative development process involving several partners (vehicle manufacturers, research institutions, suppliers and developers), with different perspectives and requirements. Finally, we deployed the system so that it is fully integrated within the project's data toolchain and usable by all the partners.

It is important to highlight that this paper focuses on the method, workflow and tools and does not discuss the actual domain-specific data, which will be the subject of another publication.

The remainder of the paper is organized as follows. Section 2 gives an overview of the related work, while Section 3 presents the methodology and the consequent specifications for the target process. Section 4 presents the design and implementation of the CDB. Section 5 discusses what we have learnt from the deployment of the system, while Section 6 draws the final conclusions.

## 2. Related Work

Beside the general overviews cited in the Introduction, there is a rich literature on privacy and risk management in projects. For instance, [9] deal with Risk Assessment in Multi-Disciplinary Engineering Projects, [10] with privacy risks when sharing data on information systems. Furthermore, [11] investigates the validity of sharing privacy-preserving versions of datasets. They propose a

Privacy-preserving Federated Data Sharing (PFDS) protocol that each agent can run locally to produce a privacy-preserving version of its original dataset. The PFDS protocol is evaluated on several standard prediction tasks and experimental results demonstrate the potential of sharing privacy-preserving datasets to produce accurate predictors. In addition, [12] provides an extensive review of data analytic applications in road traffic safety, with particular attention to crash risk modelling.

Furthermore, [13] deals with integrating diverse knowledge through boundary spanning processes, with a particular focus on multidisciplinary project teams. The concept of a Project Consortia Knowledge Base (PC-KB) is presented in [14] in an integration framework based on semantic knowledge that facilitates project-level communication as well as access to project data across tool and partner boundaries.

Commercial companies (e.g., Amazon, Microsoft, Google) have established efficient cloud ecosystems for data management providing very powerful services, but they rely on proprietary technologies, with very limited interoperability and development opportunities for third parties. However, we could not find in the literature guidelines on how to exploit these cloud technologies to support project partners in processing big data to address quantitative research questions.

In recent years, a number of field operational tests (FOTs) have been executed to test new Advanced driver-assistance systems (ADAS) in authentic traffic conditions, involving thousands of drivers (e.g., euroFOT [15,16]). With a view to ensure scientific soundness, the Field opErational teSt supporT Action (FESTA) project developed a methodology for field operational tests (FOTs), with three main focuses: user, vehicle, context [8]. This methodology is described in the FESTA Handbook which has been frequently updated according to the latest lessons learned [17]. It records lessons learned and provides best practices collected in several European FOTs in the last ten years. L3Pilot decided to adopt this methodology, as illustrated in the next section.

Several collaborative industrial research projects have been conducted in Europe addressing the first levels of automated driving. The AdaptIVe project developed several functionalities providing various levels of driver assistance, such as partial, conditional and high automation [18]. Drive C2X investigated cooperative awareness, which was enabled by periodic message exchange between vehicles and roadside infrastructure [19,20]. The FOT-Net Data project prepared the Data Sharing Framework, which provides hands-on recommendations on how to manage data sharing of data from the transportation research area [21]. The TEAM project developed an app suite for supporting collaborative road mobility [22].

## 3. Methodology

### 3.1. Overview

The RQs for all impact areas in the L3Pilot project (listed in the Introduction, see also [7]) were generated through the top-down approach recommended by the FESTA Handbook [17]. The process began with a review of the descriptions of automated driving functions (ADFs) that were going to be piloted during the project. Therefore, in the early stages, only high-level RQs (Levels 1 and 2 in Table 1 example) were defined, to meet the project objectives.

**Table 1.** An example on definition of logging requirements for a hypothesis [7].

| Item | Example |
|---|---|
| Evaluation area | Technical and traffic |
| RQ level 1 | "What is the impact of the ADF on driving behaviour?" |
| RQ level 2 | "What is the ADF impact on driven speed in different scenarios?" |
| RQ level 3 | "What is the ADF impact on driven speed in driving scenario X?" |
| Hypothesis | Example 1: "There is no difference in the driven mean speed for the ADF compared to manual driving." Example 2: "There is no difference in the standard deviation of speed for the ADF compared to manual driving." |
| Required Performance indicators (PIs) | Mean speed, standard deviation of speed, max speed, plot (speed/time) |
| Logging requirements/sensors available | CAN bus of vehicle: Ego speed in x direction |

In such a top-down approach, the generation RQs and hypotheses typically is based on theoretical understanding of the mechanisms how different impact areas are influenced. Thus, the process was started with an extensive literature review, which aimed to identify the key elements related to different impact areas. In addition, the review aimed to find knowledge gaps. The process was iterative (generate RQs, review them) to ensure that all major topics were covered and RQs were well formulated. In this step, a wide range of RQs was created, not limiting them by means of any single data collection method. The RQs were simply based on literature and the experience of the project members in previous, related work. The generation of the first (higher) level of research questions was structured according to the four L3Pilot evaluation areas. The second stage involved the development of more detailed RQs related to specific components of the higher-level questions, where appropriate. For each RQ, the underlying hypothesis is then made explicit. Table 1 provides an example in the Technical and Traffic area.

The top-down approach to setting RQs was followed by a bottom-up revision. In that phase, the RQs were cross-checked for their feasibility in terms of the data generation, a suitable experimental procedure at the pilot sites and availability of evaluation methods and tools [23]. RQs were prioritized based on test site characteristics, data coding and processing demand, ethical constraints, resources and time available in the project and importance for the research. In this phase, some of the first two levels of RQs were updated to be in line with the evaluation possibilities in the project.

In line with the FESTA Handbook, the next steps after generation of the hypotheses concerned the definition of the relevant performance indicators (we cover them in detail in the next subsections) and of the logging needs related to them. Here, we differentiated the subjective and objective data [7]. Questionnaires would collect subjective data across test participants (drivers and possible passengers), and objective data would be collected mostly from the data loggers of the test vehicles, additional cameras installed on them, and, when necessary, from external data sources (e.g., weather information, road type, etc.) [24]. Figure 1 provides an overview of the RQ definition and implementation workflow.

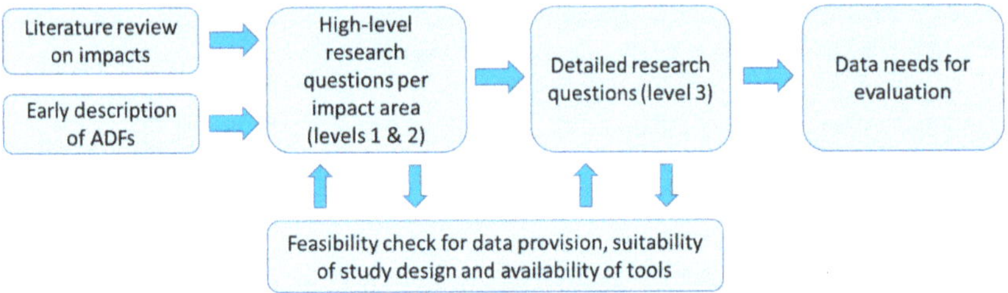


**Figure 1.** Overview of the research questions (RQ) definition and implementation workflow.

*3.2. Objective (Vehicular) Data*

The adopted methodology requires defining (i) the performance indicators (PIs) through which the RQs can be answered, (ii) the derived measures that are needed to calculate these indicators, and (iii) the actual vehicular signals that are needed to calculate these measures (see the full process in [24]). Beside the standard vehicular signals (e.g., speed, acceleration, pedal activity, etc.), source data come also from state-of-the-art automated driving sensors, such as cameras, Light Detection and Ranging (lidars) and radars. As a single PI may be derived from different alternative derived measures—and a single measure from different alternative signals—a collaboration was set up among the evaluation team (who express the data needs), the vehicle owners (who provide and share the data) and those responsible for developing the processing tools. The collaboration aimed at defining performance indicators (PIs) and derived measures for all RQs [7], as illustrated in the Table 1 example. The overall analysis defined a set of signals to be provided by all the vehicle owners that are specified in a common data format [25].

Four different types of PIs (Table 2) were defined to be computed from the vehicular signals and stored in the Consolidated database (CDB) as the factual basis for answering the RQs. PIs are typically constituted by statistical aggregations (e.g., avg/std/min/max) in significant intervals of a trip. Two PI types are computed at trip level: while Trip_PIs are general indicators synthetizing a trip, ScenarioSpecific_TripPIs are computed aggregating trip segments from a specific scenario only. The other two PI types (namely, ScenarioInstance_PI and Datapoints) are much more specific, as they are computed for each instance of a given driving scenario detected during a trip. Table 2 provides an overview of the L3Pilot vehicular PI types, with some examples. The table reports only two Datapoint types, as examples, since there is one datapoint type for each driving scenario type.

**Table 2.** An overview of the L3Pilot vehicular sensor data performance indicators (PI) types.

| PI Type | Description | Example of PIs |
|---|---|---|
| Trip PI | PIs computed at trip level | Mean (stdev) longitudinal acceleration, percentage of time elapsed per driving scenario type |
| Scenario specific Trip PI | PIs computed at trip level but only when a specific driving scenario occurs. Example of driving scenarios, described later, are: driving in a traffic jam, lane change. | Mean duration of sections with speed lower than a threshold |
| Scenario instance PI | PIs computed for each instance of a driving scenario. The same PIs are computed in each type of scenario | Mean (stdev) time headway, mean(stdev) position in lane |
| Datapoint for a Following a lead vehicle scenario | Datapoint PIs are computed for each instance of a driving scenario. Different types of scenario have a different datapoint structure. Here we report two examples. | Mean (stdev) relative velocity, Time headway at minimum time to collision |
| Datapoint for Approaching a traffic jam scenario | Datapoints are used as input for the impact assessment by either resimulating driving scenarios or constructing artificial scenarios based on statistical analyses of scenarios encountered during piloting | Vehicle speed at brake or steering onset, Longit. position of object at brake or steering onset |

Several research questions required analysing context data beyond the actual vehicular signals and questionnaire answers. Context data were useful, particularly to segment information so to allow comparisons and more focused analysis. Among context data we highlight:

- Experimental conditions. Different conditions have to be considered, such as: baseline, ADF not available, ADF off, ADF on.
- Road types. Tests are performed on various road types, such as: motorways, major urban arterials, other urban roads.
- Driving scenarios. The system has to track different types of driving scenarios, that are typical driving situations, such as uninfluenced driving, lane change, lane merge, following a lead vehicle, etc. Scenarios are computed by the L3Pilot data toolchain, processing the vehicular time series [26].

So, Trip PIs are to be computed for different segments, based on the actual experimental condition (i.e., baseline, ADF off, ADF on) and road type; and Scenario Instance PIs are to be segmented not only on the basis of the scenario type itself, as per definition, but also considering the different experimental conditions and road types.

Other metadata were mandated as well, such as driver type (professional or ordinary), temperature and speed limits, in order to better characterize the context of each PI measurement.

*3.3. Confidentiality*

There are several external factors that impact the extent to which the project RQs can be addressed. For example, when assessing the effects of the ADFs on driving (i.e., observed differences between the ADF and human driver with respect to, e.g., car-following behaviour), the maturity of the system is important to take into account and whether the test vehicle represents the driving experience of a mature product. In addition, to get the open road testing permission for an ADF, public authorities demand safety of testing. The details of the driving dynamics can also be a sensitive topic to the manufacturers. For that reason, it is important to make sure that vehicle telemetry data are kept confidential, and that the manufacturers or their tested systems cannot be identified, ranked or compared from any information shared within the project or with others, as their competitiveness at a crucial development stage would be compromised. Another challenge for gaining a good understanding of the impacts of ADFs is that the data is limited to specific test routes, speed ranges and weather conditions, which are defined by the function's Operational Design Domains (ODD). All pilot studies must also adhere to the rules of respective OEM and to national regulations on testing of automated driving, like the mandatory use of safety drivers in certain situations. Therefore, all this limits the possibilities of the project to address all of the RQs set for evaluation. To address these limitations, L3Pilot has specified a common methodology, data format and analysis toolkit, which will be used across all pilot sites and for evaluation of different ADFs [26]. Other solutions include the use of a dedicated research partner for analysis of data from a single pilot site to ensure that the access to commercially sensitive data is controlled and that data is pseudonymised before uploading to the CDB and shared among partners.

The CDB includes aggregated data from several sites in such a manner that commercially sensitive information is protected. A key requirement for being able to merge data is, that, in addition to protecting the privacy of manufacturers, the result is meaningful for the user of the result as such or those performing the following evaluation steps. In other words, the results must describe the impacts of automated driving but without compromising this privacy. The merging of data from different sites also leads to the outcome where the L3Pilot results do not represent the impact of single (OEM)-specific ADFs but the generic impacts that can be expected once these systems are introduced to the road [27].

As confidentiality is a key requirement when sharing valuable data, three main constraints were applied to data uploaded to the CDB:

- It should not be possible to identify which pilot site the data came from. For example, attention was paid not to insert metadata, such as temperature and date, that might hint to identify the location of the pilot site.
- No personal data about the driver, passengers nor other test participants.
- No possibility to characterize in detail the behaviour of ADFs. This was achieved by the fact that vehicular sensor data are not uploaded to the CDB as time series but as summarised performance indicators, which are described later.

As an ID of the trip and of the user was considered necessary to allow data owners to track their data in the CDB (also to update or delete them, if needed), an SHA-256 hashing-based pseudonymization was implemented [26]. Knowing the encrypted IDs, the data owners can track their data in the CDB, while such IDs are not decipherable by the other users.

Despite these constraints, there is no 100% guarantee that data cannot be linked to a pilot site, if malicious and highly sophisticated techniques we did not foresee are applied to the data.

*3.4. Subjective Data*

While until now we have focused on vehicular sensor data, a complete assessment of ADF functions also requires processing subjective data. One of the primary sources of data for the user and acceptance evaluation within L3Pilot is a pilot site questionnaire, which gathers subjective data from participants at the various pilot sites (for the full questionnaire see [23]). L3Pilot tests four different types of Automated Driving Functions (ADFs), including motorway, traffic jam, urban and parking. We developed a base questionnaire for background information, including questions related to sociodemographic factors, vehicle use and purchasing decisions, driving history, in-vehicle system usage, activities while driving, trip choices and mobility patterns. The data collected in the first part will be used to create different user groups for the user and acceptance evaluation. We also included questions specific to the ADF. For example, these questions assess various aspects of participants' initial reactions to using the particular ADF. To understand whether having daily access to the ADF might change any decisions or behaviours, they were reasked questions about vehicle use and purchasing decisions, driving history, in-vehicle system usage, engagement with nondriving tasks, trip choices and mobility patterns.

*3.5. Workflow Requirements*

From a process viewpoint, requirements included efficiency of workflow, and in particular, upload and download of data to/from the CDB. We highlight in particular:

- Recursive upload. The user should define the source directory and the system should automatically detect the files to be processed and do the upload of their contents to the CDB. All the subdirectories should be recursively explored.
- CSV download. The system should allow the possibility of downloading contents in either .json or .csv format, which is the typical input format for the statistical processing packages used by the analysts. The granularity of download is at the feature level. That is, a user should be able to download all the measurements of all the accessible features (Trip_PI, Scenario_Instances, Datapoints, etc.) or only some of them.
- Postediting of the performance indicators. Once the performance indicators for a trip are computed, the data provider should be able to check and edit them before the upload to the CDB.

The above requirements were typically defined during the development, in an iterative fashion, as the data providers and analysts suggested improvements based on their working experience.

## 4. Design and Implementation of the Consolidated Database (CDB)

In order to meet the above presented methodological goals, L3Pilot defined a data flowchart (Figure 2), which is the basis for the system architecture. The workflow starts with data collection at the pilot sites and ends with data analysis by impact experts. At the beginning there is a different processing for vehicular sensor data (Section 4.1) and subjective data (Section 4.2), then these data are managed seamlessly. The first, fundamental step consists in translating all the heterogeneous data sources in the Common Data Format (CDF), which has been described in [25] and made publicly available [28]. The CDF postprocessing phase is described in detail in [26], while this paper focuses on the CDB. In the CDF postprocessing, the project's analysis partners use the Derived Measures—Performance Indicator Framework (DM-PI-Framework) to enrich the vehicular signal time series from a vehicle's trip with the computed derived measures (DM) and the detected driving scenarios, which are fundamental for the computation of CDB PIs, as described in the next subsection.

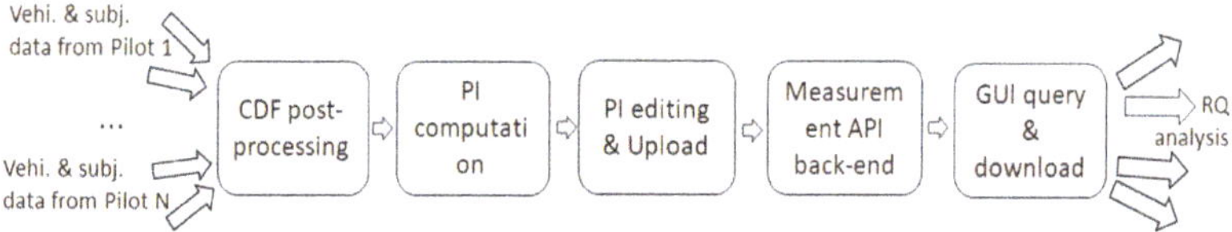


**Figure 2.** L3Pilot data workflow.

*4.1. CDB PI Computation for Vehicular Sensor Data*

The CDB PI computation step consists in synthesizing the vehicular time series so that the CDB stores only high-level information that allows tackling the project RQs, without compromising the confidentiality of the single-vehicle owner companies. This stage is undertaken by the CDB Aggregator module, which processes HDF5 files (one per each trip), containing the original time series formatted in CDF and enriched by the DM-PI-Framework's, as mentioned above. The output of the CDB Aggregator module is represented by a set of .json files storing the computed PIs. Processing an input HDF5 file, the Aggregator produces one .json file for each one of the four PI types defined in Table 3 (i.e., Trip PI, Scenario Instance PI, etc.). The .json files are ready to be uploaded to the CDB, for instance through a well-established Application Programming Interface (API) client such as Postman, or, better, through the Uploader, a dedicated module described in Section 4.3. The same information contained in the .json files is also saved in corresponding .csv files, that are more easily readable by the analysts.

**Table 3.** Measurify user roles and rights. General description and mapping in the L3Pilot case.

| Role | Description | L3Pilot Configuration/Notes |
| --- | --- | --- |
| Providers | Provider users are data owners. They can upload data and retrieve only their own data. | In L3Pilot, Providers are vehicle owners or their in-depth analysis partner for vehicular sensor data and pilot leaders for subjective data |
| Analysts | Analyst users cannot upload data but can see all the data of their typology. | In L3Pilot, analysts are the experts responding to the research questions. Utilizing the Measurify's Right resource, we have implemented three typologies, matching the type of relevant data: Technical and Traffic analysts, that access all vehicular sensor data apart from the Datapoints; Impact analysts, that access Datapoints; User analysts, that access subjective data |
| Admin | The admin configures the CDB (e.g., setting up the users and rights) and can see (only in case of need) all data entries | Given the adopted ID pseudonymization, the admin cannot resolve IDs (i.e., relating a data entry with its vehicle owner or driver). |

The CDB Aggregator module consists of a set of Matlab scripts. Figure 3 provides a high-level outlook of the programme, with three main phases: initialization, reading signals from the input HDF5 file; processing loop; and a final saving of the four types of PIs. The processing loop is the core of the programme, as it processes the time series and segments the computation of the PIs according to the context information presented in Section 3.2. First, the experimental condition is considered. Then, for each identified segment, the road type is considered. This level of segmentation leads to the computation of Trip PIs. Computation of Scenario Instance PIs and Datapoints requires further segmentation of the timeline based on the detected driving scenarios. Scenario Specific Trip PIs introduce the need for accumulating the indicator values across all the scenario instances in the trip. Similarly, the length of each scenario instance is needed for the Trip PI indicator reporting the percentage of time passed in each scenario within the trip.

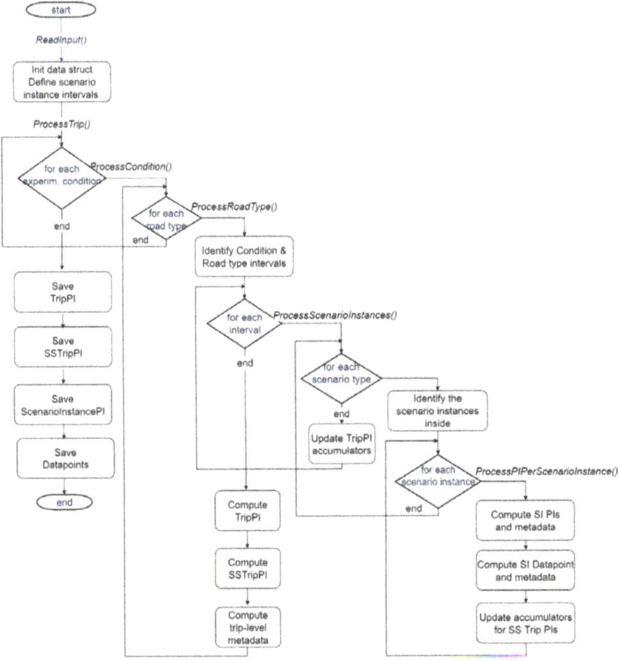


**Figure 3.** Flowchart of the vehicular sensor data processing Matlab scripts.

An example of the resulting segmentation is reported in Figure 4, where we can see eight different scenario instance PIs computed. A slice, indicated as Unrecognized 1 (U1), will not produce Scenario Instance PI, nor Datapoints, nor Scenario Specific Trip PI, as a scenario could not be detected there. However, the signal values contained in that segment will contribute to the Trip PI indicators in the ADF on condition.

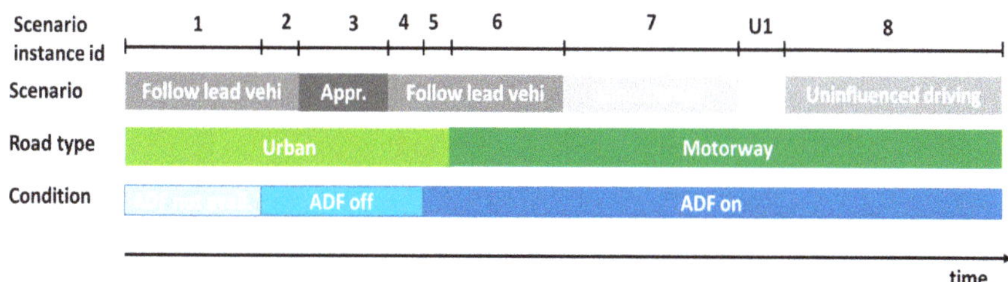


**Figure 4.** Example of scenario segmentation during a trip.

*4.2. Subjective Data Processing*

When conducting studies across multiple sites, it is essential that any data collection methodologies are applied uniformly. For example, the pilot site questionnaires are administered across all pilot sites, which vary in many respects (e.g., country language), but most relevant here is the interexperimenter variability. Therefore, to control for this variability, the questionnaire was implemented using the online tool LimeSurvey, where the only task for the pilot site staff was to transfer the translated versions of the questionnaire into the online LimeSurvey platform, enabling the administration of the questionnaire via laptops or mobile tablet computers. The use of LimeSurvey also ensures that the data output aligns with the CDF used by a consortium-wide consolidated database.

The imported surveys may then be customised and translated versions can be implemented accordingly. Pilot sites are then able to export their newly generated surveys and/or to export their results to CSV or SPSS format (Statistical Package for the Social Sciences). Note: the SPSS output of the reference version of the questionnaire is in line with the common data format requirements for the consolidated database. Although selected partners/pilot sites responsible can create, edit or view a survey, it is imperative that the questionnaire item codes are not changed, as this is the mechanics which allows tracking responses from different pilot sites. Thus, to ensure that a CDF was applied across pilot sites, instructions on the questionnaire implementation, administration and metadata were defined at the consortium level.

In terms of the administration of the questionnaire, there are differences between pilot sites regarding the length and number of drives by each participant. Therefore, the project recommendation is that the questionnaire should be completed after the last test ride, irrespective of whether a driver has multiple drives (see further details in [29]).

Following the completion of all questionnaires, each pilot site must export the test participants' response results to the SPSS file format (as illustrated in Figure 5), to ensure a common data format in the CDB for the user and acceptance evaluation area.

Quality checking has been implemented both for vehicular and subjective data. In this paper, we briefly present the procedure we set up for this second type of data.

Using the common LimeSurvey implementation should ensure that all questionnaire items and responses follow the nomenclature set out in the CDF for the official L3Pilot questionnaire. However, some partners used other implementations, and there is always the chance that some item names and codes deviate from the original. A data format map is prepared for each type of questionnaire, with all expected questions and possible answer codes and ranges. Before being uploaded to the CDB, a Matlab data quality script parses each questionnaire outcome file, searching for inconsistencies compared to the CDF.

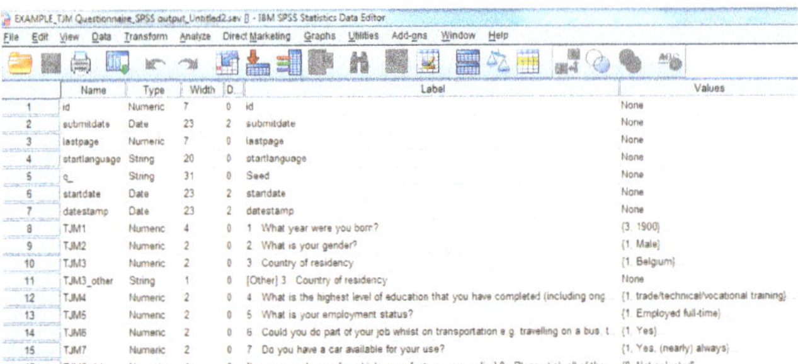

**Figure 5.** Example data output in SPSS.

A strategy is also adopted for missing data. Data may be missing because there was an error in data collection or during the data transfer process between data collection, LimeSurvey and SPSS. In order for analysts to identify cases where data is known to be missing, pilot sites are asked to fill these cells with a dummy response (1).

Should a pilot site attempt to upload a dataset including empty cells or errors from the Matlab data quality script, they will receive an error message and be asked to verify the contents of the missing cells. These steps ensure the reliability and validity of the uploaded data.

### 4.3. Uploader

The previous two steps prepare the files with the vehicular or subjective PIs to be uploaded to the CDB, via its RESTful APIs, that will be described later. For this step, a third party API development tool such as Postman can be used. However, we decided to develop an ad-hoc tool, namely the Uploader, to enhance usability, according to the specifications. This solution for uploading was preferred to a browser-based one, as it allows full access to the local file system. Given the typical On-Line Analytical Processing (OLAP) [30] pattern of usage foreseen for the CDB, we were in fact required to support an efficient upload of batches of files. The Uploader allows the user to indicate a source directory and then recursively searches in all the subdirectories all the matching .json files. Compliance is given by the name of the file, which must start with an eight-digit and end with a code indicating the type of the PI (e.g., Trip PI, Datapoint, Urban questionnaire).

Another functionality implemented by the Uploader concerns the support of postediting the csv files output by the CDB-aggregator. Postediting of PIs can be made by analysts on .csv files and the uploader offers the functionality of transforming .csv files into uploadable .json files. In addition, in this case, recursive processing of a root directory is implemented.

The Uploader is a NodeJS Command Line Interface (CLI). The user can execute some simple .json encoded command files that specify the operation to be performed on the CDB (upload, download, update, delete) or on local data (.csv to .json functionality) and the value of the corresponding parameters (e.g., source/destination directory, data override option, etc.).

### 4.4. Measurement API Back-End

For the data storage, we used Measurify (formerly, Atmosphere), an open-source RESTful API dedicated to storing measurements, typically but not exclusively from the Internet of Things [31,32]. Measurify is implemented in NodeJS and relies on MongoDB, a state-of-the-art nonrelational database, as the underlying database management system.

Measurify is a generic measurement API, which can be configured for different applications in different installations. This is achieved by inserting values for some fundamental resource collections. The most important one is the characterization of the features that describe the data type of the measurements to be uploaded and managed. This resource is used for the data integrity check at each measurement upload. For the L3Pilot installation (i.e., the CDB), we created one feature for each vehicular and subjective data type. Thus, we have four vehicular features: Trip PIs, Scenario Instance Trip PI, Datapoint (actually, since the structure of a Datapoint is scenario-dependent, we have one feature for each driving scenario type—Table 2), and Scenario Instance PI; and three subjective features: urban questionnaire, motorway questionnaire and parking questionnaire. In order to allow such a conceptually straightforward mapping, the Measurify's data structure needed to be extended so to contain data of different dimensions in each item. For instance, a measurement with a feature of type Trip PI has some scalar items (e.g., the number of harsh brakings per hour) and some vector items (e.g., percentage of time in each driving scenario type). Other configuration values concern the specification of the tags, that are strings that will be available in the user interface (UI) menus, and the specification of the constraints, that are used for specifying the relationships between tags, in order to support the automatic filling of the UI menus. Once correctly configured, the L3Pilot Measurify installation can be regularly used, i.e., for uploading and downloading the measurements.

User Roles and Data Access Rights

Measurify offers the possibility of providing three different roles to users: Providers, Analysts and Admin. Moreover, user rights can be assigned in the configuration, so to have finer control on data access for every single user. For instance, only certain features (not all) could be visible to a certain user. The mapping of user roles and rights in the L3Pilot domain is sketched in Table 3.

Data from the CDB API can be made available to the user through clients such as the Graphical User Interface (GUI), described in the next subsection, the Uploader and third parties tools, such as Postman. In this last case, there is no filter on the clients on the fields of a measurement. This would lead to a confidentiality breach, as the Trip and User IDs should not be visible to the user. Thus, the Measurify APIs have been enriched with the Fieldmask property, settable by the administrator, that allows specifying for each user what fields of a measurement are retrievable (e.g., only values and source device).

*4.5. Graphical User Interface (GUI)*

A web-based Graphical User Interface (GUI) application was created to facilitate access to the data available in the CDB. Although the data is available for download using the Uploader script described in the previous subsection, the user interface provides easier means to access, filter and download the data from a web browser.

Implementation Details

The choice of a web-based application in contrast to a native application (Windows/Linux/Mac) is motivated, firstly, by completely removing the burden of installing specific software from the user side, as the application is accessible from any web-browser, including mobile platforms. Secondly, adopting a web-based application allows maintaining a single code base, which would not be generally possible in the case of native applications since they require operating system-specific platforms. Thirdly, web-based development allows one to choose from a wide range of development frameworks, Vue.js [33] being the preferred solution for its API simplicity and flexibility, which enables fast development compared to other solutions such as AngularJS or ReactJS [34].

Architecturally, the GUI is the front-end which renders the data and resources obtained by the CDB RESTful API back-end. This interface was created envisioning a balance between user experience and software modularity. While a very flexible design may allow to easily change the GUI from the API back-end, it limits the user experience as generic forms and fields are required in this paradigm. As a compromise, we designed general Query and Download forms that are dynamically loaded with custom options according to presets in the CDB API. This design choice allows changes to scenario names, PIs and conditions to be transparent to the GUI, increasing modularity and avoiding repetition in the code base.

The GUI is deployed efficiently using Docker containers and a Docker image available in DockerHub. The CDB offers an HTTPS-enabled GUI, which is the main portal for project-wide data access.

*4.6. Functionalities*

The Query functionality is split into Vehicular and Subjective data, each with its respective page. The first is based on five query parameters: Query type (also known as a Performance indicator), Experimental condition, Road type, Driver type and Scenario type. The available options for each query parameter are dynamically loaded from the CDB API and filtered according to constraints indicating possible combinations of the query parameters. All parameters except for the first have a null choice, which ignores filtering records by that parameter.

Given choices of query parameters, the results are retrieved following a logical AND operation between all the parameter choices, i.e., the records retrieved must match all the query parameter criteria simultaneously. The results are displayed in a table which is populated asynchronously for optimal performance. The table's column's titles are obtained dynamically from the CDB, complying with the flexible design thereof. After visualizing the data, the user can also choose to download the resulting records to a CSV file using the Export button. Figure 6 illustrates a query for Scenario Instance PI with different querying parameters.

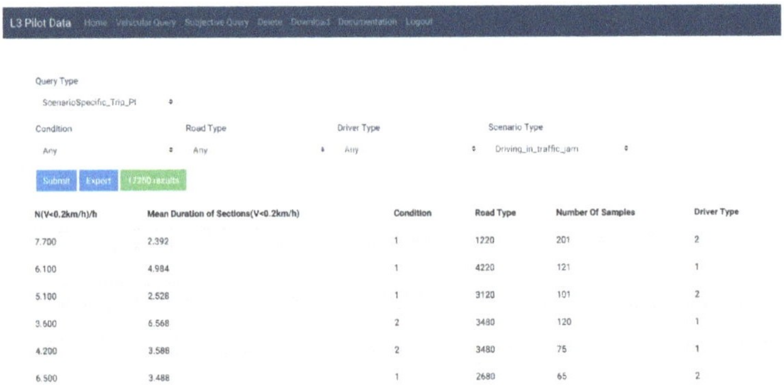

**Figure 6.** Example of vehicular query with (dummy) results displayed in a table.

Similarly, a download page allows users to download all data available in the database by compressing results for all queries into a single file. The results are saved in separate spreadsheets per query type and compressed into a single ".zip". A progress bar shows which query type is being handled at a given time. This feature has been implemented to allow data analysts to easily take frequent snapshots of all data accessible to their role.

The subjective query page offers only a query type parameter which selects a type of subjective questionnaire from the ones available. The functionality is the same as described for the vehicular query page, despite having only a single query parameter. The solution follows the modular architecture described earlier such that future requirements, such as new query parameters for subjective data, can be easily incorporated into the interface.

The delete functionality is considered in the GUI in case a data owner needs to remove a measurement no longer required (e.g., wrong measurement). The authorized user specifies the trip ID (which is known only to the Provider itself, according to the specified confidentiality rules).

## 5. Deployment at the Pilot Sites

After lab tests, the CDB was deployed in the cloud, together with the web user interface. In parallel, the Uploader was distributed to all the pilot sites. After the login to the CDB, each data row from the input files is tentatively inserted to the CDB provided that the structure of its data matches the corresponding Feature resource, which is checked for integrity assurance.

In order to make tests on their data and get familiarity with the process, the Pilot sites asked for the possibility of having own instances of the CDB, beside the cloud database, which was reserved for official data, shared among all the vehicle owners and pilot sites, in production. Thus, the Local CDB concept was defined. This was made possible thanks to the open-source release of Measurify. The installation procedure involved the configuration of Atmosphere/Measurify with the latest L3Pilot data structure and the installation of MongoDB, for the actual storage.

Given the complexity of the overall system, some partners asked for simplifying the set-up process. This led to the development of a single Docker containing the Measurify API, with automatic configuration based on Postman scripts, the MongoDB and the GUI accessible through the local host. The image is available on DockerHub.

As the internal organization rules prevented some partners from installing Docker and NodeJS (necessary for running the Uploader) on their own machines, the Information Technology (IT) experts of such organization were asked to install a local server, accessible from different machines in the local network through the web-browser GUI and the Uploader. To avoid installing NodeJS, a stand-alone version of the Uploader was set up. In one pilot site, it was preferred not to make any local installation. To this end, we set up a development installation of the CDB in the cloud.

These different architectural options allowed every partner involved in the pilot sites to get familiarity with the process, according to their different roles. Various patterns of use could be observed. Vehicle owner companies uploaded and checked their data and analysts accessed and analysed data from all the pilot sites but only concerning their specific features.

Needless to say, writing detailed instructions in the project's collaboration tool was useful to facilitate the usage of the system. Feedback on this from the users helped to improve communication and overall effectiveness, in an iterative process.

Not only did these "early" installations highlight some bugs in the code but they also enabled us to tune the overall process and suggest significant improvements, based on the experience and the analysis of the first sets of actual data. Such suggestions were discussed with the developers and then implemented. Important system functionalities have been added thanks to this collaboration. For instance, we initially considered more experimental conditions than those presented in Section 3. There was also a "Treatment" condition, aggregating a trip's measurements independent of the status of the ADF—it is sufficient that the ADF is on the vehicle, as opposed to the baseline condition. However, analysts asked to remove this condition, in order to reduce the amount of data to be processed.

A conceptual problem found at the beginning of the deployment was that if the experimental condition changes within an occurring scenario, two scenario instances are created and uploaded to the database although there is only one scenario occurring. This is fine in some scenarios. For others, however, it gives wrong results for, e.g., the duration of a lane change or the standard deviation of the speed during following a lead vehicle. The problem became apparent when looking at lane changes. During the recordings, lane changes are often not performed by the ADF but by the safety driver, or at least signs off on them. In the aggregated data this leads to three scenarios that are uploaded to the database and that are evaluated at the end, which is not the desired output. We thus introduced the concept of "partial" (uninfluenced driving, following a lead vehicle) and "complete" (all the others) scenarios. For partial scenarios, splitting them up, due to an intervening condition change (e.g., from ADF on to ADF off), is fine. For all others, the complete scenario instance is always needed, no matter the condition changes during the scenario. Moreover, all the transitions of conditions that may occur during a complete scenario need to be traced.

Another improvement concerned the maximum length of scenario instances. Since some scenario instances (particularly Uninfluenced driving and Following a lead vehicle) are quite long, in some cases, leading to a low level of meaningfulness for the PIs' statistical indicators (e.g., min/max/avg/stdev), we were asked to set a maximum length of an instance, for "limited" scenarios. Longer instances of such scenarios are split in fixed-length chunks.

We also implemented the StringMapper tool. All data on the CDB are numeric, as this format was considered best suited for automatic processing of the downloaded data by analysts using their own statistical elaboration packages. An ancillary tool (the StringMapper) was developed on request of some pilot sites, which allows mapping some columns of the downloaded .csv files from codes to the corresponding predefined string values, thus favouring data readability.

As of the end of October 2020, the CDB has been employed in 14 L3Pilot tests sites, from Italy to Sweden, both in its cloud and local versions. Vehicular sensor data are being processed by six impact analysis teams and as many traffic analysis teams, while subjective data by three teams, in order to respond to the research questions.

The cloud CDB, and the GUI as well, are hosted on an Elastic Compute Cloud (EC2) server from Amazon Web Server (AWS). A variety of hardware (CPU, memory, storage) and software (operating system) solutions can be considered, based on the project's requirements. This solution allows achieving a state-of-the-art level in terms of infrastructure performance, scalability and security. MongoDB also supports sharding, which significantly increases performance by distributing data across multiple machines. Sharding was not necessary in our application case, also because we used PIs, not the raw signal time-series.

At the time of writing, the system has been operational for eight months and the first results have already been achieved by the analysts processing the CDB data. Feedback from analysts and other project partners (vehicle manufacturers and suppliers) is a testament to the ability of the system to meet the requirements. We particularly highlight the ability of extracting and managing the indicators to answer the RQs, while preserving confidentiality, and the workflow requirements (integration with the proprietary workflows and dataflows, recursive CDB upload, .csv download and postediting of the PIs).

Feedback from all the project partners also highlights some good practices that were applied and verified during the project, implementing the workflow depicted in Figure 2:

- an extensive use of abstractions, in order to support functional extensibility and module/code reusability
- the modular approach depicted in Figure 3 for extracting PIs from signal time series revealed itself very useful to deal with a set of specification upgrades, that occurred during the project
- the development of a tool that computes the PIs from the raw data and makes them ready for sharing
- the possibility of postediting the PIs before inserting them in the shared database
- the definition of a tool for efficiently uploading files to the database.
- the development of a web-based, open-source GUI for supporting a proper user experience when querying the database
- the usage of effective, well-established data formats, such as .hdf5, .json, .csv. This was key to guarantee interoperability with different tools, particularly for data logging and data analysis, as research teams are accustomed to various tools, such as VBOX, DL2, Matlab, SPSS, etc.
- the use of state-of-the-art tools for distributed project development (e.g., for code versioning)

A key component of the system is the Measurify measurement-oriented API back-end, which has been appreciated for several reasons, such as:

- Efficient storage and sharing of complex measurements, thanks to the underlying MongoDB nonrelational database management system.
- Easy configurability by specifying the features to be supported in the specific installation (i.e., application database). In the L3Pilot CDB, the features correspond to the types of vehicular

and subjective data to be uploaded. Changes in the data structure are easily managed by simply changing the Feature records (old data are to be deleted and reinserted in the new format).

- Ability to seamlessly deal with both vehicular and subjective data
- Open source availability.
- Robustness, as the API was tested in other projects as well.
- Platform-independence, given by the use of the intrinsically platform-independent NodeJS technology and MongoDB open source tool for data storage
- Non-vendor-lockedness. Differently from the typical cloud-based data management solutions, Measurify does not depend on vendor APIs. This makes the service easily portable across cloud providers
- Ease of deployment. The CDB has been deployed in a cloud installation and locally in all the pilot sites, also on laptops.

## 6. Conclusions

This paper has proposed a methodology and implemented a workflow for quantitatively addressing RQs in a collaborative project sharing among partners sensitive big data information (both objective, from vehicular sensors, and subjective, from user questionnaires), which is still not well covered in the literature. We have applied the workflow in L3Pilot, a large-scale project aimed at assessing the impact of automated driving as a safe and efficient means of transportation on public roads. While the project is not finished and its RQs have not been answered yet, the toolchain and the CDB API/GUI have been widely used across all pilot sites. Based on our working experience, we first stress the importance of establishing a collaborative community of researchers and developers who are knowledgeable in their respective domains. This team has been vital to allow a full understanding of the requirements, development of specifications and system and proper handling of all the issues that emerged with the concrete operations in the pilot sites. Discussions between experts in different fields have been very useful to achieve quality in a reasonable timeframe.

The process of sharing data among different providers and analysts for quantitatively answering RQs is complex, as is the development of the supporting systems and tools. Thus, time and effort need to be carefully spent in order to make everything work smoothly. Iterations and flexibility/availability are needed, as specifications (also those established during the project) had to be refined based on the actual test information, also with major design implications.

Our experience has shown the key importance of a reference methodology (in our case, FESTA), to theoretically inform and coherently manage all the steps of the project, in a top-down approach, from higher-level RQs, down to quantitative PIs and the actual signals or questionnaire items needed to answer them. The methodology needs to be put in practice in the everyday work of all the involved research teams, from car manufacturers to data analysts. As we could not find a proper solution ready for the whole chain, we designed a system architecture and developed the missing tools. Particularly, we highlight the following three key components:

- the Common Data Format (CDF) [25], which allowed all partners to deal with all the data in the same format, sharing tools and knowledge, but not the proprietary data, especially those coming from advanced driving assistance systems (ADAS). Not only does the CDF cover the original signal time-series but also additional information (e.g., the driving scenarios), that are computed by the Derived Measures—Performance Indicator Framework.
- the Measurify API, a non-vendor-locked cloud system for sharing appropriate measurements among relevant partners.
- the CDB Aggregator tool, which computes the sharable information, obtained through simple statistical data processing on relevant time segments based on such factors as experimental condition, road type and driving scenario (Figures 3 and 4).

We have described the implementation of the workflow, showing the challenges to overcome to meet the expectations of a real-world SAE automation level 3 pilot project. Feedback from the various types of users of the system (data analysts, original equipment manufacturers and suppliers) has been largely positive and stresses that a proper methodology and a chain of interoperable tools to manage big data are a key factor to the success of a collaborative research project.

While the implementation is exclusively in the automotive field, we argue that the proposed methodological approach and system architecture (Figure 2) and tools are general and could be efficiently adapted and employed in different domains in order to support quantitative research analyses:

- a common data format can be defined for any application domain, if not yet available.
- the Measurify API is released open source and installations can be easily configured for different domains [31], by specifying different features (i.e., measurement types)
- the principles of the CDB Aggregator (segmentation and statistical data synthesis) are generally applicable. Different factors (experimental condition, types of context of usage of a new system to test, etc.) can be efficiently nested in the modular processing schema presented in Figure 3.

The team of automotive and traffic analysts is now processing the data to answer the project RQs, that will be published in other articles. In a longer-term view, the next steps concern the study of more advanced ADFs, also considering the further impact on safety, security, privacy and freedom. Besides this, it will be interesting to apply and verify the proposed workflow and system architecture in other application domains other than automotive.

**Author Contributions:** Conceptualization, F.B., N.O., E.H.A., G.T., S.I., T.L., H W., J.H., A.D.G., M.D, R.B.; software, F.B., N.O., E.H.A., S.M., J.H., R.B., methodology, F.B., S.I., T.L., G.T., H.W., J.H., R.B., supervision, A.D.G., M.D. All authors have read and agreed to the published version of the manuscript.

**Funding:** This project has received funding from the European Union's Horizon 2020 research and innovation programme under grant agreement No 723051. The sole responsibility of this publication lies with the authors.

**Acknowledgments:** The authors would like to thank all partners within L3Pilot for their cooperation and valuable contribution.

**Conflicts of Interest:** The authors declare no conflict of interest.

## References

1. Cummings, J.N.; Kiesler, S. Collaborative Research across Disciplinary and Organizational Boundaries. *Soc. Stud. Sci.* **2005**, *35*, 703–722. [CrossRef]
2. Vom Brocke, J.; Lippe, S. Managing collaborative research projects: A synthesis of project management literature and directives for future research. *Int. J. Proj. Manag.* **2015**, *33*, 1022–1039. [CrossRef]
3. Pérez-Padillo, J.; García Morillo, J.; Ramirez-Faz, J.; Torres Roldán, M.; Montesinos, P. Design and Implementation of a Pressure Monitoring System Based on IoT for Water Supply Networks. *Sensors* **2020**, *20*, 4247. [CrossRef]
4. Stapel, J.; Mullakkal-Babu, F.A.; Happee, R. Automated driving reduces perceived workload, but monitoring causes higher cognitive load than manual driving. *Transp. Res. Part F Traffic Psychol. Behav.* **2019**, *60*, 590–605. [CrossRef]
5. Wang, Z.; Wu, Y.; Niu, Q. Multi-Sensor Fusion in Automated Driving: A Survey. *IEEE Access* **2020**, *8*, 2847–2868. [CrossRef]
6. Ardelt, M.; Coester, C.; Kaempchen, N. Highly Automated Driving on Freeways in Real Traffic Using a Probabilistic Framework. *IEEE Trans. Intell. Transp. Syst.* **2012**, *13*, 1576–1585. [CrossRef]
7. Hibberd, D.; Louw, T.; Aittoniemi, E.; Brouwer, R.; Dotzauer, M.; Fahrenkrog, F.; Innamaa, S.; Kuisma, A.; Merat, N.; Metz, B.; et al. Deliverable D3.1 from Research Questions to Logging Requirements; Deliverable D3.1 of L3Pilot Project funded under the European Union's Horizon 2020 research and innovation programme GA No: 723051. Available online: https://l3pilot.eu/download/ (accessed on 25 November 2020)
8. Barnard, Y.; Innamaa, S.; Koskinen, S.; Gellerman, H.; Svanberg, E.; Chen, H. Methodology for Field Operational Tests of Automated Vehicles'. *Transp. Res. Procedia* **2016**, *14*, 2188–2196. [CrossRef]

9. Biffl, S.; Moser, T.; Winkler, D. Risk Assessment in Multi-Disciplinary (Software+) Engineering Projects. International Journal of Software Engineering and Knowledge Engineering (IJSEKE). *Spec. Session Risk Assess.* **2011**, *21*, 211–236.

10. Bhatia, J.; Breaux, T.D.; Friedberg, L.; Hibshi, H.; Smullen, D. Privacy Risk in Cybersecurity Data Sharing. In Proceedings of the 2016 ACM on Workshop on Information Sharing and Collaborative Security (WISCS '16), Vienna, Austria, 24–28 October 2016; pp. 57–64.

11. Fioretto, F.; Van Hentenryck, P. Privacy-Preserving Federated Data Sharing. In Proceedings of the 18th International Conference on Autonomous Agents and MultiAgent Systems (AAMAS '19), Montreal, QC, Canada, 13–17 May 2019; pp. 638–646.

12. Mehdizadeh, A.; Cai, M.; Hu, Q.; Alamdar Yazdi, M.; Mohabbati-Kalejahi, N.; Vinel, A.; Rigdon, S.; Davis, K.; Megahed, F. A Review of Data Analytic Applications in Road Traffic Safety. Part 1: Descriptive and Predictive Modeling. *Sensors* **2020**, *20*, 1107. [CrossRef] [PubMed]

13. Ratcheva, V. Integrating diverse knowledge through boundary spanning processes—The case of multidisciplinary project teams. *Int. J. Proj. Manag.* **2009**, *27*, 206–215. [CrossRef]

14. Winkler, D.; Ekaputra, F.J.; Serral, E.; Biffl, S. Efficient data integration and communication issues in distributed engineering projects and project consortia. In Proceedings of the 14th International Conference on Knowledge Technologies and Data-driven Business (i-KNOW '14), Graz, Austria, 16–19 September 2014; pp. 1–4.

15. Benmimoun, M.; Benmimoun, A. Large-Scale FOT for Analyzing the Impacts of Advanced Driver Assistance Systems. In Proceedings of the 17th ITS World Congress 2010, Busan, Korea, 25–29 October 2010.

16. Burzio, G.; Mussino, G.; Tadei, R.; Perboli, G.; Dell'Amico, M.; Guidotti, L. A subjective field test on lane departure warning function in the framework of the euroFOT project. In Proceedings of the 2nd Conference on Human System Interactions, Catania, Italy, 21–23 May 2009; pp. 608–610.

17. FOT-Net & CARTRE. FESTA Handbook. Version 7. December 2018. Available online: https://connectedautomateddriving.eu/wp-content/uploads/2019/01/FESTA-Handbook-Version-7.pdf (accessed on 25 November 2020).

18. Adaptive Project Final Report. 2017. Available online: http://www.adaptive-ip.eu/files/adaptive/content/downloads/AdaptIVe-SP1-v1-0-DL-D1-0-Final_Report.pdf (accessed on 31 July 2020).

19. Schulze, M.; Mäkinen, T.; Kessel, T.; Metzner, S.; Stoyanov, H. Final Report (D11.6) of DriveC2X. 2014. Available online: https://www.eict.de/fileadmin/redakteure/Projekte/DriveC2X/Deliverables/DRIVE_C2X_D11_6_Final_report__full_version_.pdf (accessed on 31 July 2020).

20. Boban, M.; d'Orey, P.M. Measurement-based evaluation of cooperative awareness for V2V and V2I communication. In Proceedings of the 2014 IEEE Vehicular Networking Conference (VNC), Paderborn, Germany, 3–5 December 2014; pp. 1–8.

21. Gellerman, H.; Svanberg, R.; Barnard, Y. Data Sharing of Transport Research Data. *Transp. Res. Procedia* **2016**, *14*, 2227–2236. [CrossRef]

22. Bellotti, F.; Kopetzki, S.; Berta, R.; Lytrivis, P.; Amditis, A.; Raffero, M.; Aittoniemi, E.; Basso, R.; Radusch, I.; De Gloria, A. TEAM applications for Collaborative Road Mobility. *IEEE Trans. Ind. Inf.* **2019**, *15*, 1105–1119. [CrossRef]

23. Metz, B.; Rösener, C.; Louw, T.; Aittoniemi, E.; Bjorvatn, A.; Wörle, J.; Weber, H.; Torrao, G.; Silla, A.; Innamaa, S.; et al. Deliverable D3.3 Evaluation Methods; Deliverable D3.3 of L3PilotS Project funded under the European Union's Horizon 2020 research and innovation programme GA No: 723051. Available online: https://l3pilot.eu/download/ (accessed on 25 November 2020).

24. Innamaa, S.; Aittoniemi, E.; Bjorvatn, A.; Borrack, M.; Di Lillo, L.; Fahrenkrog, F.; Gwehenberger, J.; Lehtonen, E.; Louw, T.; Malin, F.; et al. Deliverable D3.4 Evaluation Plan; Deliverable D3.4 of L3Pilot Project funded under the European Union's Horizon 2020 research and innovation programme GA No: 723051. Available online: https://l3pilot.eu/download/ (accessed on 25 November 2020).

25. Hiller, J.; Svanberg, E.; Koskinen, S.; Bellotti, F.; Osman, N. The L3Pilot Common Data Format—Enabling efficient automated driving data analysis. In Proceedings of the 26th International Technical Conference on the Enhanced Safety of Vehicles, Eindhoven, The Netherlands, 10–13 June 2019.

26. Hiller, J.; Koskinen, S.; Berta, R.; Osman, N.; Nagy, B.; Bellotti, F.; Rahman, A.; Svanberg, E.; Weber, H.; Arnold, E.H.; et al. The L3Pilot Data Management Toolchain for a Level 3 Vehicle Automation Pilot. *Electronics* **2020**, *9*, 809. [CrossRef]

27. Innamaa, S.; Merat, N.; Louw, T.; Metz, B.; Streubel, T.; Rösener, C. Methodological challenges related to real-world automated driving pilots. In Proceedings of the ITS World Congress, Singapore, 21–25 October 2019.

28. L3Pilot Common Data Format. Available online: https://github.com/l3pilot/l3pilot-cdf (accessed on 31 July 2020).

29. Louw, T.; Merat, N.; Metz, B.; Wörle, J.; Torrao, G.; Satu, I. Assessing user behaviour and acceptance in real-world automated driving: The L3Pilot project approach. In Proceedings of the 8th Transport Research Arena TRA 2020, Helsinki, Finland, 27–30 April 2020.

30. Vassiliadis, P.; Sellis, T. A survey of logical models for OLAP databases. *SIGMOD Rec.* **1999**, *28*, 64–69. [CrossRef]

31. Berta, R.; Kobeissi, A.; Bellotti, F.; De Gloria, A. Atmosphere, an Open Source Measurement-Oriented Data Framework for IoT. *IEEE Trans. Ind. Inform.* **2020**. [CrossRef]

32. Measurify. Available online: https://github.com/measurify (accessed on 31 July 2020).

33. Macrae, C. *Vue. js: Up and Running: Building Accessible and Performant Web Apps*; O'Reilly Media, Inc.: Newton, MA, USA, 2018.

34. Wohlgethan, E. Supporting Web Development Decisions by Comparing Three Major JavaScript Frameworks: Angular, React and Vue. Js. Ph.D. Thesis, Hochschule für Angewandte Wissenschaften Hamburg, Hamburg, Germany, 2018.

*Article*

# A Portable Support Attitude Sensing System for Accurate Attitude Estimation of Hydraulic Support Based on Unscented Kalman Filter

**Xuliang Lu, Zhongbin Wang *, Chao Tan, Haifeng Yan, Lei Si and Dong Wei**

School of Mechatronic Engineering, China University of Mining and Technology, Daxue Road, Xuzhou 221116, China; LuXuliang@cumt.edu.cn (X.L.); tccadcumt@126.com (C.T.); YanHaifeng@cumt.edu.cn (H.Y.); lei.si@cumt.edu.cn (L.S.); weidongcmee@cumt.edu.cn (D.W.)
* Correspondence: wangzbpaper@126.com; Tel./Fax: +86-516-8359-0758

Received: 11 August 2020; Accepted: 15 September 2020; Published: 23 September 2020

**Abstract:** To measure the support attitude of hydraulic support, a support attitude sensing system composed of an inertial measurement unit with microelectromechanical system (MEMS) was designed in this study. Yaw angle estimation with magnetometers is disturbed by the perturbed magnetic field generated by coal rock structure and high-power equipment of shearer in automatic coal mining working face. Roll and pitch angles are estimated using the MEMS gyroscope and accelerometer, and the accuracy is not reliable with time. In order to eliminate the measurement error of the sensors and obtain the high-accuracy attitude estimation of the system, an unscented Kalman filter based on quaternion according to the characteristics of complementation of the magnetometer, accelerometer and gyroscope is applied to optimize the solution of sensor data. Then the gradient descent algorithm is used to optimize the key parameter of unscented Kalman filter, namely process noise covariance, to improve the accuracy of attitude calculation. Finally, an experiment and industrial application show that the average measurement error of yaw angle is less than 2° and that of pitch angle and roll angle is less than 1°, which proves the efficiency and feasibility of the proposed system and method.

**Keywords:** support attitude; inertial measurement unit; coal mining; unscented Kalman filter; quaternion; gradient descent

---

## 1. Introduction

Hydraulic support is an important safety support equipment in the automatic coal mining working face, which is of great significance to the safety of coal miners and the normal operation of coal mining equipment. In recent years, with the continuous development of intelligent mining technology [1,2], the roboticized hydraulic support technology has been paid more and more attention by scholars and coal mining managers. In particular, the precise estimation of the support attitude of hydraulic support directly determines the intelligent ability of hydraulic support. Therefore, it is urgent to study a new attitude sensing method to obtain accurate support attitude of hydraulic support [3], especially in the special application scenarios with a large demand for the number of sensors and complex environment.

The existing studies on the support attitude sensing measurement of hydraulic support are mainly focused on the mechanism and kinematics of hydraulic support and trying to achieve the accurate estimation of support attitude by kinematic analysis of its mechanism [4]. In [5,6], Polish scholars analyzed the structural composition and empirical formula of support strength of hydraulic support, and developed a new type of support mechanism for vertical displacement of hydraulic support, but which can only approximately estimate the yaw angle of hydraulic support by experience. In [7,8], through the analysis of the four-linkage mechanism and parameters of hydraulic support, the kinematics equation is established, and then the support attitude can only be roughly estimated due to the neglected clearance between the hinge joint and the pin shaft. In [9], a laser radar detection method is used for measuring the relative attitude angle of hydraulic support based on the position of inspection robot, but the estimation result is disturbed greatly due to the influence of robot motion vibration. In [10,11], a mechanical-electrical-hydraulic coordination technique based on the dynamic response of the load is exploited to adjust the attitude of hydraulic support, but it is easily affected by the pressure of the pump station. In [12], a virtual adjustment method of support attitude under the propulsive state of hydraulic support groups is proposed and the estimation accuracy is highly dependent on the modeling precision of the virtual model, and different hydraulic supports need different virtual models, which is difficult to establish and has poor universality. In [13], an approach for monitoring the posture of hydraulic support with fiber Bragg grating tilt sensor based on gravity is presented, but it is not applied to measure yaw angle of the canopy. However, in automatic coal mining working face with the complicated geological structures, the above methods are characterized by complex solutions, difficult modelling, large estimation disturbance and lack of universality with estimating attitude or relative attitude indirectly based on kinematic parameters of mechanism, or can only estimate a single attitude angle. In this paper, we try to explore a new approach for estimating directly the support attitude of hydraulic support with comprehensive consideration of estimation accuracy, measurement method, universality and sensor size.

Inertial measurement unit (IMU) can be independent of the existing mechanism of hydraulic support to estimate support attitude directly, which is not affected by the mechanism of hydraulic support itself. In recent years, the development of microelectromechanical system technology has accelerated leaps and bounds the development process of IMU, which makes the IMUs become low in cost, low in power consumption and small in size. Additionally, it is widely applied in robotics [14,15], navigation [16,17] and virtual reality [18]. Scientific research is not limited to sensor applications [19], but also includes the performance of IMUs [20]. The performance improvement and calibration analysis of IMU provide precise attitude estimation technologies for its application [21], such as the method of improving estimation performance based of sine rotation vector; the acceleration and magnetic field measurements are transformed into the differences between the Euler angles of the measured attitude and the predicted attitude to correct the predicted attitude [22]. However, the measurement accuracy and performance of these low-cost MEMS inertial measurement units are easily limited by the complex working environment [23,24], especially the automatic coal mining working face with complex magnetic interference produced by coal structure and high-power electromechanical equipment. Therefore, it is necessary to use a filtering algorithm to improve the estimation accuracy.

The Kalman filter (KF) and particle filter (PF) are the most studied and widely used filtering algorithms. The Kalman filter for linear filtering and the extended Kalman filter (EKF) with Jacobian matrix and the unscented Kalman filter (UKF) for nonlinear filtering are frequently used to process inertial information, including object tracking [25] and rotation estimation [26]. In order to reduce yaw drift of attitude rotation estimation in indoor scenarios, an improved EKF combining INS mechanization algorithm and zero velocity update methodology is proposed to improve the accuracy of yaw estimation [27], but the Jacobian matrix is used to update the state transition matrix and the observation matrix in the linearization steps of system equation in attitude determination, which leads to poor stability and large estimation bias. Owing to the system, equations are updated based on the sigma-point approximation generated by unscented transformation in UKF [28,29], which has been demonstrated to have more attitude solution accuracy and strong robustness than EKF in estimating attitude with highly nonlinear kinematics models. In [30,31], a quaternion-based MEMS sensors fusion algorithm and heading estimation approach based on rotation matrix and principal component analysis are proposed to improve the heading estimation accuracy of indoor pedestrians, which shows that quaternion is easier to implement than other methods in sensor fusion technology. However, the particle filter is a sequential importance sampling filtering method based on Bayesian estimation [32] and its idea is based on the Monte Carlo method to represent probability with particle set, which can be used in any state space models [33,34]. In [35], a method of filtering and predicting time-varying signals under model uncertainty is proposed to realize dynamic estimation of the data. In [36], a particle metropolis Hastings algorithm driven by multiple parallel particle filters is proposed to make inference on dynamic and static variables. In any case, although the particle filter has good estimation performance in nonlinear filtering, the mathematical process of particle filters is more complex, and the algorithm implementation and calculation are difficult. Moreover, the resampling, particle degradation and calculation amount in the calculation process lead to the complexity of the implementation process and poor real-time performance. Therefore, the operation time is much longer than that of UKF.

Through the above research on attitude estimation of nonlinear systems, and considering the estimation accuracy, calculation speed and real-time performance of algorithm, UKF is proved to be a powerful technique for estimating attitude angle and a superior alternative to the EKF and PF in various nonlinear system filtering [37,38]. Unfortunately, the key parameter of UKF selected by experience, process noise covariance Q, has a great influence on the estimation accuracy in the complex working environment [39], so the research on tuning of parameter Q is very meaningful [40,41]. Therefore, the gradient descent algorithm [42] with the advantages of fast global convergence, faster operation speed and simple realization could be used to tune the process noise covariance Q to improve the estimation performance of UKF.

In this paper, a support attitude sensing system with a character of intrinsic safety is designed to measure the support attitude of hydraulic support in the special application scenarios with a large demand for the number of sensors and complex environment. The designed support attitude sensing system is a nine-axis low-cost MEMS measurement unit composed of a gyroscope, magnetometer and accelerometer. The proposed gradient descent algorithm-optimized quaternion-based unscented Kalman filter makes full use of the characteristics of complementation of the magnetometers without long-term drift, accelerometers and gyroscopes which are not affected by magnetic disturbance to improve the support attitude estimation accuracy of hydraulic support.

The remaining parts of this paper are organized as follows. In Section 2, the support attitude sensing system designed in our laboratory is described. Section 3 introduces the proposed optimized quaternion-based unscented Kalman filter based on the complementary characteristics of MEMS sensors in detail. In Section 4, an experiment is conducted and analyzed to validate the estimation performance of the proposed system and approach. Then, industrial applications and tests are carried out in Section 5. Section 6 summarizes our conclusions and future work.

## 2. Support Attitude Sensing System

### 2.1. Establishing the Coordinate System and Attitude Angle

The two coordinate systems and three attitude angles need to be established in the support attitude measurement of hydraulic support, as shown in Figure 1, which are the geographical coordinate system, carrier coordinate system, pitch angle, yaw angle and roll angle. The geographical coordinate system (frame n) is the North-East-Down (NED) orthogonal coordinate defined by the relative geoid. The Down axis (D axis) is perpendicular to the ellipsoid of the reference system and points to the interior of the earth. The North axis (N axis) points to true north (the velocity vector projection along the earth's rotation angle on the plane perpendicular to the D axis). Finally, the East axis (E axis) points horizontally east and completes the right-hand orthogonal coordinate system. The carrier coordinate system (frame b) is fixed to the carrier, which is the reference coordinate system. The $Z_b$ axis and $X_b$ axis point to the bottom and forward along the longitudinal axis of the carrier, respectively. The $Y_b$ axis points to the right wing (side) and completes the right-hand orthogonal coordinate system with the axes $X_b$ and $Z_b$.

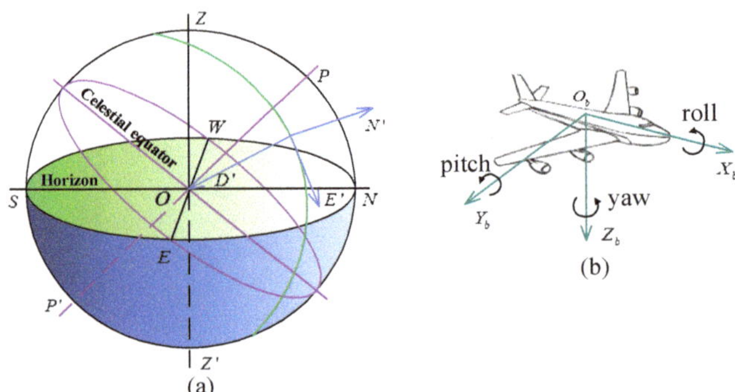



**Figure 1.** Definition of coordinate systems and carrier attitude angles: (**a**) local geographical coordinate system ($N'E'D'$) defined relative to the geoid; (**b**) carrier coordinate system ($X_bY_bZ_b$) and carrier attitude angles including roll, pitch and yaw angles.

The support attitude of hydraulic support includes yaw angle, roll angle and pitch angle. The yaw angle of rotation around $Z_b$ axis of the carrier is the angle between the N axis and the longitudinal axis $X_b$ of the carrier, measured on the horizontal plane and clockwise is positive. The pitch angle of rotation around the $Y_b$ axis of the carrier is the angle between the horizontal plane and the longitudinal axis $X_b$ of carrier, measured on the vertical plane and the direction of the carrier's head up is positive. The roll angle of rotation around the $X_b$ axis of the carrier is the angle between the horizontal plane and the cross axis $Y_b$ of the carrier, measured on the cross section, and the lifting direction of the carrier on the left side is positive.

The carrier coordinate system and NED coordinate system can coordinate transformation with each other by rotating yaw angle, pitch angle and roll angle around the X axis, Y axis and Z axis in turn, and the transformation relationship from ONED to $OX_bY_bZ_b$ is shown in Figure 2. The coordinate transformation matrix is also called direction cosine matrix, which is written as:

$$
\begin{aligned}
C_n^b &= [\phi]_{N_2}[\theta]_{E_1}[\psi]_D \\
&= \begin{bmatrix}
\cos\theta\cos\psi & \cos\theta\sin\psi & -\sin\theta \\
\cos\psi\sin\phi\sin\theta - \cos\phi\sin\psi & \sin\phi\sin\theta\sin\psi + \cos\phi\cos\psi & \sin\phi\cos\theta \\
\cos\phi\sin\theta\cos\psi + \sin\phi\sin\psi & \cos\phi\sin\theta\sin\psi - \sin\phi\cos\psi & \cos\phi\cos\theta
\end{bmatrix}
\end{aligned} \tag{1}
$$

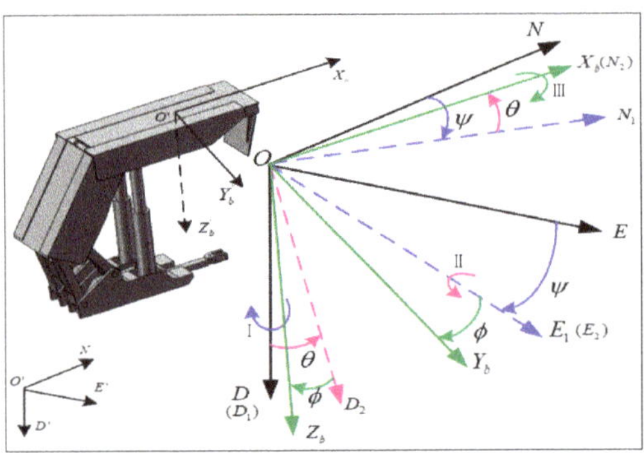



**Figure 2.** Coordinate transformation and directional cosine matrix definition of the canopy between frame b and frame n based on hydraulic support.

On the contrary, the transformation from $OX_bY_bZ_b$ to ONED is written as:

$$C_b^n = \begin{bmatrix} \cos\theta\cos\psi & \sin\phi\sin\theta\cos\psi - \cos\phi\sin\psi & \cos\phi\sin\theta\cos\psi + \sin\phi\sin\psi \\ \cos\theta\sin\psi & \sin\phi\sin\theta\sin\psi + \cos\phi\cos\psi & \cos\phi\sin\theta\sin\psi - \sin\phi\cos\psi \\ -\sin\theta & \sin\phi\cos\theta & \cos\phi\cos\theta \end{bmatrix} \tag{2}$$

where $\phi$, $\theta$ and $\psi$ are roll angle, pitch angle and yaw angle of support attitude of hydraulic support canopy, respectively.

### 2.2. Intrinsically Safe Micro Inertial Sensor

The support attitude sensing system based on the intrinsically safe micro inertial sensor is designed as a small strapdown inertial navigation system in our lab to meet the coal mine special environment navigation requirements of hydraulic support, mine inspection robot, support robot and driving robot. The intrinsically safe micro inertial sensor combining the IMU and magnetometer is composed of a three-axis gyroscope, three-axis magnetometer, three-axis accelerometer, an intrinsically safe microprocessor and a flameproof shell. The three single-axis gyroscopes and accelerometer use the ICM-42605 with the measuring range of ±125°/s and ±16 g. The three single-axis magnetometers use the LSM3030C with a measurement range of ±16 gauss magnetic full scale. STM32MP157 is selected as the microprocessor of the system and it is based on the high-performance dual-core Arm® Cortex®-A7 32-bit RISC core with an operating frequency of up to 650 MHz. The support attitude sensing system uses an I2C/SPI bus for digital output.

### 2.3. Support Attitude Estimation

According to the mining process of automatic coal mining working face, there are two working states in the support process for the canopy of hydraulic support, namely, static working state and moving working state. An accelerometer with superior static characteristics has high measurement accuracy in static state. A gyroscope with stable static performance has high measurement accuracy in moving state. A magnetometer with the measurement stability in a long time test is easily disturbed by the external perturbed magnetic sources produced by coal rock structure and high-power shearer equipment. Nevertheless, the gyroscope is not disturbed by magnetic sources from the automatic coal mining working face. Therefore, this paper proposes a new attitude estimation method for measuring the pitch angle, yaw angle and roll angle of hydraulic support based on the characteristics of complementation of accelerometers, gyroscopes and magnetometers, which eliminates the errors derived from the MEMS sensors by the quaternion-based UKF.

In addition, the tuning parameter of UKF, the process noise covariance Q, is usually a general value set by experience, which sometimes makes UKF produce large estimation errors, especially in a complex environment. Therefore, a gradient descent algorithm is used to optimize Q to improve the estimation accuracy and performance in this paper. The flow chart of accurate estimation of support attitude of hydraulic support using the quaternion-based unscented Kalman filter through tuning process noise covariance is shown in Figure 3. The other key tuning parameter, the observation noise covariance R, is related to the deviation of the magnetometer, accelerometer and gyroscope, which can be obtained by calculating the expectation of the square of deviation of the sensors.

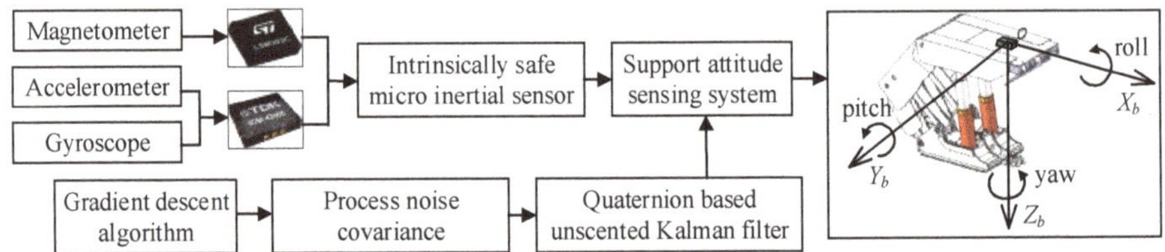



**Figure 3.** Block diagram of support attitude sensing system based on unscented Kalman filter for attitude measurement according to the complementary characteristics of the MEMS inertial sensor.

### 2.3.1. Support Attitude Estimation Based on Magnetometer

The magnetometer output is the sensitive geomagnetic field intensity. On the earth's surface, the geomagnetic field always points to the north along the magnetic induction, with components in the north direction and vertical direction of that, and no component in the east direction. Therefore, magnetic field coordinate is (N, 0, D) in the geomagnetic coordinate system. The magnetometer output is $M_n^{mag} = \begin{bmatrix} M_N & 0 & M_D \end{bmatrix}^T$ when the geomagnetic coordinate system coincides with the frame b. The magnetometer output is $M_n = \begin{bmatrix} M_x & M_y & M_z \end{bmatrix}^T$ in frame n and $m_b = \begin{bmatrix} m_x & m_y & m_z \end{bmatrix}^T$ in frame b. According to $m_b = C_n^b|_{\psi=\psi_m} \cdot M_n^{mag}$, the relationship of geomagnetic field intensity between the two coordinate systems can be obtained as:

$$\begin{bmatrix} m_x \\ m_y \\ m_z \end{bmatrix} = \begin{bmatrix} M_N \cos\theta \cos\psi_m - M_D \sin\theta \\ M_D \sin\phi \cos\theta + M_N(\sin\phi \sin\theta \cos\psi_m - \cos\phi \sin\psi_m) \\ M_D \cos\phi \cos\theta + M_N(\cos\phi \sin\theta \cos\psi_m + \sin\phi \sin\psi_m) \end{bmatrix} \tag{3}$$

Thus, the components of geomagnetic vector in a horizontal direction are:

$$\begin{cases} m_z \cos\phi \sin\theta + m_y \sin\phi \sin\theta + m_x \cos\theta = M_N \cos\psi_m \\ m_z \sin\phi - m_y \cos\phi = M_N \sin\psi_m \end{cases} \tag{4}$$

Then, the yaw angle can be expressed as:

$$\psi_m = \arctan\left(\frac{M_y}{M_x}\right) = \arctan\left(-\frac{m_y \cos\phi - m_z \sin\phi}{m_x \cos\theta + m_y \sin\theta \sin\phi + m_z \sin\theta \cos\phi}\right) \tag{5}$$

In fact, the geomagnetic North Pole and the geographical North Pole do not coincide, and the difference between them is one magnetic declination angle $\alpha$ (calculated by IGRF or WMM geomagnetic model according to the longitude and latitude height of the observation point). Therefore, the real local yaw angle of the carrier obtained based on magnetometer is:

$$\psi_{magn} = \psi_m + \alpha \tag{6}$$

### 2.3.2. Support Attitude Estimation Based on Accelerometer

The accelerometer output is the sensitive carrier acceleration, including the acceleration of the support attitude sensing system and the gravitational acceleration. For support attitude angle computation with an accelerometer, the relationship between the gravitational acceleration $\begin{bmatrix} 0 & 0 & g \end{bmatrix}^T$

in frame n and the gravitational acceleration $a_b = \begin{bmatrix} a_x & a_y & a_z \end{bmatrix}^T$ measured by the accelerometer in frame b could be expressed as:

$$\begin{bmatrix} a_x \\ a_y \\ a_z \end{bmatrix} = C_n^b \begin{bmatrix} 0 \\ 0 \\ g \end{bmatrix} \begin{bmatrix} -g\sin\theta \\ g\cos\theta\sin\phi \\ g\cos\theta\cos\phi \end{bmatrix} \tag{7}$$

According to the acceleration output of Equation (7), the roll and pitch angles obtained based on accelerometer can be calculated as follows:

$$\begin{cases} \theta = -\arctan\left(\dfrac{a_x}{\sqrt{a_y^2 + a_z^2}}\right) \\ \phi = \arctan\left(\dfrac{a_y}{a_z}\right) \end{cases} \tag{8}$$

### 2.3.3. Support Attitude Estimation Based on Gyroscope

The gyroscope output is the sensitive carrier angular acceleration. In this paper, quaternion is applied to calculate the support attitude of hydraulic support. Quaternion has the advantage of faster calculation speed, smooth interpolation, effectively avoiding the universal lock problem and small storage space. Quaternion is a vector composed of four elements, and matrix expression and normalized quaternion of that are defined as follows:

$$q = \begin{bmatrix} q_0 & q_1 & q_2 & q_3 \end{bmatrix}^T, q_0^2 + q_1^2 + q_2^2 + q_3^2 = 1 \tag{9}$$

Due to the movement of the carrier, $q$ is constantly updated. The updated dynamic model based on quaternion using the angular velocity of the gyroscope is expressed as:

$$\dot{q} = \frac{1}{2} q \otimes \omega^b \tag{10}$$

where $\omega^b = \begin{bmatrix} \omega_x & \omega_y & \omega_z \end{bmatrix}$ is the angular velocity and measured based on the gyroscope. Then, the matrix form of Equation (10) is as follows:

$$\begin{bmatrix} \dot{q}_0 \\ \dot{q}_1 \\ \dot{q}_2 \\ \dot{q}_3 \end{bmatrix} = \frac{1}{2} \begin{bmatrix} 0 & -\omega_x & -\omega_y & -\omega_z \\ \omega_x & 0 & \omega_z & -\omega_y \\ \omega_y & -\omega_z & 0 & \omega_x \\ \omega_z & \omega_y & -\omega_x & 0 \end{bmatrix} \begin{bmatrix} q_0 \\ q_1 \\ q_2 \\ q_3 \end{bmatrix} \tag{11}$$

The Picard successive approximation method of quaternion differential equation is used to solve Equation (11), and the discrete form is as follows:

$$q(t_k) = \left[ \cos\left(\frac{\Delta\alpha}{2}\right)I + \sin\left(\frac{\Delta\alpha}{2}\right)\frac{\Delta\Omega}{\Delta\alpha} \right] q(t_{k-1}) \tag{12}$$

where $\Delta\alpha$ and $\Delta\Omega$ involve the integration of angular velocity $\omega^b$ in the $k$th sampling period, $k = 1, 2, \ldots n$, $\Delta\alpha = \sqrt{\alpha_x^2 + \alpha_y^2 + \alpha_z^2}$ is the angular increment of the gyroscope in the $[t_{k-1}, t_k]$ sampling interval,

$$\alpha_i = \int_{t_{k-1}}^{t_k} \omega_i dt, i = x, y, z, \text{ and } \Delta\Omega = \int_{t_{k-1}}^{t_k} \begin{bmatrix} 0 & -\omega_x & -\omega_y & -\omega_z \\ \omega_x & 0 & \omega_z & -\omega_y \\ \omega_y & -\omega_z & 0 & \omega_x \\ \omega_z & \omega_y & -\omega_x & 0 \end{bmatrix} dt, dt \text{ is the sampling interval.}$$

The coordinate transformation matrix expressed by direction cosine, Euler angle and quaternion from carrier coordinate system to the NED coordinate system is written as:

$$
\begin{aligned}
C_b^n &= \begin{bmatrix} C_{11} & C_{12} & C_{13} \\ C_{21} & C_{22} & C_{23} \\ C_{31} & C_{32} & C_{33} \end{bmatrix} \\
&= \begin{bmatrix} \cos\theta\cos\psi & \sin\phi\sin\theta\cos\psi - \cos\phi\sin\psi & \cos\phi\sin\theta\cos\psi + \sin\phi\sin\psi \\ \cos\theta\sin\psi & \sin\phi\sin\theta\sin\psi + \cos\phi\cos\psi & \cos\phi\sin\theta\sin\psi - \sin\phi\cos\psi \\ -\sin\theta & \sin\phi\cos\theta & \cos\phi\cos\theta \end{bmatrix} \\
&= \begin{bmatrix} q_1^2 - q_2^2 - q_3^2 + q_0^2 & 2(q_1q_2 - q_0q_3) & 2(q_0q_2 + q_1q_3) \\ 2(q_0q_3 + q_1q_2) & -q_1^2 + q_2^2 - q_3^2 + q_0^2 & 2(q_2q_3 - q_0q_1) \\ 2(q_1q_3 - q_0q_2) & 2(q_0q_1 + q_2q_3) & -q_1^2 - q_2^2 + q_3^2 + q_0^2 \end{bmatrix}
\end{aligned}
\tag{13}
$$

Therefore, the support attitude angle can be expressed in the form of Euler angle represented by direction cosine, and is written as:

$$
\begin{cases} \phi = \arctan\dfrac{C_{32}}{C_{33}} \\ \theta = \arctan\dfrac{-C_{31}}{\sqrt{C_{32}^2 + C_{33}^2}} \\ \psi = \arctan\dfrac{C_{21}}{C_{11}} \end{cases}
\tag{14}
$$

The carrier yaw angle in form of Euler angle represented by quaternion is written as:

$$
\psi_{gyro} = \arctan\left(\frac{2(q_1q_2 + q_0q_3)}{q_0^2 + q_1^2 - q_2^2 - q_3^2}\right)
\tag{15}
$$

## 3. The Optimized Quaternion-Based Unscented Kalman Filter

In this section, we propose an unscented Kalman filter based on quaternion for estimating the support attitude of hydraulic support. In order to improve the performance of the support attitude estimation algorithm, the gradient descent algorithm is used for tuning the key parameter of the unscented Kalman filter, i.e., process noise covariance Q. The whole procedure of accurate estimation of support attitude using the gradient descent algorithm-optimized quaternion-based unscented Kalman filter (GD-UKF) is also presented.

### 3.1. Gradient Descent Algorithm

Gradient descent algorithm with the advantages of simple implementation is a common method to solve unconstrained optimization problems. The basic idea of the gradient descent is to select an appropriate initial value $x^{(0)}$, update the value of $x$ iteratively, and minimize the objective function until it converges. The input of gradient descent is the objective function $f(x)$, the gradient function $g(x) = \nabla f(x)$ and the calculation accuracy $\varepsilon$; the output is the minimum point $x^*$ of $f(x)$. The process of the gradient descent can be summarized in detail as follows:

Step 1.1: Initial key parameters $x^{(0)}$ and $k$.

Step 1.2: Calculate the objective function $f\left(x^{(k)}\right)$.

Step 1.3: Calculate the gradient $g_k = g(x^{(k)})$. If $\|g_k\| < \varepsilon$, stop iteration and let $x^* = x^{(k)}$. Otherwise, let $p_k = -g(x^{(k)})$ and then find $\lambda_k$ for equation $f\left(x^{(k)} + \lambda_k p_k\right) = \min\limits_{\lambda \geq 0} f(x^{(k)} + \lambda p_k)$.

Step 1.4: Set the variable $x^{(k+1)} = x^{(k)} + \lambda_k p_k$ and update the function $f\left(x^{(k+1)}\right)$. If the expression $\|f(x^{(k+1)}) - f(x^{(k)})\| < \varepsilon$ or $\|x^{(k+1)} - x^{(k)}\| < \varepsilon$, stop iteration and let $x^* = x^{(k)}$. Otherwise, let $k = k + 1$ and go back to Step 1.3 to find another better minimum.

When the objective function is a convex function, the solution of gradient descent is the global optimal solution.

### 3.2. Unscented Transformation

For unscented Kalman filter in a nonlinear system, unscented transformation (UT) is the core technique for propagating mean and covariance based on Cholesky decomposition, and can effectively approximate mean and covariance changes of the random variables when it undergoes a nonlinear transformation, including cross-correlation between state and measurement. The basic principle of UT can be considered in this way. We assume that the mean and covariance of a random variable X with n dimensions are $\bar{x}$ and $P$, respectively, and mean $\bar{x}$ and covariance $P$ propagate through a nonlinear function $y = f(x)$. In order to calculate the statistics of the variable y, 2n + 1 sigma points $X^{(i)}$ and corresponding weights $W^{(i)}$ are formed and calculated as follows:

(1)　Calculate 2n + 1 sigma points

$$\begin{cases} X^{(0)} = \bar{x}, i = 0 \\ X^{(i)} = \bar{x} + \left( \sqrt{(n+\lambda)P} \right)_i, i = 1 \sim n \\ X^{(i)} = \bar{x} - \left( \sqrt{(n+\lambda)P} \right)_i, i = n+1 \sim 2n \end{cases} \tag{16}$$

where $\left( \sqrt{p} \right)^T \left( \sqrt{p} \right) = p$, $\left( \sqrt{p} \right)_i$ is the *i*-th column of the matrix of the square root.

(2)　Calculate the corresponding weight of sigma point

$$\begin{cases} W_m^{(0)} = \frac{\lambda}{n+\lambda} \\ W_c^{(0)} = \frac{\lambda}{n+\lambda} + \left(1 - \alpha^2 + \beta\right) \\ W_m^{(i)} = W_c^{(i)} = \frac{\lambda}{2(n+\lambda)}, i = 1 \sim 2n \end{cases} \tag{17}$$

where $\lambda = \alpha^2(n + \kappa) - n$ is the scaling factor for reducing total prediction error; $\alpha$ is set to a small positive value to control the distribution of sigma points; $\kappa$ is the parameter to be selected to ensure a positive semidefinite and is usually set to 0; and $\beta$ is a nonnegative weight coefficient and is set for 2 for Gaussian distribution in this paper. The subscript $m$ and $c$ represent covariance and mean, respectively.

### 3.3. Unscented Kalman Filter Design

Generally, the dynamic system of the unscented Kalman filter can be described in two system equations: firstly, state equation; and secondly, observation equation. Considering the discrete-time nonlinear dynamic model of support attitude estimation of hydraulic support, the model state variable is $x = \begin{bmatrix} q_0 & q_1 & q_2 & q_3 \end{bmatrix}^T$, which is a vector composed of four elements of quaternion; and the model observation variable is $z = \begin{bmatrix} \phi & \theta & \psi \end{bmatrix}^T$, which is a vector composed of attitude angles. The state variable is unknown and is constantly updated according to the gyroscope; the observation variable is known and measured by the accelerometer and magnetometer. The details of the mathematical model are described with the following steps.

(1)　State equation

The state prediction is based on the previous optimal estimation, and the discrete-time nonlinear dynamic state equation of unscented Kalman filter is shown as:

$$x_k = A_{k-1}x_{k-1} + w_{k-1} \tag{18}$$

where $x_k$ and $x_{k-1}$ are the prior estimation at time $k$ and the posterior estimation at time $k-1$, respectively; $w_k$ is the process noise with covariance Q and simplified as independent Gaussian white noise. $A_k$ is the state transition matrix and can be obtained as:

$$A = \cos\left(\frac{\Delta\alpha}{2}\right)I + \sin\left(\frac{\Delta\alpha}{2}\right)\frac{\Delta\Omega}{\Delta\alpha} \tag{19}$$

(2)    Observation equation

The correction state is the essential step for refining measurement estimation. The observation equation is expressed as

$$z_k = H[x_k] + v_k \tag{20}$$

where $v_k$ is the independent Gaussian measurement noise with noise covariance R. $H_k[x_k]$ is the output function and can be calculated as follows:

$$H[x] = \begin{bmatrix} \arctan\left(2(q_2 q_3 + q_0 q_1)/\left(q_0^2 - q_1^2 - q_2^2 + q_3^2\right)\right) \\ \arcsin(2(q_0 q_2 - q_1 q_3)) \\ \arctan\left(2(q_1 q_2 + q_0 q_3)/\left(q_1^2 - q_2^2 - q_3^2 + q_0^2\right)\right) \end{bmatrix} \tag{21}$$

### 3.4. Noise Covariance of Process and Observation

The observation noise covariance and process noise covariance are the two key parameters of unscented Kalman filter, which sometimes leads to large support attitude estimation errors. However, it is essential to optimize the noise covariance of the process and observation to minimize the support attitude estimation errors of hydraulic support.

(1) Process noise covariance

In the process of hydraulic support moving and canopy supporting in the automatic coal mining working face, the uneven floor could cause random acceleration of the hydraulic support movement. Meanwhile the coupling effect of roof rock and hydraulic support can produce random impact on the hydraulic support in the moving process and static state, and the left and right adjacent hydraulic support could also produce impact vibration on hydraulic support. Therefore, the coupling effect between hydraulic support and the surrounding environment could lead to random vibration and acceleration on the support attitude sensing system in the static and moving process of hydraulic support, which leads to difficulty in determining the process covariance Q.

We assume that $\omega_x = \overline{\omega}_x + w_x$, $\omega_y = \overline{\omega}_y + w_y$, $\omega_z = \overline{\omega}_z + w_z$ where $\overline{\omega}_x, \overline{\omega}_y, \overline{\omega}_z$ are the mean of $\omega_x, \omega_y, \omega_z$; $w_x, w_y, w_z$ are the process deviations of the support attitude sensing system caused by the disturbance of coupling interaction between the hydraulic support and the surrounding environment. $\sigma_{\omega_x}^2 = E\left(w_x \cdot w_x^T\right)$, $\sigma_{\omega_y}^2 = E\left(w_y \cdot w_y^T\right)$ and $\sigma_{\omega_z}^2 = E\left(w_z \cdot w_z^T\right)$ are the variance of $w_x, w_y, w_z$. Thus, Equation (11) can be rewritten as:

$$\begin{aligned}
\begin{bmatrix} \dot{q}_0 \\ \dot{q}_1 \\ \dot{q}_2 \\ \dot{q}_3 \end{bmatrix} &= \frac{1}{2} \begin{bmatrix} 0 & -(\overline{\omega}_x + w_x) & -(\overline{\omega}_y + w_y) & -(\overline{\omega}_z + w_z) \\ (\overline{\omega}_x + w_x) & 0 & (\overline{\omega}_z + w_z) & -(\overline{\omega}_y + w_y) \\ (\overline{\omega}_y + w_y) & -(\overline{\omega}_z + w_z) & 0 & (\overline{\omega}_x + w_x) \\ (\overline{\omega}_z + w_z) & (\overline{\omega}_y + w_y) & -(\overline{\omega}_x + w_x) & 0 \end{bmatrix} \begin{bmatrix} q_0 \\ q_1 \\ q_2 \\ q_3 \end{bmatrix} \\
&= \frac{1}{2} \begin{bmatrix} 0 & -\overline{\omega}_x & -\overline{\omega}_y & -\overline{\omega}_z \\ \overline{\omega}_x & 0 & \overline{\omega}_z & -\overline{\omega}_y \\ \overline{\omega}_y & -\overline{\omega}_z & 0 & \overline{\omega}_x \\ \overline{\omega}_z & \overline{\omega}_y & -\overline{\omega}_x & 0 \end{bmatrix} \begin{bmatrix} q_0 \\ q_1 \\ q_2 \\ q_3 \end{bmatrix} + \frac{1}{2} \begin{bmatrix} -q_1 & -q_2 & -q_3 \\ q_0 & -q_3 & q_2 \\ q_3 & q_0 & -q_1 \\ q_2 & q_1 & q_0 \end{bmatrix} \begin{bmatrix} w_x \\ w_y \\ w_z \end{bmatrix}
\end{aligned} \tag{22}$$

The second term on the right of Equation (22) can be considered as the process noise

$$w = \frac{1}{2}\begin{bmatrix} -q_1 & -q_2 & -q_3 \\ q_0 & -q_3 & q_2 \\ q_3 & q_0 & -q_1 \\ -q_2 & q_1 & q_0 \end{bmatrix}\begin{bmatrix} w_x \\ w_y \\ w_z \end{bmatrix}$$ and the process noise covariance Q is:

$$Q = E(w \cdot w^T) = \frac{1}{4}\begin{bmatrix} -q_1 & -q_2 & -q_3 \\ q_0 & -q_3 & q_2 \\ q_3 & q_0 & -q_1 \\ -q_2 & q_1 & q_0 \end{bmatrix}\begin{bmatrix} \sigma_{\omega_x}^2 & 0 & 0 \\ 0 & \sigma_{\omega_y}^2 & 0 \\ 0 & 0 & \sigma_{\omega_z}^2 \end{bmatrix}\begin{bmatrix} -q_1 & -q_2 & -q_3 \\ q_0 & -q_3 & q_2 \\ q_3 & q_0 & -q_1 \\ -q_2 & q_1 & q_0 \end{bmatrix}^T \tag{23}$$

However, it can be determined from Equation (23) that the optimal Q of UKF can be obtained by optimizing variance $\sigma_{\omega_x}^2$, $\sigma_{\omega_y}^2$ and $\sigma_{\omega_z}^2$ based on the optimization algorithm.

(2) Observation noise covariance

The observation noise covariance R is determined by the measurement process and related to the characteristics of the measuring instrument, which can be obtained through long-term probability statistics of sensor measurement data. In this paper, the observation noise covariance R is determined by the measurement deviation of accelerometer and magnetometer.

We assume that $v_{a_x}$, $v_{a_y}$ and $v_{a_z}$ are the observation deviation of accelerometer three-axis output $a_x$, $a_y$ and $a_z$. Then, the first-order Taylor series expansion of trigonometric function is applied to Equation (8), and the deviation of the roll and pitch angles measured by the accelerometer can be derived as

$$\begin{bmatrix} v_\theta \\ v_\phi \end{bmatrix} = \begin{bmatrix} -\dfrac{\sqrt{a_y^2+a_z^2}}{a_x^2+a_y^2+a_z^2} & \dfrac{a_x a_y}{\left(a_x^2+a_y^2+a_z^2\right)\sqrt{a_y^2+a_z^2}} & \dfrac{a_x a_z}{\left(a_x^2+a_y^2+a_z^2\right)\sqrt{a_y^2+a_z^2}} \\ 0 & \dfrac{a_z}{a_y^2+a_z^2} & -\dfrac{a_y}{a_y^2+a_z^2} \end{bmatrix}\begin{bmatrix} v_{a_x} \\ v_{a_y} \\ v_{a_z} \end{bmatrix} = \Phi_{\theta\phi}\begin{bmatrix} v_{a_x} \\ v_{a_y} \\ v_{a_z} \end{bmatrix} \tag{24}$$

Through Equation (24), the covariance of pitch angle and roll angle based on the accelerometer is as follows:

$$\sigma_{\theta\phi}^2 = E\left(\begin{bmatrix} v_\theta & v_\phi \end{bmatrix}^T \cdot \begin{bmatrix} v_\theta & v_\phi \end{bmatrix}\right) = \Phi_{\theta\phi}\begin{bmatrix} \sigma_{a_x}^2 & 0 & 0 \\ 0 & \sigma_{a_y}^2 & 0 \\ 0 & 0 & \sigma_{a_z}^2 \end{bmatrix}\Phi_{\theta\phi}^T \tag{25}$$

where $\sigma_{a_x}^2 = E\left(v_{a_x} \cdot v_{a_x}^T\right)$, $\sigma_{a_y}^2 = E\left(v_{a_y} \cdot v_{a_y}^T\right)$ and $\sigma_{a_z}^2 = E\left(v_{a_z} \cdot v_{a_z}^T\right)$ is the variance of accelerometer three-axis output $a_x$, and $a_z$.

Similarly, through Equations (3) and (5) in the first-order Taylor series expansion of trigonometric function, the deviation of yaw angle measured by magnetometer can be derived as

$$\begin{aligned} v_{\psi_{magn}} &= \begin{bmatrix} -\dfrac{M_y}{M_x^2+M_y^2} & \dfrac{M_x}{M_x^2+M_y^2} \end{bmatrix}\begin{bmatrix} M_x \\ M_y \end{bmatrix} \\ &= \begin{bmatrix} -\dfrac{M_y}{M_x^2+M_y^2} & \dfrac{M_x}{M_x^2+M_y^2} \end{bmatrix}\begin{bmatrix} \cos\theta & \sin\theta\sin\phi & \sin\theta\cos\phi \\ 0 & -\cos\phi & \sin\phi \end{bmatrix}\begin{bmatrix} v_{m_x} \\ v_{m_y} \\ v_{m_z} \end{bmatrix} = \Phi_\psi\begin{bmatrix} v_{m_x} \\ v_{m_y} \\ v_{m_z} \end{bmatrix} \end{aligned} \tag{26}$$

where $v_{m_x}$, $v_{m_y}$ and $v_{m_z}$ are the deviation of magnetometer output $m_x$, $m_y$ and $m_z$.

Through Equation (26), the covariance of yaw angle based on magnetometer is as follows:

$$\sigma_\psi = E\left(v_{\psi_{magn}} \cdot v_{\psi_{magn}}^T\right) = \Phi_\psi\begin{bmatrix} \sigma_{m_x}^2 & 0 & 0 \\ 0 & \sigma_{m_y}^2 & 0 \\ 0 & 0 & \sigma_{m_z}^2 \end{bmatrix}\Phi_\psi^T \tag{27}$$

where $\sigma^2_{m_x} = E\left(v_{m_x} \cdot v^T_{m_x}\right)$, $\sigma^2_{m_y} = E\left(v_{m_y} \cdot v^T_{m_y}\right)$ and $\sigma^2_{m_z} = E\left(v_{m_z} \cdot v^T_{m_z}\right)$ are the variances of the magnetometer output $m_x$, $m_y$ and $m_z$, respectively.

Finally, it can be determined from Equations (25) and (27) that the observation noise covariance R is:

$$R = \begin{bmatrix} \sigma^2_{\theta\phi} & 0 \\ 0 & \sigma^2_{\psi} \end{bmatrix} \tag{28}$$

### 3.5. Unscented Kalman Filter Based on Gradient Descent

The UKF with Kalman linear filtering framework is not the traditional method of linearizing nonlinear functions, which uses unscented transformation to propagate mean and covariance. Compared with the extended Kalman filter, the UKF without deriving Jacobian matrix can make the state variable approximate the probability density distribution of nonlinear function and is more robust and accurate. However, the process noise covariance Q affects the filtering performance to a certain extent, and gradient descent can be used to improve the unscented Kalman filter. The flow of gradient descent algorithm-optimized quaternion-based unscented Kalman filter (GD-UKF) can be elaborated as below:

Step 2.1: The key parameters of the initial setup are dimension (n = 4), initial value $\hat{x}_0 = E(x_0)$, covariance initial value $P_0 = E\left((x_0 - \hat{x}_0)(x_0 - \hat{x}_0)^T\right)$ of random variable. Through Equations (16) and (17), a set of sigma points and the corresponding weights are computed by: $i = 1, 2, \cdots, 2n + 1$

$$X^{(i)}_{k-1} = \begin{bmatrix} \hat{x}_{k-1} & \hat{x}_{k-1} + \sqrt{(n+\lambda)P_{k-1}} & \hat{x}_{k-1} - \sqrt{(n+\lambda)P_{k-1}} \end{bmatrix} \tag{29}$$

Step 2.2: The predicted value and covariance matrix of system state variable calculated through the state Equation (18) based on the sigma points obtained in Step 2.1 are expressed as:

$$\begin{aligned} X^{(i)}_k &= A_{k-1} X^{(i)}_{k-1} \\ \hat{x}^-_k &= \sum_{i=0}^{2n} W^{(i)}_m X^{(i)}_k \\ P^-_k &= \sum_{i=0}^{2n} W^{(i)}_c \left[X^{(i)}_k - \hat{x}^-_k\right]\left[X^{(i)}_k - \hat{x}^-_k\right]^T + Q \end{aligned} \tag{30}$$

Step 2.3: Through using UT again for the predicted value in Step 2.2, the new sigma points are calculated as follows: $i = 1, 2, \cdots, 2n + 1$

$$X^{(i)}_k = \begin{bmatrix} \hat{x}^-_k & \hat{x}^-_k + \sqrt{(n+\lambda)P^-_k} & \hat{x}^-_k - \sqrt{(n+\lambda)P^-_k} \end{bmatrix} \tag{31}$$

Step 2.4: The predicted value and covariance matrix of observation variable calculated through the observation in Equation (20) based on the new sigma points obtained in Step 2.3 are expressed as follows: $i = 1, 2, \cdots, 2n + 1$

$$\begin{aligned} Z^{(i)}_k &= H\left[X^{(i)}_k\right] \\ \hat{z}^-_k &= \sum_{i=0}^{2n} W^{(i)}_m Z^{(i)}_k \\ P_{z_k z_k} &= \sum_{i=0}^{2n} W^{(i)}_c \left[Z^{(i)}_k - \hat{z}^-_k\right]\left[Z^{(i)}_k - \hat{z}^-_k\right]^T + R \\ P_{x_k z_k} &= \sum_{i=0}^{2n} W^{(i)}_c \left[X^{(i)}_k - \hat{z}^-_k\right]\left[Z^{(i)}_k - \hat{z}^-_k\right]^T \end{aligned} \tag{32}$$

Step 2.5: The unscented Kalman filter gain matrix $K$ is calculated as:

$$K_k = P_{x_k z_k} P^{-1}_{z_k z_k} \tag{33}$$

Step 2.6: The state update and covariance update of the unscented Kalman filter system are calculated as:

$$\hat{x}_k = \hat{x}_k^- + K_k\left[z_k - \hat{z}_k^-\right]$$
$$P_k = P_k^- - K_k P_{z_k z_k} K_k^T \tag{34}$$

Step 2.7: The process noise covariance Q is updated by calculating and updating process variance $\sigma = \begin{bmatrix} \sigma_{\omega_x}^2 & \sigma_{\omega_y}^2 & \sigma_{\omega_z}^2 \end{bmatrix}^T$ of the support attitude sensing system based on gradient descent. And the objective function of gradient descent is

$$f(\sigma) = L(m_b, a_b, \omega^b, \sigma) = RMSE(x, \hat{x}_k) = \sqrt{E\left[(x - \hat{x}_k)\cdot(x - \hat{x}_k)^T\right]}$$
$$\sigma = \arg\min_{\sigma} f(\sigma) \tag{35}$$

where $L(\cdot)$ is the loss function; $RMSE(\cdot)$ is the root mean square error of estimation results; $x$ is the true state value of the support attitude sensing system. The flowchart of the proposed GD-UKF is presented in Figure 4.

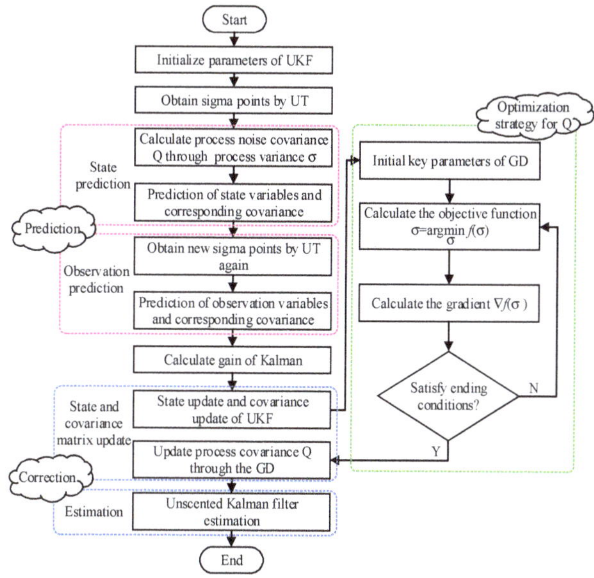


**Figure 4.** The flowchart of the proposed gradient descent algorithm-optimized quaternion-based unscented Kalman filter.

## 4. Experiments and Analysis

In this section, in order to evaluate measurement accuracy and performance of the designed support attitude sensing system based on GD-UKF, the static test and dynamic test experiment are carried out on the two-axis experiment platform. Then, the experimental results are contrasted with the Dutch XSENS high-precision attitude measurement system to prove the out-performing and practicability of the designed system.

### 4.1. Experiment on Two-Axis Turntable

The two-axis turntable experimental platform with U-frame structure is a high-precision electric precision machinery and is illustrated in Figure 5. The experiment platform is used for static test with an accuracy of 0.05° and dynamic test with an accuracy of 0.1°. The intrinsically safe micro inertial sensor is mounted on the precision electric rotary table.

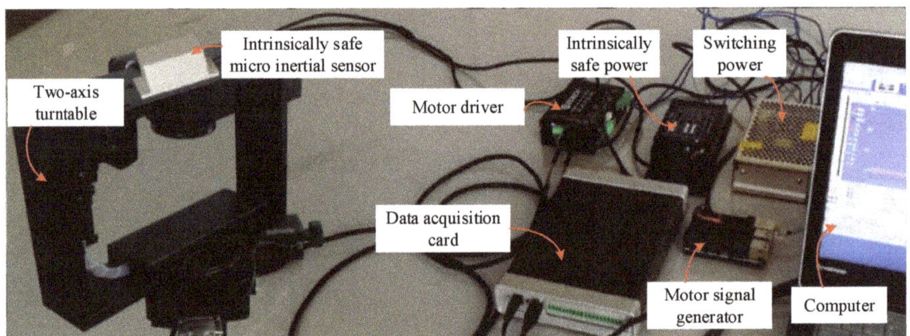


**Figure 5.** The experimental platform for testing the support attitude sensing system.

In the actual work of hydraulic support in the automatic coal mining working face, the hydraulic supports are closely arranged to support the surrounding rock. According to the coupling relationship between the hydraulic support and surrounding rock, the hydraulic support can appear yaw, pitch and tilt in the movement process, and the allowable range of the canopy pitch is ±15°. Additionally, the roll angle of hydraulic support, that is, the inclination of the coal mining working face, is nonnegative and the maximum value is 40°. The coupling relationship between adjacent hydraulic supports limits the allowable ranges of hydraulic support yaw to ±10°. Therefore, in the experiment setup, the test ranges of pitch, roll and yaw of the experimental platform are set to ±15°, 0° to 40° and ±10°, respectively, which is consistent with the actual working state of hydraulic support. Meanwhile, the attitude angle estimation performance of the support attitude sensing system based on Kalman filter, unscented Kalman filter and the proposed gradient descent algorithm-optimized quaternion-based unscented Kalman filter are tested and compared, respectively. The dynamic test of pitch angle is set from −15° to +15° in 0 s to 200 s, and the angle is maintained at 15° for static test in 200 s to 500 s. The test experiment of roll angle is designed as a dynamic test from 0° to 40° in 0 s to 250 s and static test at 40° in 250 s to 500 s. The dynamic test from −10° to 10° in 0 s to 250 s and the static test at 10° in 250 s to 500 s of yaw angle are set.

*4.2. Result Analysis*

The experimental results of the support attitude sensing system-based KF, UKF and GD-UKF on the two-axis experimental platform are shown in Figures 6–8, in which Figures 6, 7 and 8a show three attitude angles measured by the three methods, and the attitude calculation results of the three methods are consistent with the variational trend set by the experimental platform. In this paper, the root mean square error is used to evaluate the estimation accuracy of the algorithms, and the estimation errors of the three attitude angles are shown in Figures 6b, 7 and 8. The estimation errors (shown in the red curve) of pitch angle and roll angle based on Kalman filter are relatively large, and the maximum errors are greater than 3°, as shown in Figures 6b and 7b. Compared with the Kalman filter, the unscented Kalman filter improves the estimation accuracy to a certain extent, the blue curve is closer to the green curve than the red curve, but the estimation errors with maximum error greater than 2.4° are still relatively large. The reason for this is that the covariance Q, the key parameter of the algorithm, is set to the empirical value. In Figures 6 and 7, the estimation value (the pink curve) based on the proposed GD-UKF, is rather close to the theoretical value (the green curve), the dynamic and static estimation errors are less than 1° and 0.5°, respectively. The reason is that Q is tuned by gradient descent, and the variances $\sigma_{\omega_x}^2$, $\sigma_{\omega_y}^2$ and $\sigma_{\omega_z}^2$ in the optimized Q matrix are infinitely close to the true process noise covariance value caused by the inherent deviation of the experimental platform, as shown in Table 1, which greatly improves the estimation performance of UKF. Therefore, the estimation errors of pitch angle and roll angle based on GD-UKF are distinctly smaller than those of UKF and KF methods, and the accuracies of dynamic estimation and static estimation are better than 1° and 0.5°, respectively.

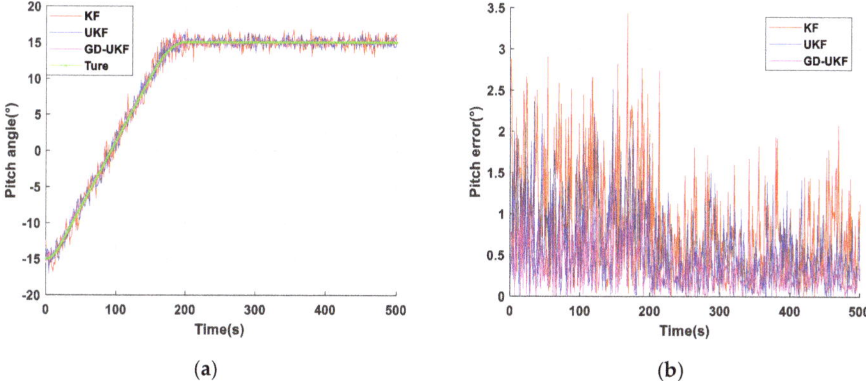

**Figure 6.** Pitch angle test of support attitude sensing system: (**a**) pitch angle estimation based on three methods; (**b**) pitch error of three methods.

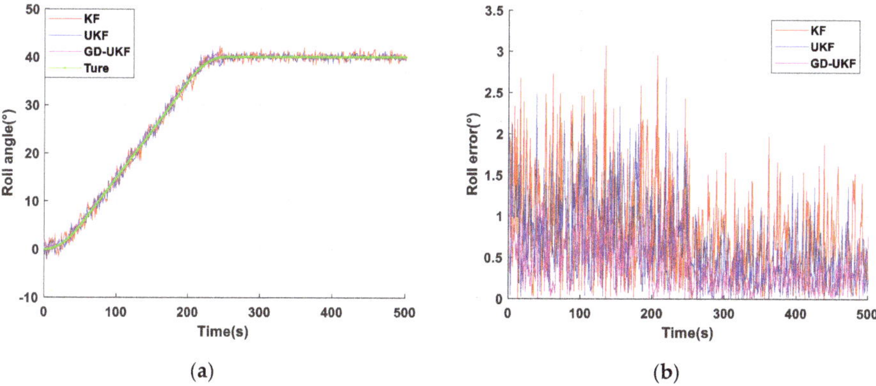

**Figure 7.** Roll angle test of support attitude sensing system: (**a**) roll angle estimation based on three methods; (**b**) roll error of three methods.

**Table 1.** Process noise covariance.

| | Inherent Covariance of Experimental Platform | | | Covariance Optimized by Gradient Descent | | |
|---|---|---|---|---|---|---|
| Static test | 0.0025 | 0 | 0 | 0.0046 | 0 | 0 |
| | 0 | 0.0025 | 0 | 0 | 0.0053 | 0 |
| | 0 | 0 | 0.0025 | 0 | 0 | 0.0032 |
| Dynamic test | 0.0100 | 0 | 0 | 0.0196 | 0 | 0 |
| | 0 | 0.0100 | 0 | 0 | 0.0182 | 0 |
| | 0 | 0 | 0.0100 | 0 | 0 | 0.0134 |

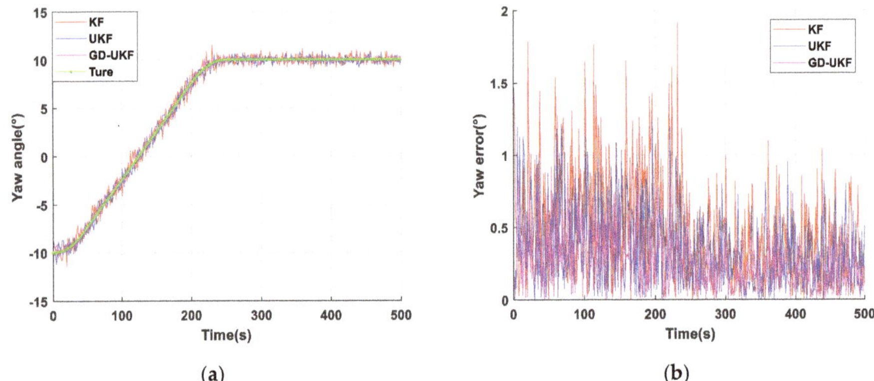

**Figure 8.** Yaw angle test of support attitude sensing system: (**a**) yaw angle estimation based on three methods; (**b**) yaw error of three methods.

As shown in Figure 8a, the fluctuation of yaw angle estimation results of three methods are stable. The maximum estimation errors based on KF and UKF are 1.8° and 1.3°, respectively, as shown in Figure 8b, the average estimation error of yaw angle of GD-UKF is 0.2° to 0.4° smaller than that of the other two algorithms, which is not obviously demonstrated to have the superior estimation performance of the proposed algorithm. The reason is that there is no magnetic field interference such as high-power equipment in the laboratory, and the estimations of yaw angle by the three methods are almost the same in the environment without magnetic noise interference. For the estimation performance of yaw angle based on GD-UKF, an industrial application needs to be operated in the automatic coal mining working face to validate the superiority and practicability of the proposed algorithm in yaw angle estimation.

The accuracy of our support attitude sensing system, with the static measurement accuracy better than 0.5° and dynamic measurement accuracy better than 1°, is lower than that of MTi-630 AHRS based on industrial-grade MEMS sensor from XSENS [43], but the support attitude sensing system is much cheaper than MTi-630 AHRS and its size is also smaller, which can meet the support accuracy requirements of hydraulic support. Moreover, the support attitude sensing system with 200 mW power consumption has lower power consumption than MTi-630 AHRS with 345 mW power consumption and especially meets the technical requirements of an intrinsically safe system [44], which can be directly applied to the automatic coal mining working face in explosive atmospheres. Considering the support accuracy requirement, the total price and power consumption of sensors and intrinsically safe technology requirements in the automatic coal mining working face, we argue that the designed support attitude sensing system is more suitable for the special application scenarios with a large demand for the number of sensors and complex environment. Therefore, it can be directly applied to measure the support attitude of hydraulic support canopy, and is also more suitable for the coal mine special environment navigation applications of other mobile equipment like mine inspection robots, support robots, driving robots, and so on.

## 5. Industrial Experiment and Application

In this section, our support attitude sensing system based on the proposed GD-UKF is applied in the automatic coal mining working face to test practical performance, as shown in Figure 9. The industrial experiment and application were tested at the 13230 coal mining working face of Gengcun Mine of Yima Coal Industrial Group Co., Ltd. (Yima, China). The intrinsically safe micro inertial sensor was installed on the canopy of hydraulic support to measure support attitude of hydraulic support. The attitude angle information was transmitted to the remote monitoring center by the network switch, and then the support attitude of hydraulic support was adjusted by the support attitude controller to satisfy the support requirements of coal mining.

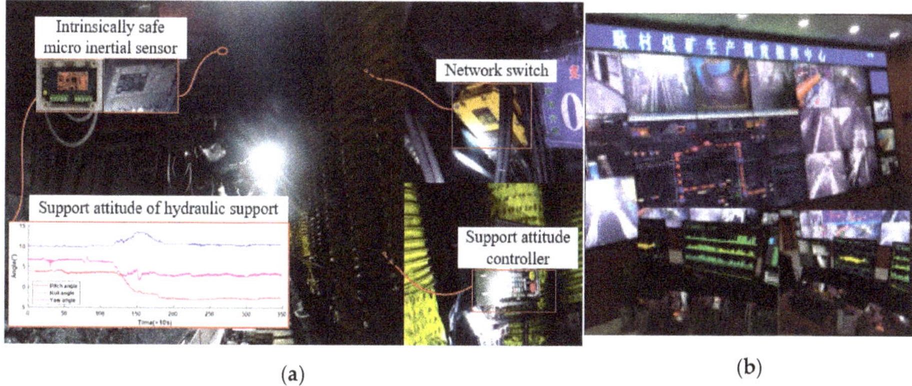


**Figure 9.** Industrial application of the support attitude sensing system: (**a**) application environment; (**b**) remote monitoring center.

The initial pitch angle of the hydraulic support canopy was 3.5°, then the angle was adjusted to −2.6° and remained stationary. In this process, the roll angle fluctuated and finally returned to the initial angle of 10° which was the inclination of the 13,230 coal mining working face, and the yaw angle was also changed from 6.9° to 3.9°. The industrial application results using the support attitude sensing system are shown in Figure 10. The pitch and roll estimation errors of support sensing system base GD-UDF were less than 1°, which could meet the requirements of the automatic coal mining working face. In order to verify the feasibility and superiority of the proposed algorithm for magnetic disturbances filtering, an industrial experiment of yaw angle was carried out, and the test results were shown in Figure 11. However, we could obviously find out that the yaw angle had a larger variation range than the other two angles and the yaw estimation error was relatively large, as shown in Figure 11a, the measurement error of original data was about 5° and that of yaw angle based on GD-UKF was less than 2°. The reason is that the high-power equipment, such as a shearer in the automatic coal mining working face, can generate a certain complex external magnetic field interference, which has a great influence on the yaw estimation of hydraulic support and is difficult to eliminate.

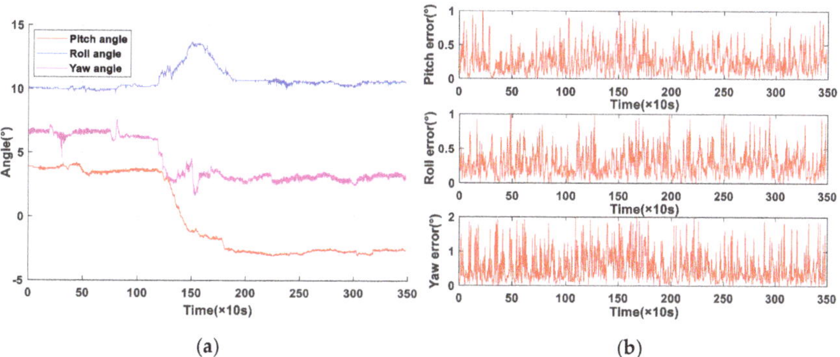


**Figure 10.** Industrial application results of support attitude sensing system-based GD-UKF: (**a**) attitude angle estimation; (**b**) estimation error.

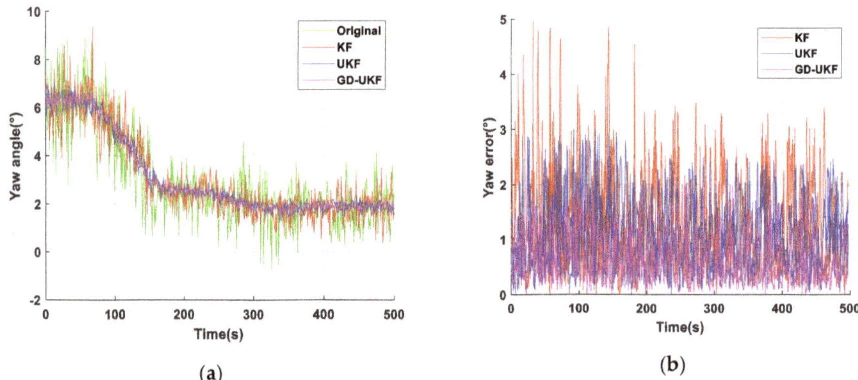

**Figure 11.** Estimation results of yaw angle tested in industrial experiment: (**a**) yaw angle estimation based on three methods; (**b**) estimation error of three methods.

The industrial application has verified that the support attitude estimation of hydraulic support with our support attitude sensing system using gradient descent algorithm-optimized quaternion-based unscented Kalman filter for attitude solution has high measurement accuracy, but the movement conditions of the automatic coal mining working face are complex, especially the magnetic disturbance. Therefore, the experiments for improving the yaw angle estimation algorithm need to be carried out under more complicated conditions.

## 6. Conclusions and Future Work

In order to tackle the problem of support attitude measurement of hydraulic support, this paper proposes a support attitude sensing system with the feature of intrinsic safety based on the low-cost MEMS-IMU and STM32MP157. The proposed gradient descent algorithm-optimized quaternion-based unscented Kalman filter makes full use of the characteristics of complementation of gyroscope, accelerometer and magnetometer to improve the accuracy of attitude estimation. In the proposed method, the gradient descent algorithm is employed to tune the process noise covariance for optimizing the state forecasting estimation. To verify the attitude estimation performance of the designed system and the proposed approach, an experiment was carried out and some analysis and comparisons were conducted. The experiment analysis results show that the support attitude sensing system based on the proposed approach has accurate attitude estimation performance with a dynamic measurement precision better than 1° and a static measurement precision better than 0.5°. Finally, the industrial experiment and application for estimating support attitude of hydraulic support in the automatic coal mining working face were performed to validate the practicability and feasibility of the designed support attitude sensing system.

However, the error elimination of support attitude sensing system is still a challenge, especially eliminating the estimation error of yaw angle in the environment of complex magnetic disturbance. In future studies, the authors will focus on algorithm research on error elimination in the complex magnetic perturbation environment. These studies may include updating the covariance of the algorithm in combination with other potential interference factors affecting support attitude estimation to further improve the estimation accuracy. In addition, the tuning parameter intelligent algorithm based on the maximum likelihood estimation of mathematic statistics is also worth further research, and the covariance is approximately solved based on the statistical model trained by machine learning with existing data.

**Author Contributions:** Methodology, X.L.; software, D.W.; formal analysis, X.L.; data curation, C.T. and H.Y.; writing—original draft preparation, X.L.; writing—review and editing, Z.W. and L.S.; visualization, L.S.; supervision, Z.W. and C.T. All authors have read and agreed to the published version of the manuscript.

**Funding:** The supports of China Postdoctoral Science Foundation (No.2019M661974) and the Priority Academic Program Development (PAPD) of Jiangsu Higher Education Institutions in carrying out this research are gratefully acknowledged.

**Conflicts of Interest:** The authors declare no conflict of interest.

## References

1. Wang, G.; Liu, F.; Pang, Y.; Ren, H.; Ma, Y. Coal mine intellectualization: The core technology of high quality development. *J. China Coal. Soc.* **2019**, *44*, 349–357. [CrossRef]
2. Hu, S.; Tang, C.; Yu, R.; Liu, F.; Wang, X. Intelligent coal mine monitoring system based on the Internet of Things. In Proceedings of the 2013 3rd International Conference on Consumer Electronics, Communications and Networks, Xianning, China, 20–22 November 2013; pp. 380–384.
3. Wang, J.; Huang, Z. The recent technological development of intelligent mining. *Engineering* **2017**, *3*, 444. [CrossRef]
4. Guan, E.; Miao, H.; Li, P.; Liu, J.; Zhao, Y. Dynamic model analysis of hydraulic support. *Adv. Mech. Eng.* **2019**, *11*, 1687814018820143. [CrossRef]
5. Gwiazda, A. Design of the roof support with strait-Line mechanism. *Adv. Mater. Res.* **2014**, *1036*, 553–558. [CrossRef]
6. Wang, G.; Wang, L.; Guo, Y. Determining the support capacity based on roof and coal wall control. *J. China. Coal. Soc.* **2014**, *39*, 1619–1624.
7. Zhang, Y.; Zhang, H.; Gao, K.; Xu, W.; Zeng, Q. New method and experiment for detecting relative position and posture of the hydraulic support. *IEEE Access* **2019**, *7*, 181842–181854. [CrossRef]
8. Cao, L.; Sun, S.; Zhang, Y.; Guo, H.; Zhang, Z. The research on characteristics of hydraulic support advancing control system in coal mining face. *Wirel. Pers. Commun.* **2018**, *102*, 2667–2680. [CrossRef]
9. Yang, X.; Wang, R.; Wang, H.; Yang, Y. A novel method for measuring pose of hydraulic supports relative to inspection robot using LiDAR. *Measurement* **2020**, *154*. [CrossRef]
10. Meng, Z.; Zeng, Q.; Wan, L.; Liu, P. Pose adjusting simulation of hydraulic support based on mechanical-electrical-hydraulic coordination. *Teh. Vjesn.* **2018**, *25*, 1110–1118.
11. Zeng, X.; Meng, G.; Zhou, J. Analysis on the pose and dynamic response of hydraulic support under dual impact loads. *Int. J. Simul. Model.* **2018**, *17*, 69–80. [CrossRef]
12. Ge, X.; Xie, J.; Wang, X.; Liu, Y.; Shi, H. A virtual adjustment method and experimental study of the support attitude of hydraulic support groups in propulsion state. *Measurement* **2020**, *158*, 107743. [CrossRef]
13. Liang, M.; Fang, X.; Li, S.; Wu, G.; Ma, M.; Zhang, Y. A fiber Bragg grating tilt sensor for posture monitoring of hydraulic supports in coal mine working face. *Measurement* **2019**, *138*, 305–313. [CrossRef]
14. Pareek, S.; Manjunath, H.; Esfahani, E.T.; Kesavadas, T. MyoTrack: Realtime estimation of subject participation in robotic Rehabilitation using sEMG and IMU. *IEEE Access* **2019**, *7*, 76030–76041. [CrossRef]
15. Xu, Z.; Lihua, D.; Zhong, S.; Ning, L. Study of the navigation method for a snake robot based on the kinematics model with MEMS IMU. *Sensors* **2018**, *18*, 879.
16. Wen, K.; Yu, K.; Li, Y.; Zhang, S.; Zhang, W. A New quaternion Kalman Filter based foot-mounted IMU and UWB tightly-coupled method for indoor pedestrian navigation. *IEEE Trans. Veh. Technol.* **2020**, *69*, 4340–4352. [CrossRef]
17. Hu, G.; Zhang, W.; Wan, H.; Li, X. Improving the heading accuracy in indoor pedestrian navigation based on a decision tree and Kalman Filter. *Sensors* **2020**, *20*, 1578. [CrossRef]
18. Hachaj, T.; Piekarczyk, M. Evaluation of pattern recognition methods for head gesture-based interface of a virtual reality helmet equipped with a single IMU sensor. *Sensors* **2019**, *19*, 5408. [CrossRef]
19. Chen, Z.; Zhu, Q.; Soh, Y.C. Smartphone inertial sensor-based indoor localization and tracking with iBeacon corrections. *IEEE Trans. Ind. Inform.* **2016**, *12*, 1540–1549. [CrossRef]
20. Zhong, M.; Guo, J.; Yang, Z. On real time performance evaluation of the inertial sensors for INS/GPS integrated systems. *IEEE Sens. J.* **2016**, *16*, 6652–6661. [CrossRef]

21. Wu, Y.; Luo, S. On misalignment between magnetometer and inertial sensors. *IEEE Sens. J.* **2016**, *16*, 6288–6297. [CrossRef]

22. Ko, N.; Jeong, S.; Bae, Y. Sine rotation vector method for attitude estimation of an underwater robot. *Sensors* **2016**, *16*, 1213. [CrossRef]

23. Chung, H.; Ojeda, L.; Borenstein, J. Accurate mobile robot dead-reckoning with a precision-calibrated fiber-optic gyroscope. *Robot. Autom. IEEE Trans.* **2001**, *17*, 80–84. [CrossRef]

24. Hong, S.K.; Park, S. Minimal-drift heading measurement using a MEMS gyro for indoor mobile robots. *Sensors* **2008**, *8*, 7287–7299. [CrossRef] [PubMed]

25. Hamid, K.R.; Talukder, A.; Islam, A.K.M.E. Implementation of fuzzy aided Kalman filter for tracking a moving object in two-dimensional. *Int. J. Fuzzy Log. Intell. Syst.* **2018**, *18*, 85–96. [CrossRef]

26. Hu, P.; Xu, P.; Chen, B.; Wu, Q. A self-calibration method for the installation errors of rotation axes based on the asynchronous rotation of rotational inertial navigation systems. *IEEE Trans. Ind. Electron.* **2018**, *65*, 3550–3558. [CrossRef]

27. Jiménez, A.R.; Seco, F.; Prieto, J.C.; Guevara, J. Indoor pedestrian navigation using an INS/EKF framework for yaw drift reduction and a foot-mounted IMU. In Proceedings of the 7th Workshop on Positioning, Navigation and Communication (WPNC 2010), Dresden, Germany, 11–12 March 2010; pp. 135–143.

28. Wang, G.; Li, N.; Zhang, Y. Hybrid consensus sigma point approximation nonlinear filter using statistical linearization. *Trans. Inst. Meas. Control* **2017**, *40*, 2517–2525. [CrossRef]

29. Zhao, J.; Mili, L. A robust generalized-maximum likelihood unscented Kalman Filter for power system dynamic state estimation. *IEEE J. Sel. Top. Signal Process.* **2018**, *12*, 578–592. [CrossRef]

30. Deng, Z.; Wang, G.; Hu, Y.; Wu, D. Heading estimation for indoor pedestrian navigation using a smartphone in the pocket. *Sensors* **2015**, *15*, 21518–21536. [CrossRef]

31. Abdelrahman, A.; Naser, E.S. Low-cost MEMS-based pedestrian navigation technique for GPS-denied areas. *J. Sens.* **2013**, 572–575.

32. Chopin, N.; Jacob, P.E.; Papaspiliopoulos, O. SMC 2: An efficient algorithm for sequential analysis of state space models. *J. R. Stat. Soc.* **2013**, *75*, 397–426. [CrossRef]

33. Carvalho, C.M.; Johannes, M.S.; Lopes, H.F.; Polson, N.G. Particle learning and smoothing. *Stat. Sci.* **2010**, *25*, 88–106. [CrossRef]

34. Martino, L.; Elvira, V.; Camps-Valls, G. Group importance sampling for particle filtering and MCMC. *Digit. Signal Process.* **2018**, *82*, 133–151. [CrossRef]

35. Lingo Urteaga, M.; Bugallo, F.; Djuric, P.M. Sequential Monte Carlo methods under model uncertainty. In Proceedings of the IEEE statistical Signal Processing Workshop, Palma de Mallorca, Spain, 26–29 June 2016.

36. Martino, L.; Elvira, V.; Camps-Valls, G. Distributed particle metropolis-Hastings schemes. In Proceedings of the IEEE Statistical Signal Processing Workshop, Freiburg, Germany, 10–19 June 2018.

37. Yang, C.; Shi, W.; Chen, W. Comparison of unscented and extended Kalman Filters with application in vehicle navigation. *J. Navig.* **2017**, *70*, 411–431. [CrossRef]

38. Yuan, X.; Yu, S.; Zhang, S.; Wang, G.; Liu, S. Quaternion-based unscented Kalman Filter for accurate indoor heading estimation using wearable multi-sensor system. *Sensors* **2015**, *15*, 10872–10890. [CrossRef] [PubMed]

39. Zhang, L.; Sidoti, D.; Bienkowski, A.; Pattipati, K.R.; Kleinman, D.L. On the identification of noise covariances and adaptive Kalman Filtering: A new look at a 50 year-old problem. *IEEE Access* **2020**, *8*, 59362–59388. [CrossRef]

40. Zhang, X.; Li, P.; Tu, R.; Lu, X.; Ge, M.; Schuh, H. Automatic calibration of process noise matrix and measurement noise covariance for multi-GNSS precise point positioning. *Mathematics* **2020**, *8*, 502. [CrossRef]

41. Poddar, S.; Kottath, R.; Kumar, V.; Kumar, A. Adaptive sliding Kalman filter using nonparametric change point detection. *Measurement* **2016**, *82*, 410–420. [CrossRef]

42. Murray, R.; Swenson, B.; Kar, S. Revisiting normalized gradient descent: Fast evasion of saddle points. *IEEE Trans. Autom. Control* **2019**, *64*, 4818–4824. [CrossRef]

43. MTi-630 AHRS. Available online: https://www.xsens.com/products/mti-600-series (accessed on 9 July 2020).
44. Explosive Atmospheres. *Part 18: Intrinsically Safe System*; GB 3836.18-2010/IEC 60079-25: 2003; General Administration of Quality Supervision, Inspection and Quarantine of the People's Republic of China: Beijing, China; Standardization Administration of the People's Republic of China: Beijing, China, 2010.

*Article*

# Analysis and Comparison of Rad-Hard Ring and LC-Tank Controlled Oscillators in 65 nm for SpaceFibre Applications [†]

**Danilo Monda, Gabriele Ciarpi and Sergio Saponara ***

Department of Information Engineering (DII), University of Pisa, 56126 Pisa, Italy;
danilo.monda@phd.unipi.it (D.M.); gabriele.ciarpi@ing.unipi.it (G.C.)
* Correspondence: sergio.saponara@iet.unipi.it
† This article is the extended version of the conference papers "Analysis and Comparison of Ring and LC-tank Oscillators for 65 nm Integration of Rad-Hard VCO for SpaceFibre Applications" published in the Proceedings of International Conference on Applications in Electronics Pervading Industry, Environment and Society, Pisa, Italy, 11–13 September 2019.

Received: 28 June 2020; Accepted: 14 August 2020; Published: 17 August 2020

**Abstract:** This work presented a comparison between two Voltage Controlled Oscillators (VCOs) designed in 65 nm CMOS technology. The first architecture based on a Ring Oscillator (RO) was designed using three Current Mode Logic (CML) stages connected in a loop, while the second one was based on an LC-tank resonator. This analysis aimed to choose a VCO architecture able to be integrated into a rad-hard Phase Locked Loop. It had to meet the requirements of the SpaceFibre protocol, which supports frequencies up to 6.25 GHz, for space applications. The full custom schematic and layout designs are shown, and Single Event Effect simulations results, performed with a double exponential current pulses generator, are presented in detail for both VCOs. Although the RO-VCO performances in terms of technology scaling and high-integration density were attractive, the simulations on the process variations demonstrated its inability to generate the target frequency in harsh operating conditions. Instead, the LC-VCO highlighted a lower influence through Process-Voltage-Temperature simulations on the oscillation frequency. Both architectures were biased with a supply voltage of 1.2 V. The achieved results for the second architecture analyzed were attractive to address the requirements of the new SpaceFibre aerospace standard.

**Keywords:** ring oscillator; LC-tank oscillator; SpaceFibre; rad-hard circuits; radiation effects; high-speed data transfer

---

## 1. Introduction

Several thousand launch activities have been performed during the last half-century, and with the rapid development of technology, satellites are playing an important role in human society. These systems are widely used for navigation, communication, and earth observation. One of the first communication experiments with laser was conducted between two Low Earth Orbit (LEO) satellites and a geostationary satellite ARTEMIS. The experiment was performed with a data rate of up to 50 Mbps. Then, other experiments followed with an increased data rate to achieve inter-satellite communication links. Today, current trends in satellites show a rapid increase in data traffic and digital processing. The throughput of next-generation satellites for digital telecom applications, as well as scientific missions, surveillance, and remote sensing, will exceed terabits per second of data that must be processed on board. For instance, the high-resolution cameras and synthetic aperture radars need high-speed communications between the instruments and the on-board data storage system [1]. The optical technology, thanks to its high bandwidth-length product, the lightweight cabling, and electromagnetic hardness, can potentially be the solution for data-rate increment in satellites. In this direction, the European Space Agency (ESA) has recently released the new SpaceFibre standard for on-board satellite communication up to 6.25 Gbps [2,3]. The standard describes the very high-speed serial link and network technology, and it was designed specifically for use on-board spacecraft and satellites. This protocol provides a coherent quality of service mechanism able to support bandwidth reserved, scheduled, and priority-based qualities of service. SpaceFibre provides robust, long-distance communications for launcher applications and supports avionics applications with deterministic delivery constraints using virtual channels. Communication performances are strongly related to the ability to synchronize the receiver with the transmitter. This issue is typically fixed with a Clock Data Recovery (CDR), and the key block used for its synchronization is the Phase Locked Loop (PLL). The Voltage Controlled Oscillator (VCO) is the core system, inside the PLL, able to generate the required frequency of 6.25 GHz to be compliant with the SpaceFibre protocol. Although the required Total Ionizing Dose (TID) level is lower than 1 Mrad for space applications [4], the main problems are due to Single Event Effects (SEEs) that temporarily disturb the typical operation of the circuit. This work targets, as implementation technology, a commercial 65 nm CMOS from TSMC (Taiwan Semiconductor Manufacturing Company). This technology, thanks to its thin gate-oxide thickness, could be considered radiation hard up to few hundred Mrad TID levels, as proved in [5], and by us in previous designs of other high-speed circuits in [6–8]. To the best of the authors' knowledge, in literature and market, there are not examples of rad-hard VCOs able to work at 6.25 GHz. The paper [9] showed the design of a PLL in the range from 0.2 GHz to 1.2 GHz, designed in 65 nm STMicroelectronics space technology. This system was irradiated up to 300 krad TID level, and its behavior was verified with different protons. In [10], a comparison between Ring Oscillator (RO) and LC-tank VCO for PLL was made for Large Hadron Collider's (LHC) applications. Both were designed for a working frequency from 2.2 GHz to 3.2 GHz, and the SEE test performed with heavy-ions showed that the LC-VCO had a larger cross-section than the RO-VCO. Varactors have been identified as the most sensitive part of LC-tank architectures, and Triple Modular Redundancy (TMR) technique has been adopted to face SEEs in the design of the phase frequency divider. The goal of this work was to compare the performances of the widely used RO and LC controlled oscillators in radiation environments and to contribute with new approaches for exploiting the characteristics that have made these systems the most implemented.

This work is an extension of the preliminary work presented by us at the conference [11]. With respect to the conference presentation, this work presented the complete full custom design of schematic and layout for both the RO and the LC-tank controlled oscillator (respectively reported in Sections 2 and 3). Moreover, this work in Section 4 provides transient and SEE simulations results, missing in [11]. Section 5 compares this work vs. The state-of-the-art. Conclusions are drawn in Section 6.

## 2. Ring Oscillator Based on a Cascade of Three Current Mode Logic (CML) Buffer

### 2.1. Ring Oscillator Schematic Design

The RO-VCO presented in this work is composed of a cascade of inverting amplifiers in closed-loop, as shown in Figure 1. The transconductance $g_m$ is the gain of the single amplifier, while R and C are the equivalent output resistance and the equivalent input capacitance, respectively, of previous and following stages. According to Figure 1, the open-loop gain of the system composed of N generic stages is expressed as

$$H(j\omega) = \left(-\frac{g_m R}{1 + j\omega RC}\right)^N \tag{1}$$

For the Barkhausen oscillation criterion [12], the module of the transfer function has to be higher than one for the start-up condition and then equal to one to sustain the oscillation, while the transfer function phase has to be an integer multiple of $2\pi$.

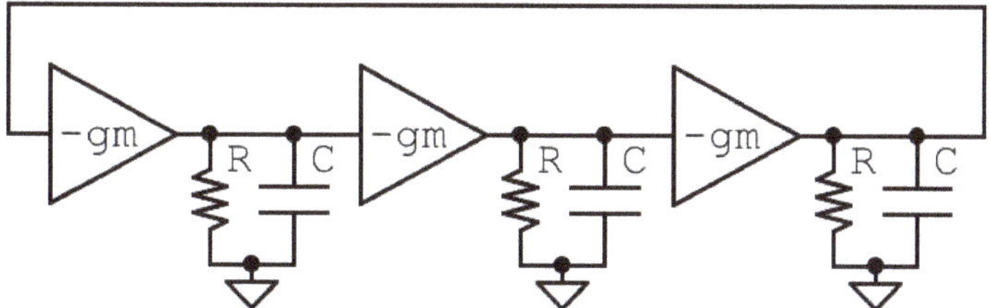

**Figure 1.** Ring oscillator modalized using inverting stage amplifiers.

Applying this criterion at the model in Figure 1, we obtain the oscillation condition in terms of design parameters, expressed as

$$g_m R \geq \frac{1}{cos\theta} \tag{2}$$

where $\theta$ is the phase shift introduced by each RC load, which for the Barkhausen oscillation criterion must be an integer multiple of $\pi/N$. In a ring oscillator, the frequency $f_0 = 1/2N\tau_D$, where $\tau_D$ is the delay of a single stage, and N is the number of stages in the loop. In order to limit power consumption and to reduce the silicon area to decrease the number of collisions caused by ionizing particles, $N = 3$ was chosen for the RO-VCO design. Although two stages ring oscillator provides a quadrature clock, as demonstrated in [13], a three stages oscillator is conventionally used for differential architecture [14]. Moreover, a smaller value of N provides a better phase noise [15] and a higher value of the working frequency $f_0$. With this choice, in accordance with Equation (2), the following condition is extracted as the main design guideline

$$g_m R \geq 2 \tag{3}$$

Although CMOS architectures are largely used for their low static-power dissipation and high integration density, the designed RO-VCO is composed of three CML stages. The current mode logic architecture, based only on n-MOSFETs and resistors, is more suitable for high-frequency applications, thanks to their lower voltage swing and lower output impedance than a standard CMOS approach [16,17]. Moreover, the use of a differential structure allows obtaining higher common-mode disturb immunity than the use of a single-ended structure, as in classic CMOS circuits [18]. Guard rings and deep n-well are also used for the design of MOSFETs devices to prevent Single Event Latch-up (SEL) and to mitigate SEEs [19,20]. The single CML stage, shown in Figure 2, is made by a source coupled pair with a resistive load, a simple current mirror, and accumulation-mode MOSFETs varactors. Active components M1 and M2 are designed with the minimum channel length allowed by technology, and the transistor width is chosen in order to ensure, in the worst case, a $g_m{}^*R$ value of 4, which is two times higher than the critical value expressed in Equation (3). The supply voltage for this technology is 1.2 V, and the value chosen for resistors shifts the output common-mode voltage level at 0.9 V. The RO-VCO bias current is controlled by the external generator $I_0$ through the simple current mirror M3 and M4 with a unity current gain. These MOSFETs are designed with the maximum MOSFET length allowed by the RF-device model to increase the output resistance. A current of 4 mA feeds the controlled oscillator, and the post-layout simulated power consumption is 18 mW. In order to take control of the oscillation frequency, a couple of varactors are added at the output of each stage [21,22].

The frequency tuning is made, thanks to accumulation-mode MOSFETs devices. A single varactor is designed by 40 fingers divided into 2 groups, and each finger is designed with the minimum finger length of 200 nm and a finger width of 550 nm. They can assume the value in the range from 69.53 fF to 34.93 fF, respectively, for the minimum and maximum value of the control voltage in the typical case. As shown in Figure 3a, the variation of the capacitance value through the corner cases is lower than 5%.

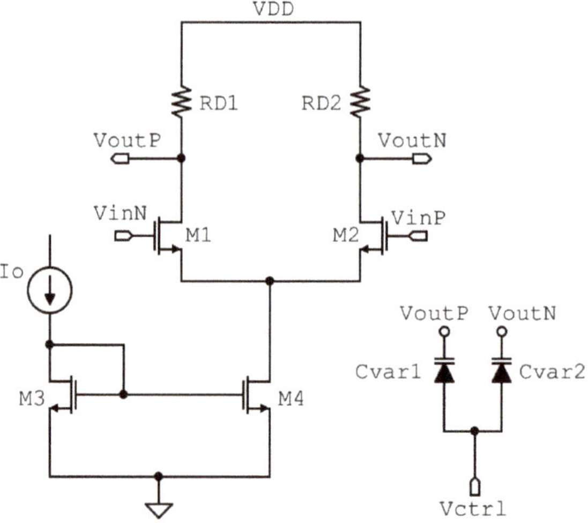

**Figure 2.** Circuit schematic of the single-stage, based on a Current Mode Logic (CML) buffer, of the ring oscillator and a couple of varactors connected at the two outputs.

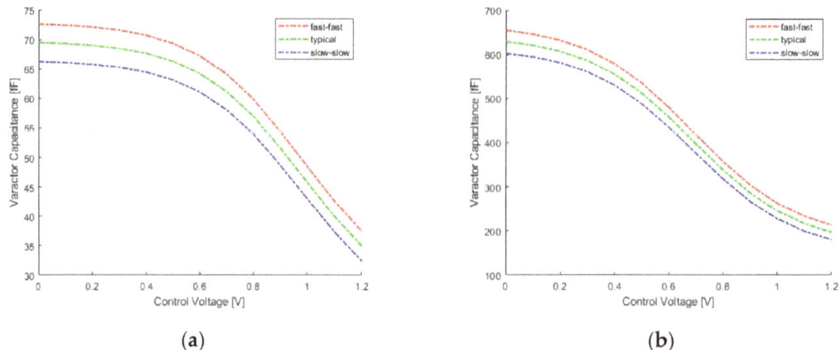

(**a**)                                        (**b**)

**Figure 3.** Varactor capacitance vs. control voltage for RO-VCO (**a**) and LC-VCO (**b**), different corner cases. RO, Ring Oscillator; VCO, Voltage Controlled Oscillator.

The oscillation frequency of the RO-VCO based on a CML architecture is closely related to the value of the gate capacitance [23], and it is expressed by the relation $f_0 = 1/2\pi RC_T$, where R is the parallel between the pull-up CML resistive load and the output MOSFET resistance, while $C_T$ is the cumulative capacitance due by varactors and the gate capacitance of the following stage.

## 2.2. Ring Oscillator Layout Design

The complete layout of the RO-VCO designed in 65 nm CMOS bulk-silicon technology is shown in Figure 4. The simple current mirror, in the bottom side, and the three source-coupled pairs are designed, adopting the common centroid technique to increase matching. All the gate terminals are turned in the same way so that the current flows in the same direction, and the space between instances is the minimum allowed by technology rules. A trade-off between metal width and length is made to prevent the electro-migration phenomena due to high current density. Moreover, alternate layers perpendicular to each other are drawn to minimize parasitic capacitances that lead to a frequency reduction. The total layout area of the proposed RO-VCO is $249 \times 86$ μm$^2$.

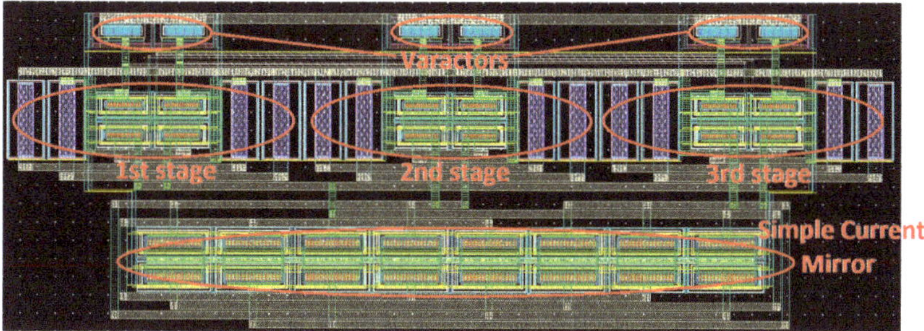



**Figure 4.** Full custom layout of the RO-VCO based on the CML buffer designed with Cadence Virtuoso [24].

## 3. LC-Tank Oscillator

### 3.1. LC-Tank Schematic Design

The second architecture designed is based on an LC-tank resonator. This architecture bases its oscillation frequency on the filtering effect of an LC-tank, leaving for active components only the role of setting the feedback gain [25] and compensate for the loss of the inductor.

The design guideline to respect Barkhausen oscillation criterion must be

$$g_m > 1/R_P \tag{4}$$

where $g_m$ is the value of the transconductance of the n-MOSFETs devices inside the cross-coupled cell, and $R_P$ is the parasitic resistance of the inductor [26]. Figure 5 shows the schematic of the LC-VCO designed to generate the target 6.25 GHz frequency. A polysilicon resistor is used to shift the output common-mode level at $V_{DD}/2$, preventing the damaging or lifetime reduction of the low-voltage MOSFETs used for the cross-coupled pair.

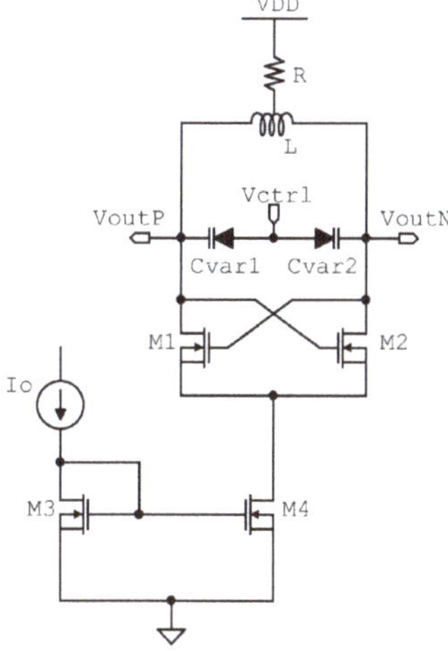



**Figure 5.** LC-tank VCO circuit schematic and a couple of varactors connected at the outputs.

This resistor is connected to the center tap of the symmetrical inductor chosen for its lower layout area than that of two separate inductors. In order to achieve the best frequency performance of this technology, the cross-coupled pair is sized using minimum length MOSFETs and a MOSFET width of 3.6 µm to guarantee a cell gain of at least 6 dB for start-up condition. The VCO bias current is controlled by the external current Io through the simple current mirror M3 and M4 with a current gain of 5, and the power consumption is less than 3 mW. The oscillation frequency of the LC-VCO is set by $f_0 = 1/\left(2\pi \sqrt{L(C + Cvar)}\right)$ [26], where C is the equivalent capacitance due to the cross-coupled cell and the first stage of the output buffer, and Cvar is the capacitance of the accumulation-mode MOSFETs varactors connected at the controlled oscillator outputs. The Tuning Range (TR) is made with the control voltage Vctrl in the range from 0 V to $V_{DD}$, and varactors assume, respectively, the value in the range from 629.6 fF to 197.6 fF, as shown in Figure 3b. A single varactor is composed of 120 fingers divided into 6 groups, and each finger is designed with 300 nm finger length and 1.2 µm finger width.

Figure 6 shows the simulated frequency response of the VCO for the two extreme values of the control voltage, and a minimum cell gain of about 10 dB for the minimum value of the control voltage, allowing to achieve a robust start-up condition for the oscillator.

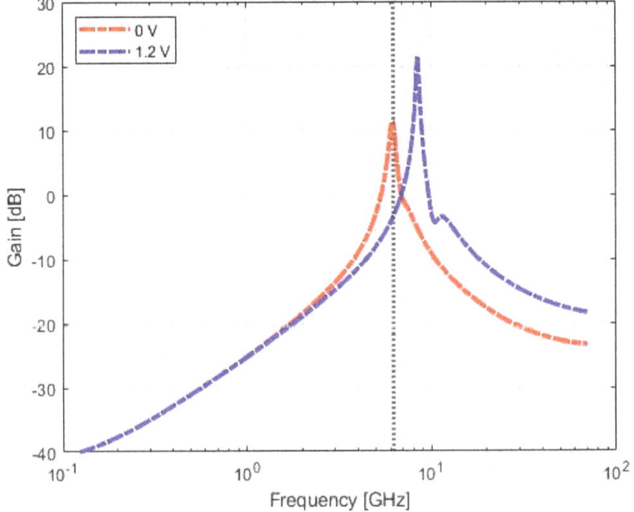


**Figure 6.** Frequency response simulated for minimum (**red line**) and maximum (**blue line**) values of the control voltage. The vertical marker indicates the target frequency of 6.25 GHz.

### 3.2. LC-Tank Layout Design

The complete layout of the LC-VCO is shown in Figure 7, and it is composed of the simple current tail mirror, varactors, cross-coupled cell, inductor, and poly-silicon resistance from bottom to the top.

The current mirror is designed as a single strip, and a common centroid technique is adopted for the cross-coupled cell. Moreover, the minimum space allowed by technology rules is used, helping to increase matching. About 85% of the total area is occupied by the differential inductor ($177 \times 198$ µm$^2$) that has a quality factor of 20. It has been chosen with an odd number of turns because the two output terminals are on the same side of the cell, thus making the routing shorter with MOSFETs devices. Moreover, the single resistor connected to the center tap helps to reduce the metal connection length between the inductor and the cross-coupled cell.

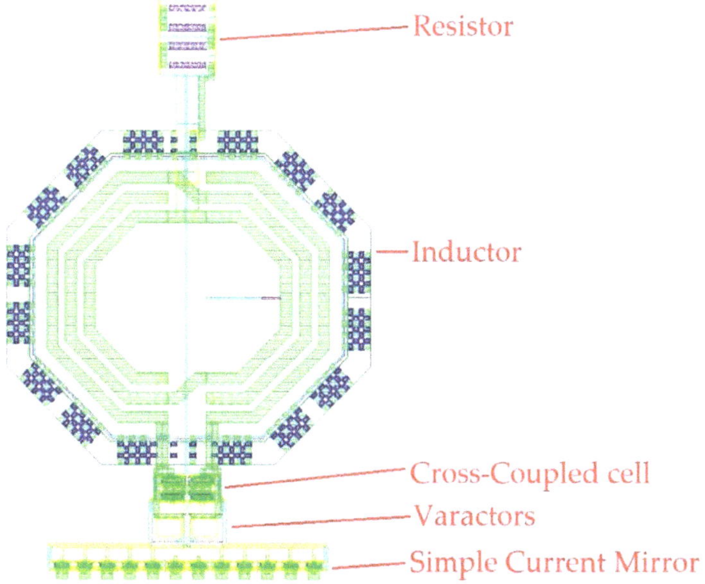


**Figure 7.** Full custom layout of the LC-tank VCO designed with Cadence Virtuoso [21].

The oscillator is designed to work properly in the temperature range −55 °C, +125 °C with 10% variations of current bias and voltage supply. The total layout area of the proposed LC-VCO is $308 \times 198 \ \mu m^2$.

## 4. Simulations Results

### 4.1. Design Simulations

The small length size n-MOSFETs allowed to achieve high-frequency performance, but on the other hand, this choice increased the deviation of the device parameters from the typical condition. Although the frequency tuning was made with the use of accumulation-mode varactors, the frequency shift due to the technology simulations was so high that it could not be compensated using the control voltage. Figure 8 shows a post-layout simulation of the free-running oscillation frequency of the RO-VCO for the only three corners process. The frequency values were plotted versus an increasing value of the control voltage from the minimum to the maximum values. The oscillation frequency in the slow-slow corner case did not reach the 6.25 GHz frequency value required by the SpaceFibre standard, even using the maximum value of the control voltage. In the fast-fast corner case, the frequency was higher than the targeted frequency, even with the minimum value of the control voltage. Although the RO-VCO resulted as strongly dependent on the device parameters, in space applications, the best components should be selected.

Although n-MOSFETs devices in the cross-coupled cell were designed with the minimum MOSFET length, the frequency shift in the LC-VCO, due to the technology simulations, could be recovered with the use of varactors and the control voltage. This can be seen by the curves in Figure 9, showing LC-tank VCO post-layout simulation of the free running-frequency versus control voltage in fast-fast (red line), typical (green line), and slow-slow (blue line) technology corner cases.

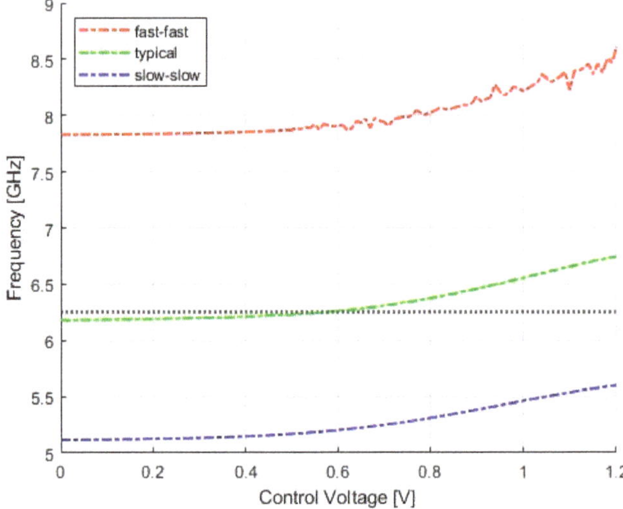

**Figure 8.** RO-VCO post-layout simulation of the free-running frequency versus control voltage in fast-fast (**red line**), typical (**green line**), and slow-slow (**blue line**) technology corner cases. The horizontal marker indicates the target frequency.

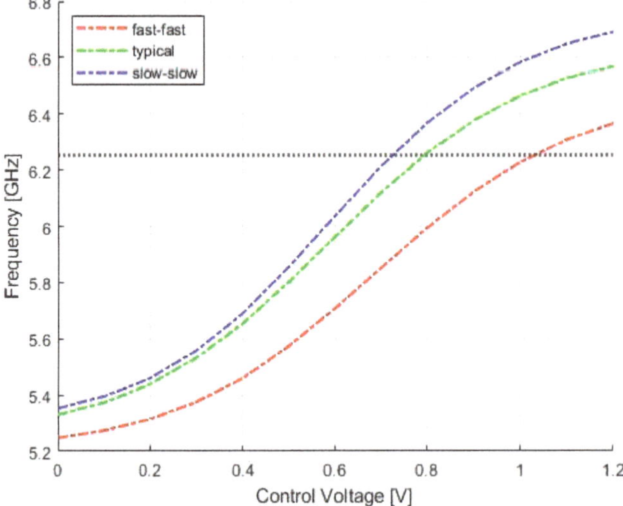

**Figure 9.** LC-tank VCO post-layout simulation of the free-running frequency versus control voltage in fast-fast (**red line**), typical (**green line**), and slow-slow (**blue line**) technology corner cases. The horizontal marker indicates the target frequency.

In addition to technology simulations, thus increasing the simulation realism, PVT (Process-Voltage-Temperature) simulations were performed by also changing temperature and supply voltage for both architectures. The SpaceFibre standard required the system to properly work under harsh conditions. In Table 1, the process and fabrication results are listed, respectively, in the third and fourth columns. The frequency variations were calculated as a variation from the nominal condition for temperature, supply voltage, and polarization current in each technology corner. The variations were obtained for temperature variations in the range −55 °C, 125 °C, and for ±10% supply voltage and polarization current deviations. Fabrication results were expressed as the frequency standard deviation σ, and data were obtained from the Monte Carlo simulations. Monte Carlo simulations were performed with 200 simulations in the nominal condition for each corner case considering process and mismatch variations.

**Table 1.** Frequency variations and standard frequency deviation, respectively, for PVT (Process-Voltage-Temperature) and Monte Carlo simulations.

| Oscillator | Technology Corners | Frequency Variations | Standard Deviation σ (Hz) |
|---|---|---|---|
| RO | slow-slow | 31.46% | 0.63 |
| | typical | 28.34% | 0.89 |
| | fast-fast | 15.72% | 3.34 |
| LC | slow-slow | 9.01% | 99.69 |
| | typical | 9.07% | 578.5 |
| | fast-fast | 7.04% | 132.2 |

Figure 10 shows the simulated phase noise of the two architectures with the harmonic balance simulation in post-layout. Both VCOs were simulated at the same frequency, and the LC-VCO exhibited a better phase noise of about 30 dB than the other architecture (at 1 MHz offset, in Figure 9, there was a phase noise of −110 dBc/Hz for the LC-VCO vs. The −82 dBc/Hz for the RO-VCO). Device noise was considered in every simulation for both oscillator architecture and for all simulations performed in this work.

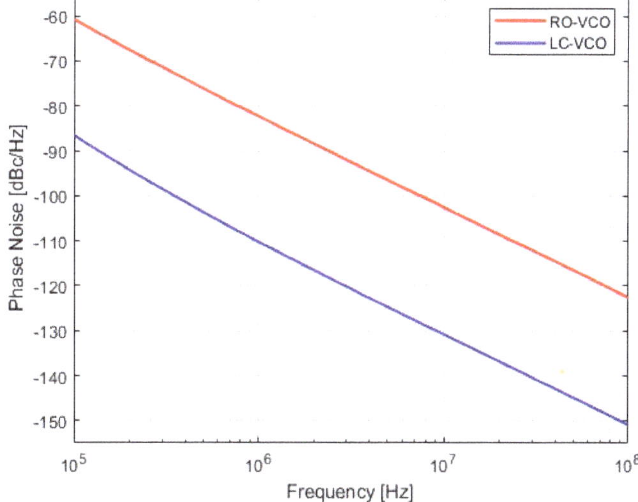


**Figure 10.** Phase noise simulated for RO-VCO (**red line**) and LC-VCO (**blue line**) at post-layout in typical conditions at the same working frequency of 6.25 GHz. These simulations were performed using the Cadence environment.

The integrated RMS jitter was calculated from Figure 10 in the bandwidth from 100 kHz to 10 MHz. The RMS jitter obtained was 9.51 ps and 0.44 ps for RO-VCO and LC-VCO, respectively. Moreover, the RO was more sensitive to temperature variations than the LC-VCO. The time-domain VCO stability was made with the use of the Allan variance [27,28], or two-sample variance, defined as

$$\sigma_x^2(\tau) = \frac{1}{2}E\left[(\overline{x_2} - \overline{x_1})^2\right] \tag{5}$$

Figure 11 shows the Allan deviation, or $\sigma$-$\tau$ plot, calculated as the square root of Equation (5) when the VCO was in the steady-state oscillation. In particular, Figure 11a,b show the comparison between the Allan deviation in frequency and in amplitude, respectively, for both architectures. LC-VCO exhibited lower variations in frequency and amplitude than the RO-VCO.

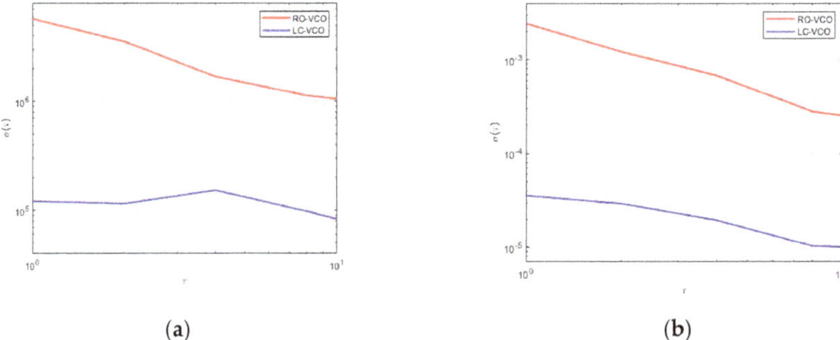

(a)       (b)

**Figure 11.** Allan deviation of (**a**) frequency and (**b**) amplitude for RO-VCO (**red line**) and LC-VCO (**blue line**) calculated in typical condition at the same oscillation frequency.

### 4.2. Single Event Effect Simulations

The model used for SEE simulations and widely accepted by the scientific community [29–31] is shown in Equation (6), where *tinj* is the injection instant, *ta* is the collection time constant of the junction, *tb* is the ion track establishment time constant, and $Q$ is the critical charge.

$$ISET = \frac{Q}{ta - tb}\left[e^{-\left(\frac{t-tinj}{ta}\right)} - e^{-\left(\frac{t-tinj}{tb}\right)}\right]$$ (6)

SEEs were modeled as double exponential current pulses at sensitivity nodes, and two different sets of values, with a Linear Energy Transfer (LET) of 5 and 60 MeV×cm$^2$/mg, were used [32]. The two sets of values were expressed for different time constants versus critical charge Q and LET. The strike of an ionizing particle could be modeled by inserting a current pulse on each P-N junction, with the direction of the injected current depending on the device type [33], as shown in Figure 12. Moreover, the effects generated by the injected currents were strongly sensitive to the circuit conditions, requiring the analysis of the system in different states.

Both VCOs based their frequency tuning on accumulation-mode MOSFETs varactors, and the output nodes resulted as the most sensitive nodes in the whole architectures. Indeed, the strike of an ionizing particle generated a voltage variation in the node that was then translated in a frequency deviation by varactors. In this subsection, SEE simulations results are discussed, respectively, for RO and LC controlled oscillators.

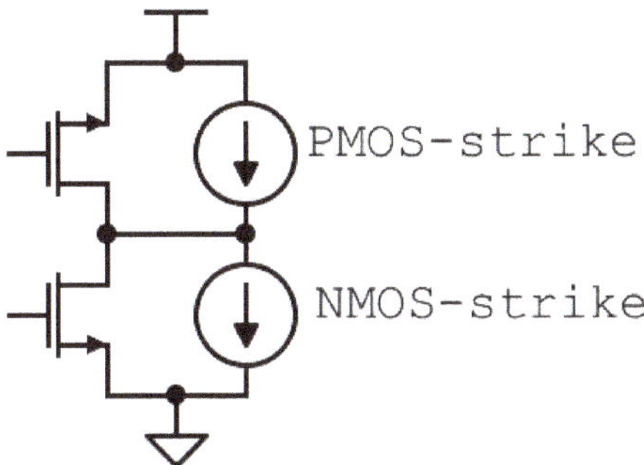


**Figure 12.** The model used for the correct stimulation of the P-N junction with the double exponential current pulses generators.

Figures 13 and 14 show the effects generated by a particle strike for the two values of LET provided for the model in Equation (5). Particles with 5 and 60 MeV×cm$^2$/mg are representant in the following figures with the label hit$_1$ and hit$_2$, respectively. The two exponential current generators excited sensitive nodes of RO-VCO at 25 ns and 30 ns, and the LC-VCO ones at 10 ns and 15 ns.

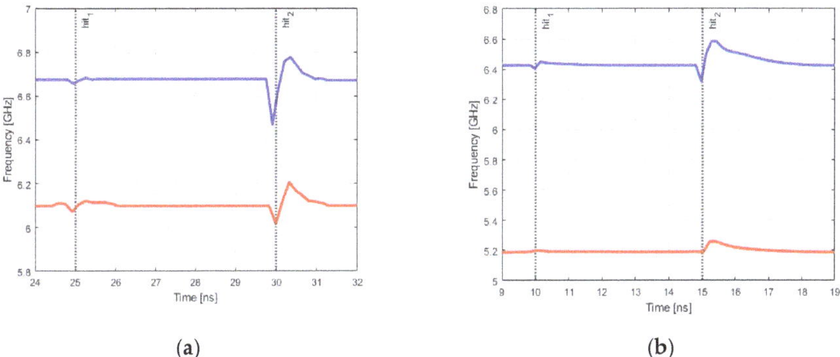

(a)  (b)

**Figure 13.** Two sets of values with a Linear Energy Transfer (LET) of 5 and 60 MeV×cm$^2$/mg were used for Single Event Effect (SEE) simulations, and the free-running frequency was plotted for the control voltage value equal to 0 V (**red line**) and $V_{DD}$ (**blue line**). (**a**) shows the free-running frequency in the RO-VCO, while (**b**) shows the free-running frequency in the LC-VCO.

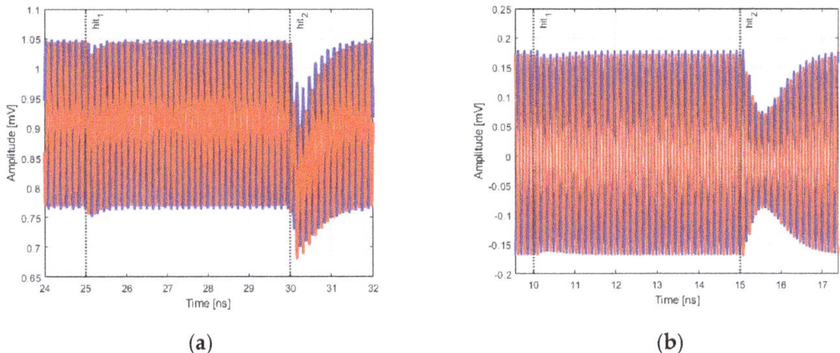

(a)  (b)

**Figure 14.** Differential amplitude variations due to the hit of the two LET values provided for the model. (**a**) shows the differential amplitude in the RO-VCO, while (**b**) shows the differential amplitude in the LC-VCO.

In Figure 13, the free-running frequency versus time is plotted for minimum (red line) and maximum (blue line) values of control voltage, and in Figure 14, the differential output amplitudes for the maximum value of the control voltage, respectively, for RO-VCO and LC-VCO are shown.

In Table 2, the data extracted from Figures 13 and 14 are listed, where the column called clock cycles shows the number of clock cycles in which the frequency assumes different values respect to the nominal due to the strike of the particle. In the last two columns, the maximum variations for frequency and amplitude are reported.

**Table 2.** Variations due to a SEE for RO-VCO and LC-VCO.

| Oscillator | Hit | Control Voltage | Clock Cycles | Frequency Variations | Amplitude Variations |
|---|---|---|---|---|---|
| RO | $hit_1$ | 0 | 6 | 0.61% | −1.65% |
|    |         | $V_{DD}$ | 3 | 0.27% | −2.93% |
|    | $hit_2$ | 0 | 9 | 1.53% | −15.38% |
|    |         | $V_{DD}$ | 9 | 3.12% | −16.27% |
| LC | $hit_1$ | 0 | 6 | 0.11% | 1.55% |
|    |         | $V_{DD}$ | 13 | 0.46% | 3.17% |
|    | $hit_2$ | 0 | 15 | 1.24% | 43.80% |
|    |         | $V_{DD}$ | 24 | 2.45% | 57.49% |

As it is shown in the last two columns of Table 2, when the VCOs were hit by the ionizing particle with a LET of 5 MeV×cm$^2$/mg (called $hit_1$ in Table 2), both architectures showed nearly the same amplitude and frequency variations. Instead, when a particle with higher LET (called $hit_2$ in Table 2) did hit the two VCOs, the amplitude variations of the LC-tank were greater than that of the RO, while the frequency variations were lower in the LC-tank-based architecture. This was despite that the LC architecture used one order greater varactor capacitances than RO one. This greater frequency deviation in RO-VCO was due to the frequency relationship with capacitance, which was 1/C for the RO-VCO and $1/\sqrt{C}$ for the LC-VCO. This attested that a small capacitance variation generated a huge frequency variation in the RO-VCO, as highlighted in Table 2.

n-MOSFETs devices in both architectures were designed with the minimum channel length targeting high-frequency applications, but a maximum number of fingers and an oversized MOSFET width were used to increase the parasitic capacitance of devices. Although this SEE mitigation technique increased the silicon area and reduced the tuning range, it increased the SEE tolerant property of both VCOs. Indeed, following the simple rule V = Q/C, if the capacitance value was increased for a fixed value of charge, then a lower variation of the voltage occurred. Moreover, guard rings and deep n-wells were adopted to isolate the devices by the charge generated in the substrate during a particle strike. Indeed, if an ionizing particle passed through the device, electron-hole pairs could be generated, which, thank to guard rings and deep n-wells, were rapidly collected, avoiding the activation of parasitic latch-up.

## 5. Comparison vs. The State-of-the-Art

A state-of-the-art comparison of voltage controlled oscillator designed in 65 nm technology is made in Table 3. In works [9,10], the PLLs were based on a RO and LC-VCO. They were irradiated up to 300 krad TID level compliant with SpaceFibre protocol and tested with different protons. Their working frequency did not meet that required by the SpaceFibre standard, and the aim of this work was to investigate the behavior of these two well-known architectures at a higher frequency. In [34], a VCO based on LC tank was optimized against SEEs, and in [35], a three stages ring oscillator was designed targeting Bluetooth front end applications, but no process simulations were performed. Another solution presented in [36] was based on a Colpitts architecture for mm-wave applications.

**Table 3.** State-of-the-art comparison.

| Ref. | Frequency (GHz) | Power Consumption (mW) | SEE Tolerant | Architecture | Area (mm$^2$) |
|---|---|---|---|---|---|
| [9] | 0.2–1.2 | n.a. | tested | Ring | n.a. |
| [10] | 2.2–3.2 | 6 | tested | Ring and LC | n.a. |
| [34] | 2.25–2.88 | 1.8 | tested | LC tank | n.a. |
| [35] | 3.7–6.5 | 2 | no | Ring | 0.011 |
| [36] | 23.8–29.1 | 2.3 | no | Colpitts | 0.221 |
| This work | 6.20–6.75 | 18 | simulated | Ring | 0.021 |
| This work | 5.35–6.55 | <3 | simulated | LC tank | 0.061 |

## 6. Conclusions

In this work, the comparison between two VCO architectures designed in a commercial 65 nm technology was made. Targeting high-frequency space applications, as the SpaceFibre protocol, a CML approach was adopted for the design of the RO-VCO. CML architecture was preferred, targeting high frequency, thanks to its lower voltage swing than a CMOS. The RO-VCO was an appetible VCO configuration in terms of technology scaling, high-integration density, and area occupancy, which was about 35% of the total silicon area required for the LC-VCO. Although the RO-VCO resulted as strongly dependent on the device parameters, in space applications, the best components should be selected. To overcome the effects of the device parameters deviation on the oscillation frequency, an LC-tank VCO architecture was designed. This architecture, despite its large area, mainly occupied by the inductor, presented promising performances in terms of the frequency range, covering the 5.35 GHz to 6.55 GHz range, in the typical case, with a control voltage swing of $V_{DD}$. SEE simulation results highlighted the output nodes as the most sensitive nodes for both VCOs, for the effects due to the varactors. Although the LC-tank VCO used one order greater varactor values than RO, and the ionizing particle hits generated higher amplitude variation on its output signals, the frequency variations of this VCO were lower than that showed by the RO architecture, thanks to the different relationship between frequency and capacitance. In the literature, VCOs based on Colpitts architecture for space applications are not available because of their large silicon area. The LC system, whose layout is shown in Figure 7, would be integrated into a 1 mm$^2$ chip containing a SERDES (Serializer-Deserializer) to test system-level performance. The whole chip would be electrically tested in standard condition, then it would be exposed to X-rays to achieve the 300 krad TID and to heavy ions for SEE characterization.

**Author Contributions:** D.M.: investigation, methodology, conceptualization, software, writing-review and editing; G.C.: investigation, methodology, conceptualization, software, writing-review and editing; S.S.: investigation, methodology, conceptualization, writing-review and editing; All authors contributed to the present paper with the same effort in finding available literature resources, conducting simulations, as well as writing the paper. All authors have read and agreed to the published version of the manuscript.

**Funding:** This work was supported by PHOS4BRAIN and ISHTAR Projects by INFN and the University of Pisa, as well as by the Dipartimento di Eccellenza Crosslab project by MIUR.

**Acknowledgments:** The authors would like to thank L. Berti, G. Mangraviti, and all the other guys from IMEC (Leuven) for useful discussions and support on the preliminary VCO design. Discussions with F. Palla and G. Magazzù are also acknowledged.

**Conflicts of Interest:** The authors declare no conflict of interest.

## References

1. Xie, L.; Wei, L. Research on Vehicle Detection in High Resolution Satellite Images. In Proceedings of the IEEE Fourth Global Congress on Intelligent Systems, Hong Kong, China, 3–4 December 2013.
2. European Space Agency for the members of ECSS. *ECSS E-ST-50-11C, Space Engineering, SpaceFibre-Very High-Speed Serial Link*; European Cooperation for Space Standardization: Noordwijk, the Netherlands, 2019.
3. Parkers, S.; Ferrer, A. *SpaceFibre Specification Draft K1*; University of Dundee: Dundee, UK, 2017.

4. Ciarpi, G.; Magazzù, G.; Palla, F.; Saponara, S. Design of Radiation-Hard MZM drivers. In Proceedings of the 20th Italian National Conference on Photonic Technologies, Lecce, Italy, 23–25 May 2018; Volume 26, pp. 1–4.

5. Menouni, M.; Barbero, M.; Bompard, F.; Bonacini, S.; Fougeron, D.; Gaglione, R.; Rozanov, A.; Valerio, P.; Wang, A. 1-Grad total dose evaluation of 65 nm CMOS technology for the HL-LHC upgrades. In Proceedings of the Topical Workshop on Electronics for Particle Physics, Aix en Provence, France, 22–26 September 2014; pp. 22–26.

6. Ciarpi, G.; Saponara, S.; Magazzù, G.; Palla, F. Radiation Hardness by Design techniques for 1 Grad TID Rad-Hard System in 65 nm standard CMOS technologies. In *Proceedings of the International Conference on Application in Electronics Pervading Industry, Environment and Society, Pisa, Italy, 11–13 September 2019*; Springer LNEE: Cham, Switzerland; Volume 627, pp. 269–276.

7. Palla, F.; Ciarpi, G.; Magazzù, G.; Saponara, S. Design of a high radiation-hard driver for Mach–Zehnder Modulators based high-speed links for hadron collider applications. *Nucl. Instrum. Methods Phys. Res. Sect. A* **2019**, *936*, 303–304. [CrossRef]

8. Ciarpi, G.; Magazzù, G.; Palla, F.; Saponara, S. Design, Implementation, and Experimental Verification of 5 Gbps, 800 Mrad TID and SEU-Tolerant Optical Modulators Drivers. *IEEE Trans. Circuits Syst. I Regul. Pap.* **2020**, *67*, 829–838. [CrossRef]

9. Malou, F.; Gasiot, G.; Chevallier, R.; Dugoujon, L.; Roche, P. TID and SEE Characterization of Rad-Hardened 1.2GHz PLL IP from New ST CMOS 65nm Space Technology. In Proceedings of the 2014 IEEE Radiation Effects Data Workshop (REDW), Paris, France, 14–18 July 2014; pp. 1–8.

10. Prinzie, J.; Christiansen, J.; Moreira, P.; Steyaert, M.; Leroux, P. Comparison of a 65 nm CMOS Ring- and LC-Oscillator Based PLL in Terms of TID and SEU Sensitivity. *IEEE Trans. Nucl. Sci.* **2017**, *64*, 245–252. [CrossRef]

11. Monda, D.; Ciarpi, G.; Mangraviti, G.; Berti, L.; Saponara, S. Analysis and Comparison of Ring and LC-tank Oscillators for 65 nm Integration of Rad-Hard VCO for SpaceFibre Applications. In *Applications in Electronics Pervading Industry, Environment and Society*; Springer LNEE: Berlin/Heidelberg, Germany, 2020; Volume 627, pp. 25–32.

12. Voinigescu, S. *High-Frequency Integrated Circuits*; Cambridge University Press: Cambridge, UK, 2013.

13. Bae, W.; Ju, H.; Park, K.; Cho, S.; Jeong, D. A 7.6 mW, 214-fs RMS jitter 10-GHz phase-locked loop for 40-Gb/s serial link transmitter based on two-stage ring oscillator in 65-nm CMOS. In Proceedings of the 2015 IEEE Asian Solid-State Circuits Conference (A-SSCC), Xiamen, China, 9–11 November 2015; pp. 1–4.

14. Razavi, B. *Design of Integrated Circuits for Optical Communications*; McGraw-Hill: New York, NY, USA, 2002.

15. Hajimiri, A.; Limotyrakis, S.; Lee, T.H. Jitter and phase noise in ring oscillators. *IEEE J. Solid-State Circuits* **1999**, *34*, 790–804. [CrossRef]

16. Heydari, P. Design and Analysis of Low-Voltage Current-Mode Logic Buffers. In Proceedings of the Fourth International Symposium on Quality Electronic Design, San Jose, CA, USA, 24–26 March 2003.

17. Tlelo-Cautle, E.; Castañeda-Aviña, P.; Trejo-Guerra, R.; Carbajal-Gómez, V. Design of a Wide-Band Voltage-Controlled Ring Oscillator Implemented in 180 nm CMOS Technology. *Electronics* **2019**, *8*, 1156. [CrossRef]

18. Gupta, K.; Pandey, N.; Gupta, M. *Model and Design of Improved Current Mode Logic Gates: Differential and Single-Ended*; Springer Nature: Berlin/Heidelberg, Germany, 2019.

19. Snoeys, W.; Faccio, F.; Burns, M.; Campbell, M.; Cantatore, E.; Carrer, N.; Casagrande, L.; Cavagnoli, A.; Dachs, C.; Di Liberto, S.; et al. Layout techniques to enhance the radiation tolerance of standard CMOS technologies demonstrated on a pixel detector readout chip. *Nucl. Instrum. Methods Phys. Res. Sect. A Accel. Spectrum. Detect. Assoc. Equip.* **2000**, *439*, 349–360. [CrossRef]

20. Camplani, A.; Shojaii, S.; Shrimali, H.; Stabile, A.; Liberali, V. CMOS IC radiation hardeningby design. *Facta Univ. Ser. Electron. Energ.* **2014**, *27*, 251–258. [CrossRef]

21. Zhang, Z.; Chen, L.; Djahanshahi, H. A SEE Insensitive CML Voltage Controlled Oscillator in 65nm CMOS. In Proceedings of the 2018 IEEE Canadian Conference on Electrical & Computer Engineering, Quebec City, QC, Canada, 13–16 May 2018; pp. 1–4.

22. Andreani, P.; Mattisson, S. On the use of MOS varactors in RF VCOs. *IEEE J. Solid-State Circuits* **2000**, *35*, 905–910. [CrossRef]

23. Farahabadi, P.M.; Miar-Naimi, H.; Ebrahimzadeh, A. Closed-Form Analytical Equations for Amplitude and Frequency of High-Frequency CMOS Ring Oscillators. *IEEE Trans. Circuits Syst. I Regul. Pap.* **2009**, *56*, 2669–2677. [CrossRef]
24. Cadence® Design Systems, Virtuoso®. Available online: https://www.cadence.com/en_US/home/tools/custom-ic-analog-rf-design/circuit-design/virtuoso-analog-design-environment.html (accessed on 28 June 2020).
25. Razavi, B. A Study of Phase Noise in CMOS Oscillators. *IEEE J. Solid-State Circuits* **1996**, *31*, 331–343. [CrossRef]
26. Razavi, B. *RF Microelectronics*; Prentice Hall: Upper Saddle River, NJ, USA, 1998; Volume 1.
27. Land, D.V.; Levick, A.P.; Hand, J.W. The use of the Allan deviation for the measurement of the noise and drift performance of microwave radiometers. *Meas. Sci. Technol.* **2007**, *18*, 1917–1928. [CrossRef]
28. Galton, I.; Weltin-Wu, C. Understanding Phase Error and Jitter: Definitions, Implications, Simulations, and Measurement. *IEEE Trans. Circuits Syst. I Regul. Pap.* **2019**, *66*, 1–19. [CrossRef]
29. Rathore, P.; Nakhate, S. Development of Radiation Hardened by Design (RHBD) of NAND gate to mitigate the effects of single event transients (SET). In Proceedings of the IEEE 1st International Conference on Power Electronics, Intelligent Control and Energy Systems (ICPEICES), Delhi, India, 4–6 July 2016; pp. 1–6.
30. Bailey, S.; Keller, B.; Der-Khachadourian, G. Project UPSET: Understanding and Protecting Against Single Event Transients. Available online: https://people.eecs.berkeley.edu/~{}stevo.bailey/documents/upset_final.pdf (accessed on 28 June 2020).
31. Wirth, G.I.; Vieira, M.G.; Neto, H.E.; Kastensmidth, F.L. Modeling the sensitivity of CMOS circuits to radiation induced single event transients. *Microelectron. Reliab.* **2008**, *48*, 29–36. [CrossRef]
32. Díez-Acereda, V.; Khemchandani, S.L.; Pino, J.D.; Mateos-Angulo, S. RHBD Techniques to Mitigate SEU and SET in CMOS Frequency Synthesizers. *Electronics* **2019**, *8*, 690. [CrossRef]
33. Prinzie, J.; Smedt, V.D. Single Event Transients in CMOS Ring Oscillators. *Electronics* **2019**, *8*, 618. [CrossRef]
34. Prinzie, J.; Christiansen, J.; Moreira, P.; Steyaert, M.; Leroux, P. A 2.56-GHz SEU Radiation Hard LCLC-Tank VCO for High-Speed Communication Links in 65-nm CMOS Technology. *IEEE Trans. Nucl. Sci.* **2018**, *65*, 407–412. [CrossRef]
35. Yen, C.; Nasrollahpour, M.; Hamedi-Hagh, S. Low-power and high-frequency ring oscillator design in 65 nm CMOS technology. In Proceedings of the IEEE 12th International Conference on ASIC, Guiyang, China, 25–28 October 2017; pp. 533–536.
36. Kashani, M.H.; Molavi, R.; Mirabbasi, S. A 2.3-mW 26.3-GHz $G_m$-Boosted Differential Colpitts VCO with 20% Tuning Range in 65-nm CMOS. *IEEE Trans. Microw. Theory Tech.* **2019**, *67*, 1556–1565. [CrossRef]

*Article*

# Analysis and Design of Integrated Blocks for a 6.25 GHz Spacefibre PLL

**Marco Mestice *, Bruno Neri, Gabriele Ciarpi and Sergio Saponara**

Department of Information Engineering (DII), University of Pisa; via G. Caruso 16, 56122 Pisa, Italy;
bruno.neri@iet.unipi.it (B.N.); gabriele.ciarpi@ing.unipi.it (G.C.); sergio.saponara@iet.unipi.it (S.S.)
* Correspondence: m.mestice@studenti.unipi.it

Received: 24 June 2020; Accepted: 17 July 2020; Published: 19 July 2020

**Abstract:** The design of a Phase-Locked Loop (PLL) to generate the clock reference for the new Spacefibre standard is presented in this paper. Spacefibre has been recently released by the European Space Agency (ESA) and supports up to 6.25 Gbps for on-board satellite communications. Taking as a starting point a rad-hard 6.25 GHz Voltage Controlled Oscillator in 65 nm technology, this work presents the design of the key blocks for an integrated PLL: a Triple Modular Redundancy Phase/Frequency Detector, a Charge Pump, and a passive Loop Filter. The modeling activities carried out in an Advanced Design System have proven that the proposed PLL can be completely integrated on-chip, with a Loop Filter area consumption of only 6000 $\mu m^2$ (considering the 65 nm technology). The design of active circuits has been carried out at the transistor level in a Cadence Virtuoso environment, implementing both system and layout rad-hard techniques, and different solutions are discussed in this paper. As a result, a compact (0.09 $mm^2$), low power (10.24 mW), dead zone free and rad-hard PLL is obtained with a Phase Noise below −80 dBc/Hz @ 1 MHz. A preliminary block view and floor plan of the test chip is also proposed.

**Keywords:** rad-hard; PLL (phase-locked loop); SEE (single event effects); Spacefibre; TID (total ionization dose); charge pump; phase/frequency detector; frequency divider

---

## 1. Introduction

Recently, the new Spacefibre standard [1] has been released by the European Space Agency (ESA). To cope with the data transfer of high bandwidth sensors, required in scientific, surveillance, and telecom satellite applications, Spacefibre supports up to 6.25 Gbps. The clock reference generator is a key block for the Spacefibre implementation. It must be hardened against SEE (Single Event Effects) [2] and TID (Total Ionization Dose) [3], it should work in a −55 to 125 °C temperature range, and it should sustain up to 6.25 GHz. Moreover, output frequency divided by 2 and by 4 should be supported. In aerospace applications, the radiation issues are mainly related to SEE because the TID levels are from several dozen to a few hundred krad, even lower if aluminum shields are used. As discussed in [4] with a 5 mm aluminum shielding, the estimated total TID received by three satellites for earth observation and environmental data collection, RazakSAT-1, SCD-2, and ALOS are 2.30 rads, 170 rads, and 24200 rads, respectively. Therefore, rad-hard techniques should be employed to mitigate SEEs, whose effect in PLLs (Phase-Locked Loops) has been widely analyzed and demonstrated [5,6]. Although several solutions have been proposed in the literature [7–9], the state-of-the-art rad-hard PLLs are limited to frequencies below 6 GHz in normal conditions(e.g. less than 3 GHz is achieved in [10–12]) but, in order to reach 6.25 GHz in all PVT (Process–Voltage–Temperature) corners and in environments pervaded by radiations, it is necessary to work at even higher frequencies in nominal conditions.

An issue that needs to be dealt with is that radiation hardness comes at the cost of power and area consumption, since redundancy techniques are usually implemented. Currently, the Triple Modular

Redundancy (TMR) technique is one of the most effective techniques for digital blocks, such as the PFD (Phase Frequency Detector) [13]. It consists of triplicating a cell and adding a voter to choose between the outputs following a majority approach. For the Charge Pump (CP), instead, voltage-switching Charge Pumps (V-CP) have been demonstrated to be a good choice in terms of SEE mitigation, but they lead to a great degradation in terms of noise performance compared to CPs [14].

Among the state-of-the-art designs, the 4.8–6 GHz PLL proposed in [15] represents a good tradeoff in terms of power consumption, area, and noise performance, while really low-noise performances are achieved in [13] with a 2.2–3.2 GHz PLL.

In this paper, we present a low-power, low-area consumption and reliable 6.25 GHz PLL for aerospace environments. Here, a complete project flow in 65 nm TSMC technology is developed starting from an already designed Voltage Controlled Oscillator (VCO) [16], from an already designed Frequency Divider (FD), and from a preliminary system level analysis performed by us in the conference work [17], of which this work is an extension. With respect to [17], in which the modeling activities and the preliminary system level analysis and design were presented, in this work, beside this, the complete development at the circuit and layout level of the blocks that compose the PLL in 65 nm TSMC technology is illustrated, together with the implemented rad-hard techniques, and with comparisons among different solutions.

The Loop Filter's components and the Charge Pump's current have been chosen through a system-level analysis carried out in an ADS (Advanced Design System) environment, while the design of the Charge Pump and the Phase/Frequency Detector has been carried out in a *Cadence Virtuoso* environment. A preliminary block view and floor plan of the test chip is also proposed.

Hereafter, Section 2 presents the PLL system-level analysis and the design of the passive Loop Filter circuit. Section 3 deals with the transistor-level design of the main circuit blocks. Section 4 shows the results of the SEE simulation and post-layout simulations of the whole PLL. In Section 5, a starting point for a test chip is proposed, while the conclusions are drawn in Section 6.

## 2. System-Level Design of PLL and Passive Filter Sizing

The target architecture of this work is a CP-PLL [18]. As shown in Figure 1, the blocks that compose the proposed PLL are a PFD, a CP circuit, a passive Loop Filter, a VCO, whose output corresponds to one of the PLL outputs, and an FD with an integer divide ratio, which generates the other three frequencies required by the application. The input signals of the PFD are the reference signal (REF in Figure 1) and the FD output, while the outputs related to the input phase/frequency difference are named UP and DOWN in the figure. These signals are converted by the CP in a charging or discharging current needed to control the VCO through the Loop Filter. The PLL proposed in this work is designed starting from an LC-tank VCO macrocell, which is integrated by the University of Pisa in 65 nm CMOS technology [16] and has the following characteristics: a frequency tuning range between 5.9 and 7.5 GHz, obtained through integrated varactors; the control voltage leads to a gain in the center of the frequency range of 2 GHz/V; the Phase Noise is below $-100$ dBc/Hz at 1 MHz from the carrier.

Starting from the system-level analysis and design, the Loop Filter's components, the Charge Pump's current ($I_{cp}$), and the Frequency Divider ratio (N) have been chosen. Thanks to a divider ratio of 40 and with 156.25 MHz as the reference frequency, the output at 6.25 GHz is obtained. According to the Spacefibre standard, the PLL has also to generate the frequencies 3.125 GHz and 1.5625 GHz, which are obtained using an integer FD with the following ratios: one-half, one-fourth, and 1/40th.

Moving onto the Loop Filter, as shown in Figure 2, it is composed of a capacitor C1, a resistor R1, which is needed to stabilize the loop, and a capacitor C2, whose aim is to reduce the spurious tones at multiples of the reference frequency and whose value, according to [19], should be no more than C1/5. Thanks to the modeling and simulation activity in the ADS environment in Section 2.1, a bandwidth of 6 MHz has been targeted as a tradeoff between the noise behavior and Loop Filter integrability. A completely integrated filter allows the avoidance of all the problems deriving from the parasitic effects of external components. A higher loop bandwidth, obtained with small passive devices, provides a better reduction of the VCO and Loop Filter contribution to Phase Noise. Instead, a lower bandwidth leads to lower CP and reference noise contributions, but it can be achieved only with large capacitors. A lower CP noise contribution can also be obtained thanks to a higher $I_{cp}$, at the cost of larger capacitors for a given bandwidth. Given all these reasons, an $I_{cp}$ of 40 µA has been chosen, and consequently, the values of C1, C2, and R1 have been derived: 8 pF, 1 pF, and 12 kΩ, respectively.

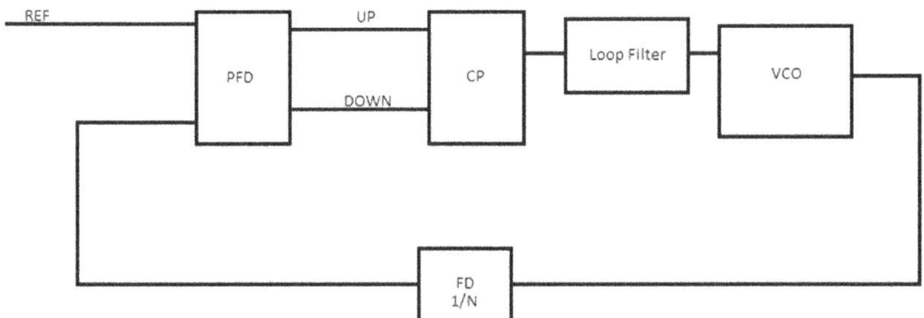

**Figure 1.** Block schematic of the Charge Pump Phase-Locked Loop (CP-PLL).

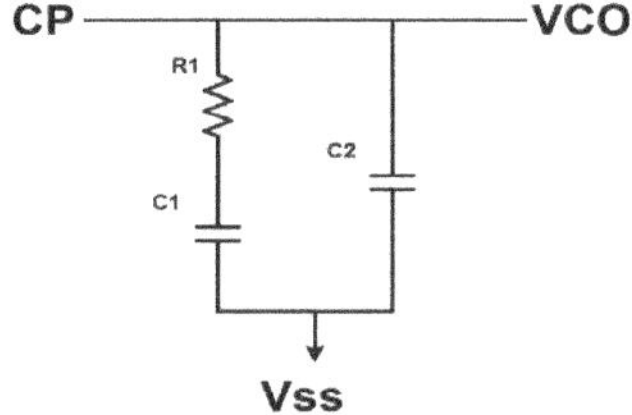

**Figure 2.** Loop Filter's schematic view.

*2.1. PLL Modeling in "Advanced Design System" Environment*

A phase domain model has been firstly built to analyze the PLL bandwidth and stability. There, all the blocks are linearized: the PFD/CP and the FD have constant gains of $I_{cp}/2\pi$ and $1/N$, respectively, while the VCO is modeled as an integrator with gain $K_{vco}$. The transfer functions for the open, Equation (1), and closed, Equation (2), loop models are the following:

$$H_{ol} = \frac{I_{cp}}{2\pi} Z(s) \frac{K_{vco}}{s} \frac{1}{N}, \tag{1}$$

$$H_{cl} = \frac{\frac{I_{cp}}{2\pi} Z(s) \frac{K_{vco}}{s}}{1 + \frac{I_{cp}}{2\pi} Z(s) \frac{K_{vco}}{s} \frac{1}{N}}, \tag{2}$$

with an $I_{cp}$ of 40 µA, $N$ of 40, and:

$$K_{vco} = 12.57 * 10^9 \ rad/(V*s), \tag{3}$$

$$Z(s) = \frac{1}{s(C1+C2)} \frac{1+sR1C1}{1+sR1\frac{C1C2}{C1+C2}}. \tag{4}$$

Figure 3 shows the results obtained using an AC simulation. As expected, R1, introducing a zero, increases the stability of the loop. In contrast, C2, introducing a pole, reduces the Phase Margin; therefore, its value has been selected to maximize the latter. The main results of the simulations are a unity gain frequency of 3.31 MHz with 50.9° of Phase Margin, and a 5.37 MHz closed loop bandwidth.

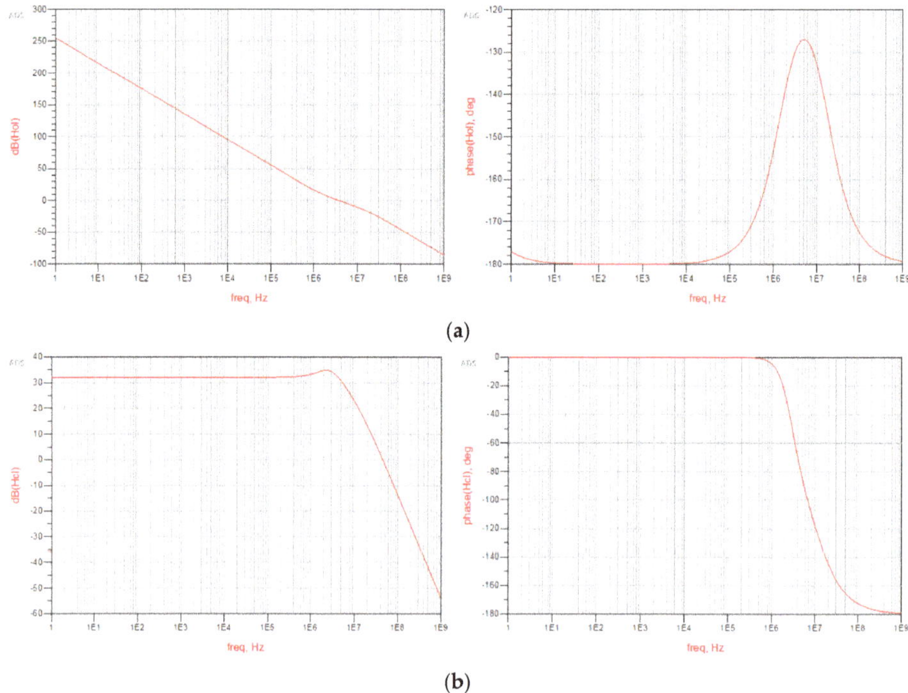

**Figure 3.** Results of the AC analysis performed on the phase domain models: magnitude on the left and phase on the right of (**a**) the open loop model and (**b**) the closed loop model.

Secondly, to evaluate the noise and lock performances, a behavioral model of the closed loop PLL, as shown in Figure 4 together with the model of the VCO alone, has been built to perform simulations in time and frequency domains. Since all the blocks of the model are noise-free, the block labeled NoiseVCO is added to consider the VCO contribution to phase noise. It is a voltage noise source that approximates with a piecewise linear curve the simulated phase noise of the VCO designed in [16], and starting from it, it generates the equivalent input noise. Regarding the Loop Filter, the noise model provided by ADS has been used for the analysis. Figure 5a shows the PLL lock behavior, which presents a lock time of 555.6 ns for a locking error below 0.01%. The peaks in Figure 5a, due to the resistor of the Loop Filter, are not seen when the PLL is locked because the CP model is ideal. However, since the real CP will be affected by non-idealities, the second capacitor is needed to attenuate these peaks. Instead, Figure 5b shows that the phase noise is well below −80 dBc/Hz. However, the reference and CP contributions have not yet been considered in this simulation, since this represents a preliminary analysis (the PFD and CP contributions to Phase Noise are analyzed in Section 3.1.3). As expected from the theory [19], the loop has a band-pass response with respect to the Loop Filter noise and a high-pass response with respect to the VCO noise.

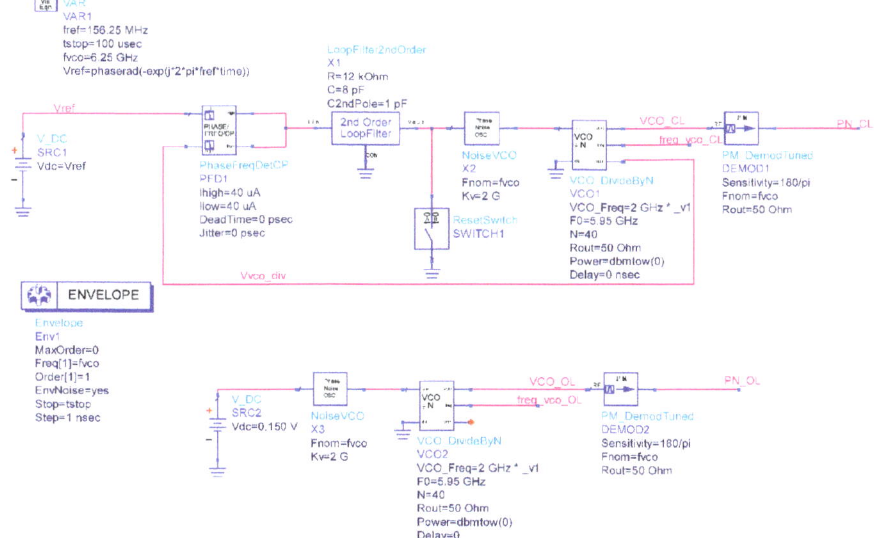

**Figure 4.** Closed-loop PLL (TOP) and Voltage Controlled Oscillator (VCO, BOTTOM) models in the time and frequency domains.

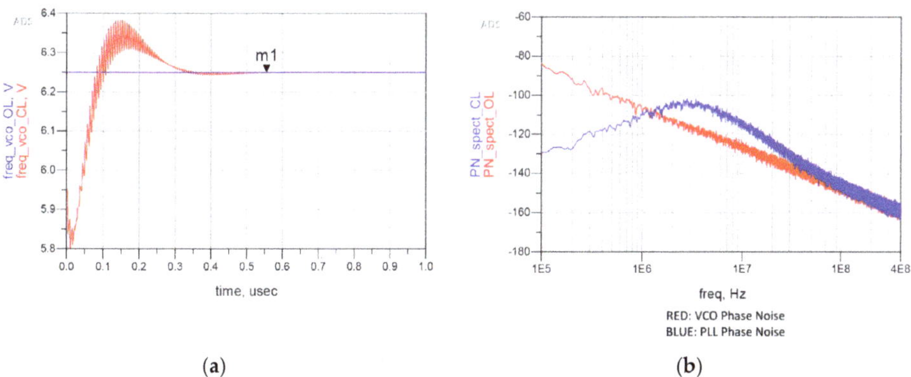



(**a**)            (**b**)

**Figure 5.** Results of the envelope analysis performed on the model of Figure 4: (**a**) locking process of the PLL; (**b**) comparison between the phase noise of the VCO and the phase noise of the closed loop PLL, considering the noise contribution of the VCO and the Loop Filter.

## 2.2. Second-Order Loop Filter's Layout

Choosing to realize C1 and C2 as Metal–Insulator–Metal (MIM) capacitors and R1 as an N-well under OD resistor, in order to minimize the area consumption, have led to the Loop Filter's layout shown in Figure 6, in which C1 can be recognized on the left, C2 can be recognized on the right, and R1 can be recognized in the lower right corner, with an area consumption of only 6000 $\mu m^2$.

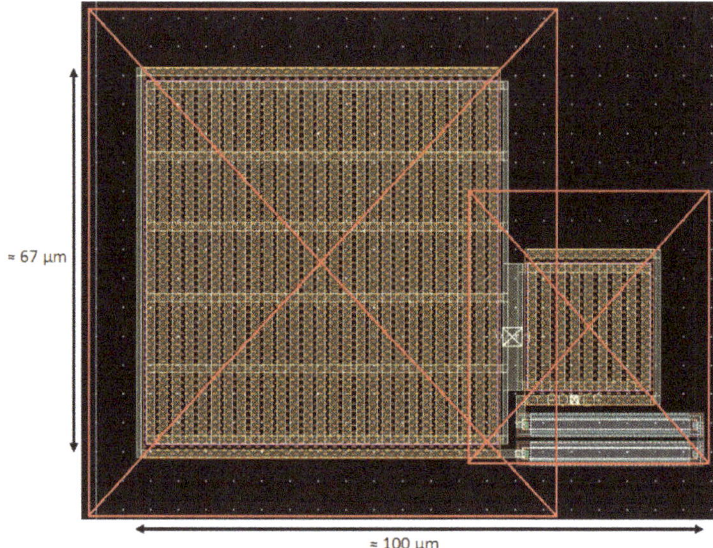

**Figure 6.** Layout in 65 nm technology of the Loop Filter in Figure 2.

## 3. Transistor Level Design of the PLL

### 3.1. Charge Pump and Phase/Frequency Detector

### 3.1.1. Charge Pump

Ideally, a Charge Pump is a current sink or source, depending on its inputs. To obtain this behavior, there are three main topologies of CP in the literature:

- Drain Switching,
- Source Switching,
- Gate Switching.

Therefore, firstly, three basic Charge Pumps [20] have been designed and compared, one for each topology: the conventional CP's Drain Switching architecture is shown in Figure 7a. It consists of two mirrors, one type n (M0 master) with two slaves (M1 and M2) and one type p (M4 master, M5 slave). The switch M3 is realized with an NMOS transistor, while the switch M6 is realized with a PMOS transistor. M0, M1, M2, M4, and M5 have no minimum length to obtain a higher output impedance and the necessary width to have the drain current of M3 (sink current) and M6 (source current) of 40 µA and the output voltage of 520 mV when the output is left open. 520 mV is the nominal control voltage needed to have 6.25 GHz as the output frequency of the VCO in [16]. Therefore, the CP has been sized to have near-zero current mismatch and an $I_{cp}$ of 40 µA when the control voltage of the VCO (Vctrl) is 520 mV. Instead, the switch transistors are wide and short to minimize their $V_{DS}$ (drain-source voltage drop).

Instead, the Source Switching-based architecture is shown in Figure 7b. Its operation is very similar to the previous CP, and therefore, it is sized with the same criteria: enhance the output impedance and reduce the current mismatch near 520 mV of Vctrl.

Finally, the Gate switching architecture is shown in Figure 7c. In contrast with the other two architectures, in this one, it is necessary to add another type p (M6, M7) and type n (M3, M4) current mirror, since causing the main type n mirror (M0, M1, M2) in the OFF state to switch off the sink current would lead to the shutdown of the source current, too. Given that, the circuit in Figure 7c has been sized with the same criteria of the other two architectures.

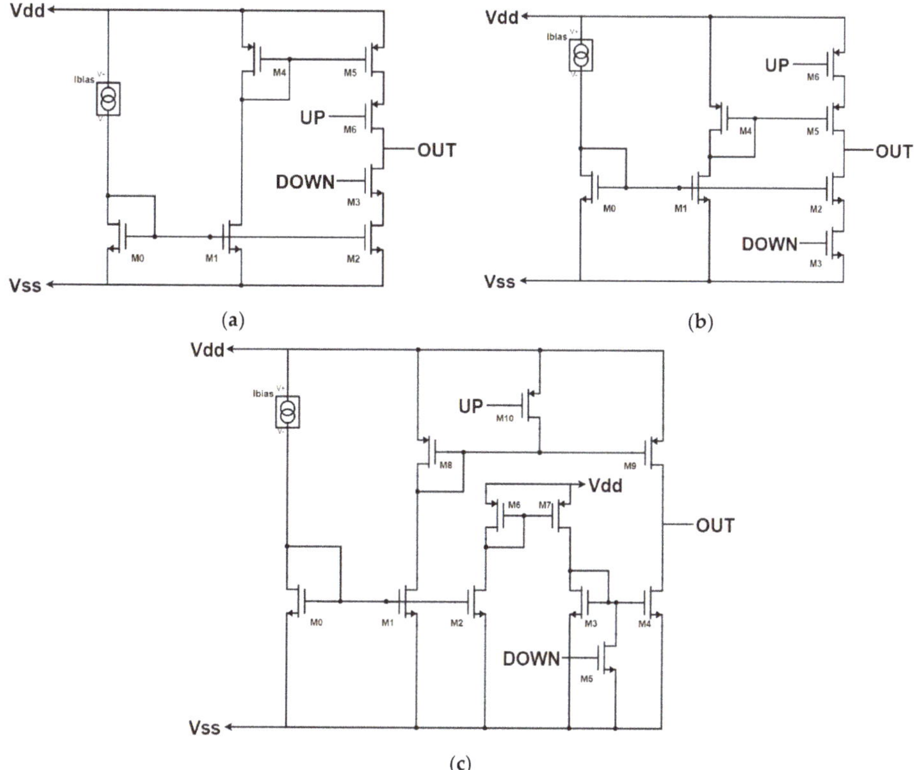


**Figure 7.** Charge Pump's architectures: (**a**) Drain Switching architecture, (**b**) Source Switching architecture, (**c**) Gate Switching architecture.

To analyze the output impedance of the three CP circuits in Figure 7, imagine placing an ideal voltage source on the output node of each CP circuit. Then, we measure the output current and its derivative when UP is low and DOWN is high for the Source and Drain Switching, and when UP is high, and DOWN is low for the Gate Switching. The three CP circuits, because of their similarity, show almost the same DC characteristics: an output range between 0.15 and 0.9 V, but quite a low output impedance of about 30 kΩ. Instead, regarding the transient behavior shown in Figure 8, the three CP architectures show different characteristics. In this case, the simulations were performed forcing the output node at 520 mV with an ideal voltage source, and the source and sink currents were measured. Since the ON time of the PFD's outputs during the reset period depends on several factors, such as the PVT conditions and layout, which were unknown at this design level, the simulations were performed considering two different values of ON time: 3.2 ns to let the two currents exhaust the transient and therefore analyze all their transient behavior; and 500 ps to approximate a more realistic reset time of the PFD and then analyze the current's behavior in an approximation of the locked state of the PLL. The Drain Switching CP in Figure 7a is the fastest one, while the Source and Gate Switching CP ones are too slow to turn ON in less than 500 ps. The Source Switching CP in Figure 7b is the slowest one, as can be seen in Figure 8b. Both the Source and Drain Switching CP architectures show peaks during the switching period, and this leads to a degradation of the transient matching. Instead, the Gate Switching CP architecture does not show any peak current during the switching period, and therefore, although the Drain Switching is the fastest architecture and all of them can be enhanced, increasing the complexity, it has been chosen as the target topology for the whole PLL design. However, it is important to notice that this architecture is intrinsically more complex because there is the need to separate the bias circuits for the UP (M8, M9, M10) and DOWN (M3, M4, M5) branches.

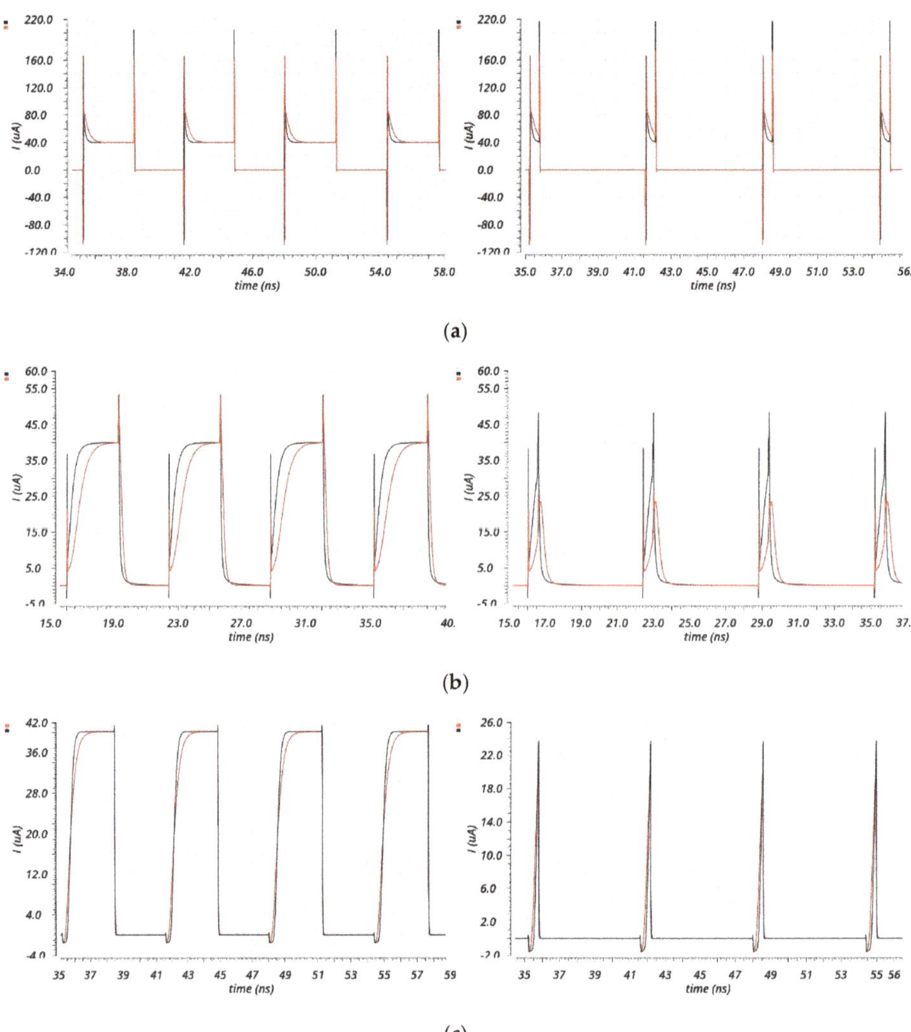

**Figure 8.** Transient behavior of the three CPs of Figure 7. The source current is represented in red, while the sink current is represented in black for two different values of ON time of the input signals (UP and DOWN): 3.2 ns on the left, which corresponds to a duty cycle of 50%, and 500 ps on the right, which corresponds to a duty cycle of 7.8125%. The Drain Switching CP results are represented in (**a**); the Source Switching CP results are represented in (**b**); the Gate Switching CP results are represented in (**c**).

Once the CP topology has been chosen, the target architecture has been modified to obtain better performances, as shown in Figure 9 versus Figure 7c. Firstly, the output simple current mirrors have been replaced with high swing current mirrors to enhance the output impedance without degrading the output range too much. This change also allowed the reduction of the sizes of the output transistors, leading to a faster CP. Moreover, both the UP and DOWN branches have their bias circuitry in order to reduce the effect of the UP signal on the sink current and of the DOWN signal on the source current. This has led to an output impedance of 250 kΩ, with an output range of 0.3–0.9 V. Instead, in terms of DC current mismatch, the worst case of process–temperature corners is 1.454 μA. Regarding the transient characteristics shown in Figure 10, the designed architecture is peak-free, and it is able to start sinking or sourcing the current in less than 500 ps.

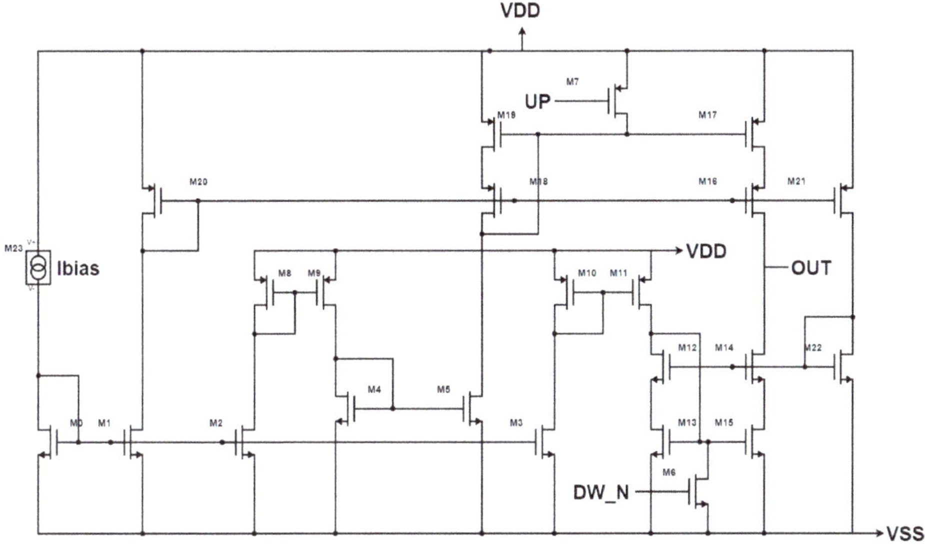


**Figure 9.** Enhanced CP's Gate Switching architecture with CMOS standard inputs.

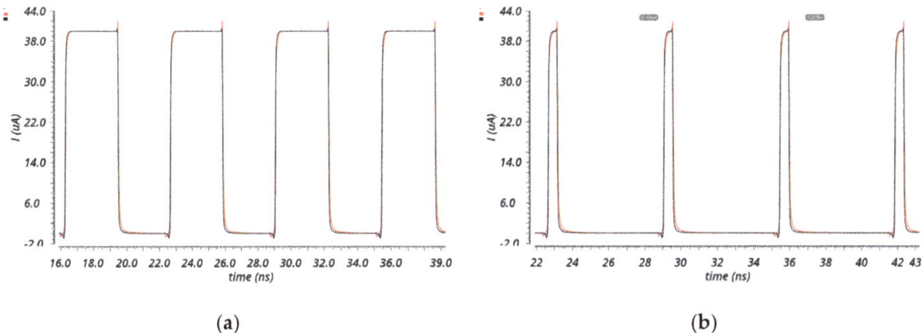


(**a**)  (**b**)

**Figure 10.** Transient behavior of the CP of Figure 9. The source current is in red, and the sink current is in black for two different values of ON time of the input signals (UP and DW_N): (**a**) 3.2 ns, (**b**) 500 ps.

A differential input Gate Switching CP has also been developed. Indeed, a differential PFD/CP guarantees a better current matching during the switching period, thanks to the symmetric input load of the CP and to the lack of necessity of adding an inverter between one of the PFD outputs and the corresponding CP's input. Hence, the switch transistors have been replaced with differential pairs

(M6–M7 and M8–M9), and the mirror M22–M23 has been added to enhance the switching speed of the source current during the ON-to-OFF switch [21], as shown in Figure 11. The DC characteristics are similar to the previous architecture: an output range of about 0.3–0.9 V, an output impedance of about 250 kΩ, and a DC current mismatch in the worst case of all the temperature–process variations of 1.66 μA. Instead, regarding the transient behavior, it is peak-free and shows a better transient current matching, as can be seen from Figure 12.

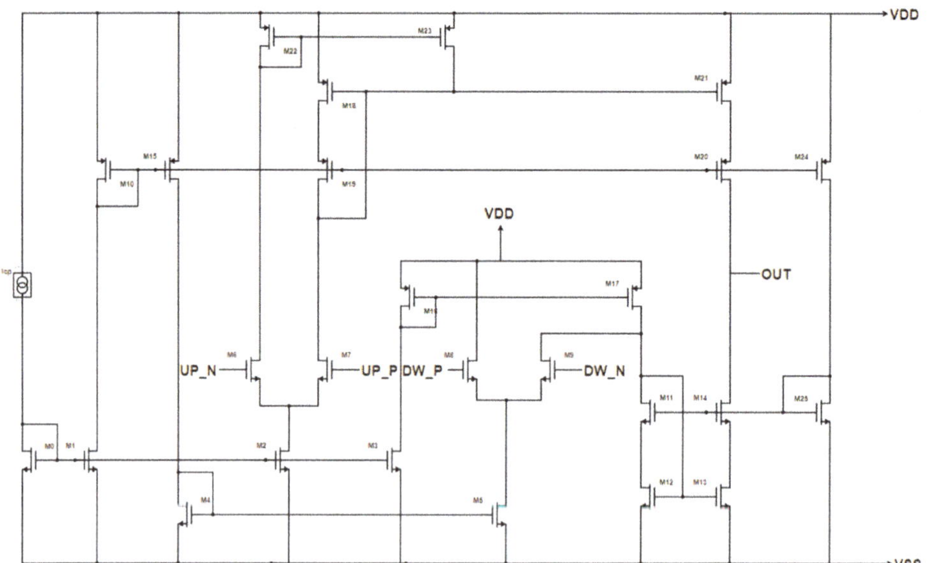


**Figure 11.** Enhanced CP's Gate Switching architecture with differential inputs.

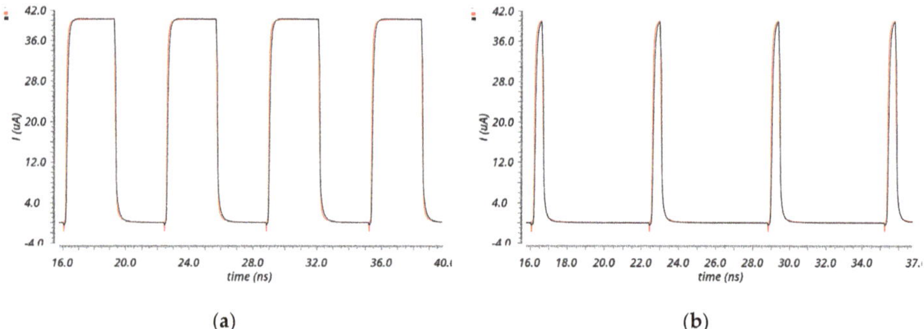

(a)

(b)

**Figure 12.** Transient behavior of the CP of Figure 11. The source current is in red and the sink current is in black for two different values of ON time of the input signals (UP_P-UP_N and DW_P-DW_N): (a) 3.2 ns, (b) 500 ps.

### 3.1.2. Phase/Frequency Detector

Since the PFD is essentially a digital block, the easiest and most effective way of hardening it against radiation is the TMR [13]. The designed architecture is shown in Figure 13a. The block-labeled PFDs in Figure 13a consist of the conventional PFD architecture shown in Figure 13b, and every PFD has its own reset voter that decides between all the output resets. In this way, the only possible loss of lock for the PLL happens when an SEE occurs on the UP and/or DOWN voters. However, this error does not lead the PFD into a wrong state causing cycle slipping, i.e., one of the input's loss (or gain) of one cycle with respect to the other input. This architecture has been developed in both CMOS logic and CML (Current Mode Logic), in order to be compatible with the designed CPs.

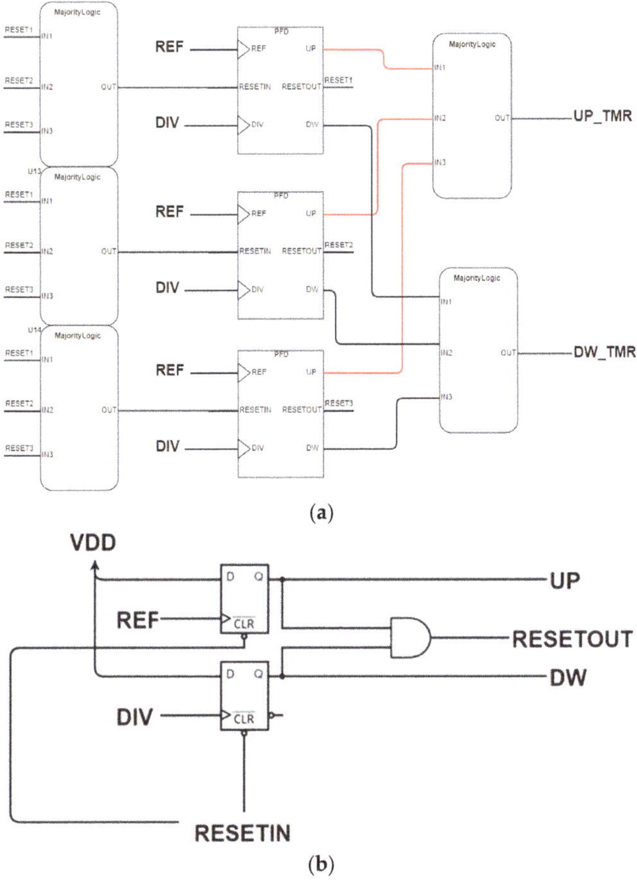

**Figure 13.** Triple modular redundant PFD architecture in (**a**) and simple PFD architecture in (**b**).

### 3.1.3. Comparison between the Two PFD/CP Architectures

In order to compare the two different solutions (CMOS and CML versions of the PFD/CP), two PLL testbenches have been built in Cadence. In these testbenches, the VCO model with parasitic effects is considered, while the CP's and PFD's schematic view has been used to reduce the simulation time. Finally, a Verilog-A frequency divider, which can be found in the Cadence's *rf-library*, has been used. For the Phase Noise comparison, because of the non-linear nature of the PLL, a direct time domain noise analysis was necessary. This has been done exploiting the Transient Noise Analysis option embedded in the Spectre RF's Transient Analysis. Since this type of simulation is time consuming, a resolution bandwidth of 500 KHz to reduce the simulation time has been set. In Figure 14, the Phase Noise results are shown: the two solutions have almost the same behavior in terms of noise, but the CMOS solution shows higher peaks at multiples of the reference frequency because of the worse current matching during the switch of the CP currents from ON to OFF and vice versa. Instead, in Table 1, the comparison in terms of DC current matching of the CP and power consumption between the CML and CMOS solutions is summarized. The DC current mismatch has been measured performing a DC simulation on the CP circuits and measuring the output current when the output node is forced at 520 mV with an ideal voltage source, and the inputs are such that both the source and sink current are ON. Considering that (1) the noise performance difference is negligible, (2) the CMOS solution has a better current matching in the worst case, and (3) the CML solution shows 25 times higher power consumption, then the CMOS solution has been chosen and has been developed at layout level for the whole PLL design.

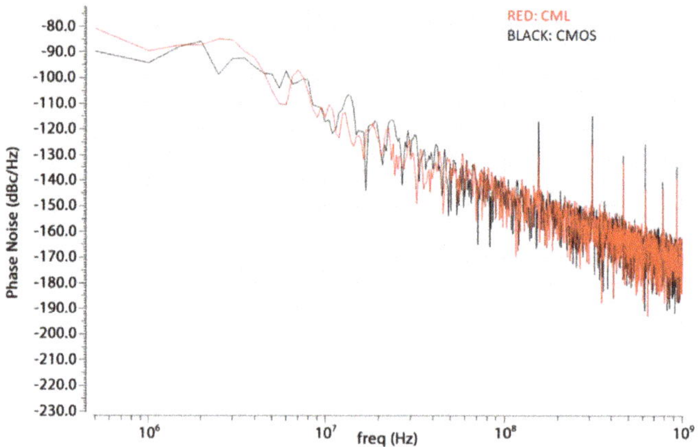


**Figure 14.** Comparison between the two CP/PFD (Phase Frequency Detector) architectures in terms of phase noise: Current Mode Logic (CML) architecture's results in red, CMOS architecture's results in black.

**Table 1.** Comparison between the two CP/PFD architectures in terms of CP's current matching and CP+PFD's power consumption.

|  | CMOS | CML |
|---|---|---|
| Charge Pump DC current mismatch (worst case) | 1.454 μA | 1.66 μA |
| Power Consumption Charge Pump + PFD | ≈200 μW | ≈5 mW |

### 3.1.4. PFD/CP Layout

The CP's layout is shown in Figure 15. The output mirrors, since their slaves are composed of two MOS in parallel, are interleaved with the master transistor to enhance the technology process matching. Moreover, every mirror and the switch transistors are surrounded by guard rings connected to the voltage supply. These rings have two main functions:

1. They reduce the possibility of SEL (Single Event Latch-Up) and Latch-Up in general;
2. They reduce the drift current generated after an SEE in the sensitive nodes near the hit node [22].

Regarding the PFD, as for the CP, guard rings have been placed around the transistor to avoid latch-up and to reduce the drift current deriving from an SEE. Moreover, the triplicated cells have been placed at a distance of 10 μm from each other to avoid MBUs (Multi Bit Upsets). The TMR technique would become useless if an SEE causes an SEU (Single Event Upset) in more than one PFD each time. In Figure 16, the complete layout of the TMR PFD is shown: the three PFDs with their reset majority logic can be recognized on the left, while the output majority logic can be recognized on the right.

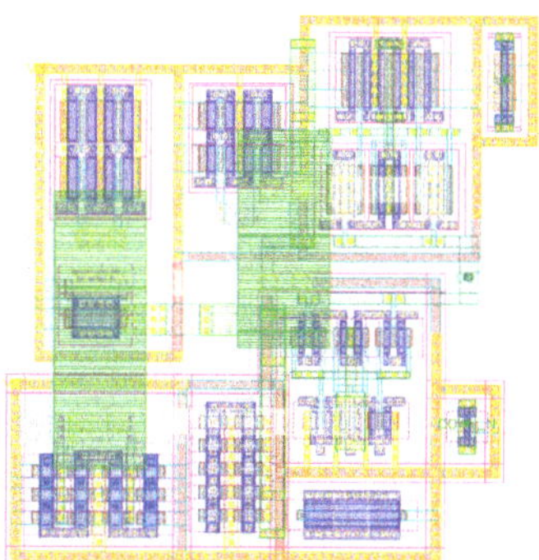

**Figure 15.** Charge Pump's layout.

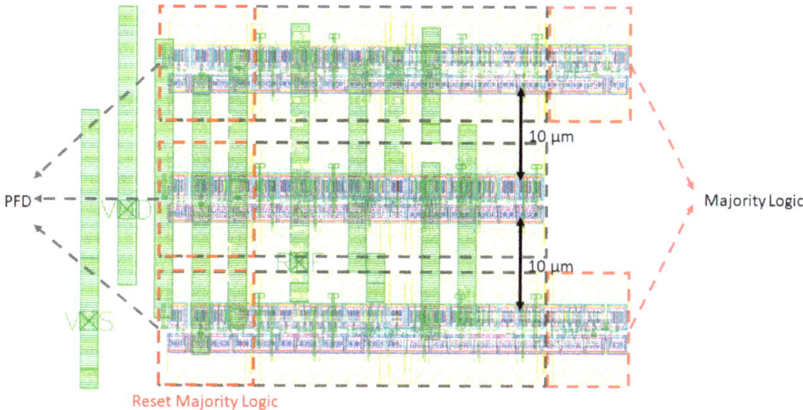


**Figure 16.** Phase/Frequency Detector's layout.

## 4. Simulation's Results

### 4.1. PFD/CP Characterization

In Figure 17, the average CP current is represented as function of the phase difference between the inputs for different technology corners (Figure 17a) and different temperatures (Figure 17b). As can be seen from the figure, the PFD/CP is dead zone-free thanks to the presence of the RESET state in the PFD.

### 4.2. Single Event Effect Simulations on the CP

The current caused on a sensible node by an SEE can be modeled as a double exponential function [23] in Equation (5). In this work, the parameter model of Equation (6), which was taken from a previous high-frequency design we have done in the same technology [24], has been used:

$$I(t) = Q \frac{e^{-\frac{t}{\tau_1}} - e^{-\frac{t}{\tau_2}}}{\tau_1 - \tau_2} . \tag{5}$$

$$\tau_1 = 200 \div 400 \, ps \,, \quad \tau_2 = 50 \div 100 \, ps \,, \quad Q = 67 \div 800 \, fC, \tag{6}$$

where $\tau_1$ and $\tau_2$ are the constant times of the double exponential shape and $Q$ is the charge released in the silicon substrate during a particle strike, which corresponds to a particle LET (Linear Energy Transfer) between 5 and 60 Mev·cm$^2$/mg.

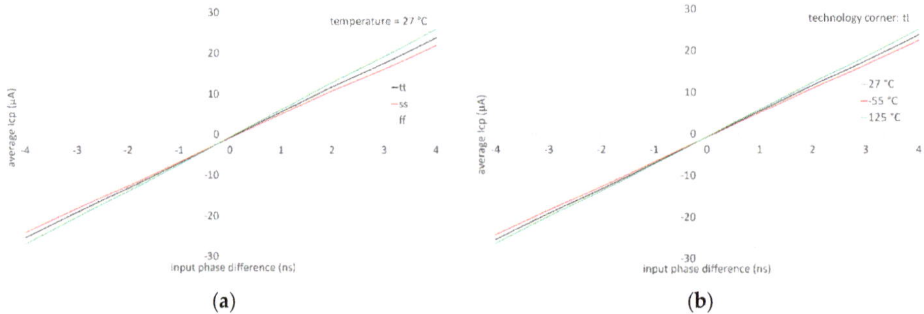

(a)       (b)

**Figure 17.** PFD/CP characteristic for different technology corners at 27 °C (**a**) and different temperatures in typical case (**b**).

The SEE analysis is focused on the most critical block, the CP, since the PFD adopts a TMR mitigation approach, the passive Loop Filter avoids the use of feedback loops and of active circuits, while the VCO was inherited from a previous SEE-tested designed block in [16]. Hence, double exponential current generators from the Cadence's analogLib library have been placed on every sensitive node of the CP (all the drains and sources of the MOSFETs that were not connected to the supply voltage), taking care of the direction of the current. Then these generators have been delayed to analyze their effect separately, and the output current has been measured in order to see how an SEE on each node of the CP affects the output current. Then, these results have been reported to the ADS behavioral model of the PLL to see how the PLL reacts to Single Event Transients (SETs) on the CP (see Figure 18).

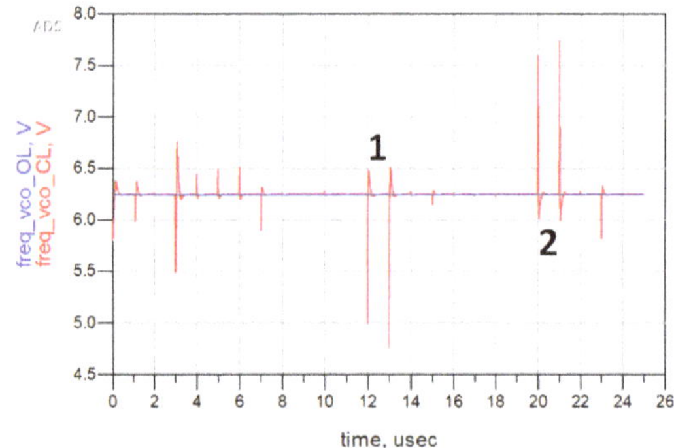

**Figure 18.** Output frequency of the ADS PLL model as function of time, for Single Event Transients (SETs) spaced 1 μs apart, hitting every sensitive node of the CP and with a Linear Energy Transfer (LET) of 60 MeV·cm$^2$/mg.

As shown in Figure 18, the PLL, which loses lock after an SET, is able to recover the locked state in less than 600 ns. The largest peaks, which are indicated in Figure 18 with the numbers 1 and 2, are the ones derived by an SET on the output nodes, as highlighted in Figure 19 with the corresponding numbers, because the charge injected into these nodes goes directly through the output node, charging or discharging the Loop Filter and then modifying the control voltage of the VCO.

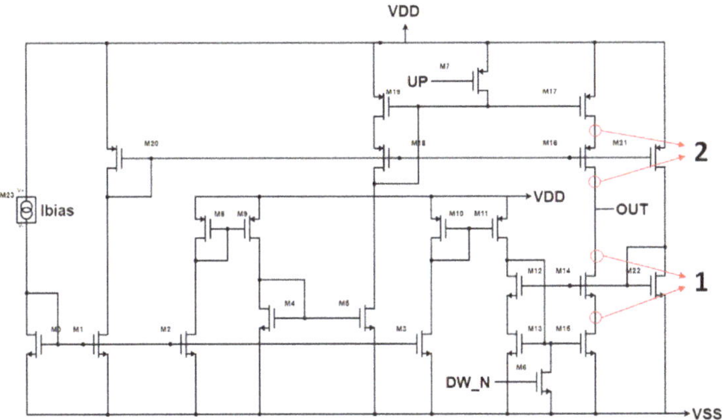

**Figure 19.** Highlight of the output nodes of the CP.

*4.3. PLL Testbench*

Once all the blocks that compose the PLL have been designed, a testbench has been built. Here, the netlist generated by the layout parasitic extraction has been used for all the blocks (CP, PFD, VCO, Loop Filter, and FD), while the connections between the blocks are still ideal. On this testbench, a Transient simulation has been performed to analyze the locking behavior and the Phase Noise. As for the simulations presented in Section 3.1.3, the Transient Noise Analysis Option has been used for the Phase Noise evaluation.

4.3.1. Locking Process

The locking process in the typical technology corner at 27 °C is shown in Figure 20. The lock time is 1 µs, which is twice that expected from the ADS behavioral model. This is mainly due to two factors: (1) the non-linearity of the VCO's characteristic, and (2) the cycle slipping. In the ADS model, the VCO was approximated to have a linear characteristic with 2 GHz/V gain, but this is true only in a small range of frequencies. For the other frequencies, a smaller gain and consequently, a smaller loop bandwidth and damping factor results, which leads to a higher lock time. Moreover, as can be seen from the locking process, at the beginning of the process, two cycle slips are seen. This is due to the RESET state of the PFD necessary to cancel the dead zone. When the phase difference between the two PFD's inputs is high enough that the new edge of the input ahead comes while the PFD is in its RESET state, this edge is not seen, and the PFD goes into the wrong state with respect to the ideal one. From that, it recovers the correct state when the cycle slip happens, but this leads to an enlargement of the lock time.

4.3.2. Noise Simulations

In Figure 21, the resulting post-layout Phase Noise for the whole PLL is shown. It is below −85 dBc/Hz, which is in line with that of the VCO. The largest contribution is due to the CP/PFD at low and middle frequencies, while the VCO's noise is predominant at higher frequencies, as expected from the theory [19].

Since in digital design, a temporal characterization of noise is usually preferred, the absolute Jitter has been measured, resulting in a peak-to-peak value of 8.8 ps and an RMS (root mean square) value of 2.03 ps (typical corner, considering 12,500 cycles, which in terms of Phase Noise means a bandwidth between 500 kHz and 1 GHz). Absolute Jitter is really important for communications systems, since it measures how much an edge of the clock is in a different position from the ideal one. For other applications, Period Jitter is of most interest, since it measures the difference between a clock period and the average one. Therefore, Period Jitter has been also measured, in the typical corner as well, resulting in a peak-to-peak of 96 fs and an RMS value of 14.744 fs.

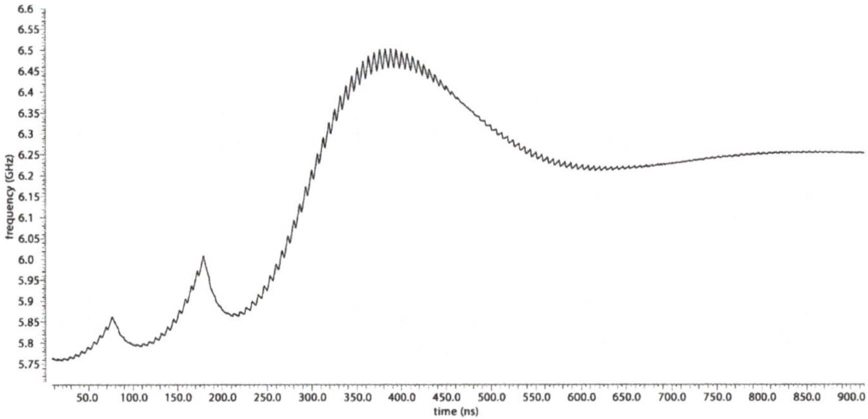

**Figure 20.** Post-layout locking process.

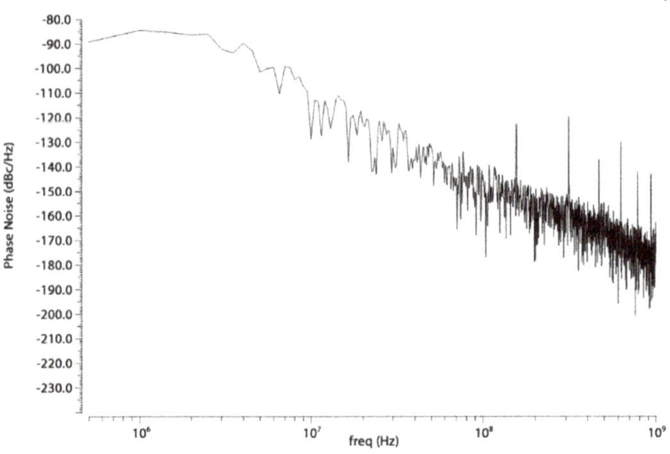

**Figure 21.** Post-layout phase noise.

## 5. Test Chip

To prototype the designed PLL, a block view of a possible test chip has been developed and is shown in Figure 22. As can be seen, it consists of a whole PLL, which can be recognized at the bottom, and a duplicate of the PFD and CP. The main idea is to test every block separately: there is the possibility of giving the inputs from an external source, and all the outputs are outputs of the test chip. Since the PFD/CP's area consumption is low, duplicating these blocks gives the possibility of testing them separately, without adding multiplexer and buffer inside the loop, disturbing the PLL's operation. The FD is not duplicated because it consists of three stages that need bias currents. This means that at least 4 pads are necessary for it and, therefore, duplicating it would lead to a too high requirement in terms of pads. Moreover, in Figure 23, a draft of the floor plan is shown: the chip's area to be fabricated using a Europractice Multi Project Wafer is 1 mm$^2$. As can be seen from the figure, the whole PLL occupies an area of about 0.09 mm$^2$ (the remaining area and pads are used for other designs not discussed in this paper). It is to be noted that the design of the pads is inherited from a previous design, where they were tested also with respect to resistance to radiation effects [24].

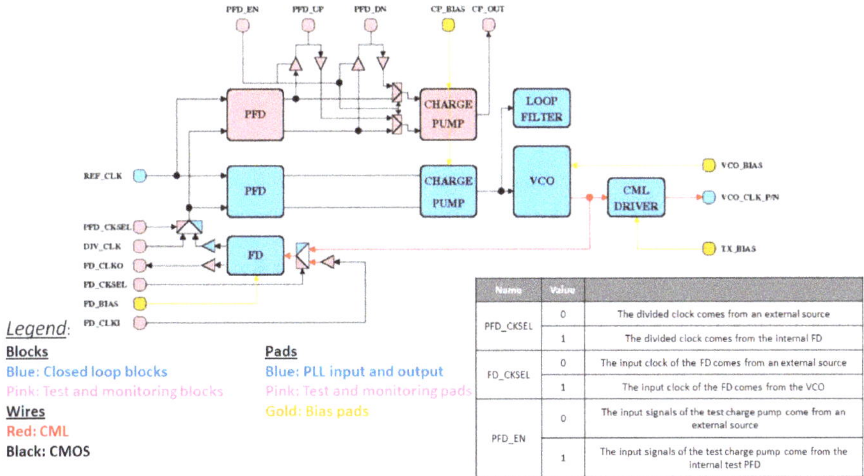



| Name | Value | |
|---|---|---|
| PFD_CKSEL | 0 | The divided clock comes from an external source |
| | 1 | The divided clock comes from the internal FD |
| FD_CKSEL | 0 | The input clock of the FD comes from an external source |
| | 1 | The input clock of the FD comes from the VCO |
| PFD_EN | 0 | The input signals of the test charge pump come from an external source |
| | 1 | The input signals of the test charge pump come from the internal test PFD |

**Figure 22.** Block schematic of the test chip.

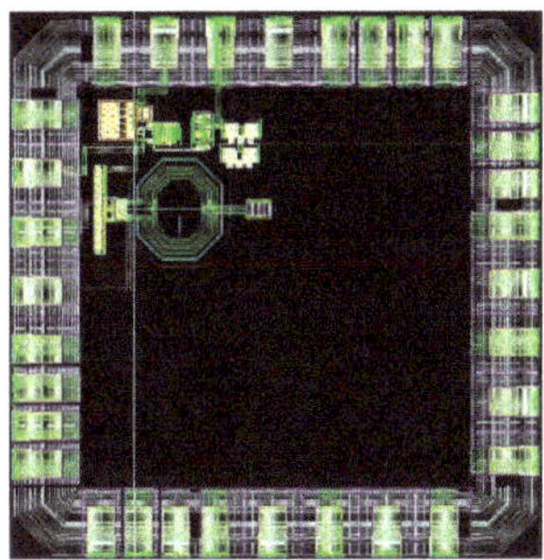

**Figure 23.** Draft floor plan of the test chip.

Although the figures and tables proposed in this paper refer to post-layout simulation results, the experience of our research group from previous designs in the same 65 nm TSMC technology, using the CAD (Computer-Aided Design) design environment and targeting similar operating frequencies, has proven that there is a coherent alignment between post-layout simulations and experimental measurements. In the next step, when the measures of the prototyped chip will be performed, a complete assessment of the results accuracy will be done.

## 6. Conclusions and Future Work

The design of a PLL for the new ESA Spacefibre standard has been presented in this paper. The work was carried out in ADS for the modeling activity and Cadence Virtuoso for the design activity. In particular, the design of a TMR PFD, a CP, and a passive Loop Filter has been presented, starting from an already designed 6.25 GHz rad-hard VCO in 65 nm technology. The modeling activity has shown that the PLL can be completely integrated on-chip, with a Loop Filter area consumption of only 6000 $\mu m^2$. The PLL is able to generate three output signals at 6.25, 3.125, and 1.5625 GHz with a gain margin of 86 dB and phase margins of 50°. The Phase Noise of the PLL, considering the higher frequency output, is below −85 dBc/Hz @ 1 MHz, and hence, it is in line with that of the VCO. The PLL power consumption in the locked state is about 10.24 mW. The PFD/CP is dead-zone free and shows a current matching below 2 µA (5% of the nominal CP's current value) in the worst case of process–temperature corners. Moreover, the PLL is highly immune to SEE on the PFD, while it is able to relock in less than 600 ns for an SEE on the CP. Finally, an area estimation has been done considering also the VCO, resulting in a total area of 0.09 mm$^2$.

In Table 2, a comparison with the state-of-the-art rad-hard PLLs is performed. In [15], a CP-PLL in 65 nm technology for high-energy physics is presented. It generates a tunable output frequency in the range of 4.8–6 GHz, which is slightly lower the SpaceFibre standard. As reported in [15], the CP-PLL was not designed to be SEE tolerant; indeed, a simple PFD was used in the PLL loop. Instead, in this work, a TMR PFD was implemented to increase the hardness against SEE. In [13], all the digital logic of the CP-PLL is implemented using a TMR approach, but its low-frequency range is not suitable for SpaceFibre application. In [25], the design of a rad-hard CP-PLL that is able to generate an output frequency in the range 1.17-3.16 GHz is reported. It bases its radiation hardness on the use of a Silicon On Sapphire (SOS) substrate. Indeed, the SOS having a resistive sapphire in the substrate reduces the creation of electron–hole pairs and their migration in the device bulk in the case of energetic ion strikes. An example of radiation hardness improvement achieved using a different technology is presented also in [26], where a high-frequency CP-PLL is designed in 250 nm SiGe technology. Instead, this work presents the design of a PLL that is suitable for space applications in standard silicon CMOS technology, whose radiation hardness is achieved by TMR and layout techniques.

**Table 2.** Comparison with the state-of-the-art Rad-Hard PLLs. SOS: Silicon On Sapphire.

|  | This Work | [13] * | [15] * | [25] * | [26] * |
|---|---|---|---|---|---|
| Technology | 65 nm CMOS | 65 nm CMOS | 65 nm CMOS | 250 nm SOS | 250 nm SiGe |
| Frequency Range (GHz) | 5.2–6.4 | 2.2–3.2 | 4.8–6 | 1.17–3.16 | 17.5–18.9 |
| Power Consumption (mW) | 10.24 | 11.7 | 18 | 102.5 | - |
| Area (mm$^2$) | 0.09 | - | 0.124 | 0.52 | 5 |
| Absolute Jitter (ps) (RMS) | 2.03 | 0.345 | 3.23 | - | - |
| Period Jitter (fs) (RMS) | 14.74 | - | 3550 | - | - |
| Phase noise @ 1MHz (dBc/Hz) | −85 | - | - | −100 | −110 |

* measured.

The next step is test chip fabrication to prototype the proposed PLL solution. The design has been submitted to tape out through MPW (Multi Project Wafer), and hence, the experimental verification of the fabricated prototype is part of the future roadmap.

**Author Contributions:** M.M.: conceptualization, investigation, methodology, software, writing—original draft, writing—review & editing; B.N.: conceptualization, methodology, supervision, writing-review & editing; G.C.: conceptualization, investigation, methodology, software, writing-review & editing; S.S.: conceptualization, investigation, methodology, supervision, writing-review & editing. All authors have read and agreed to the published version of the manuscript.

**Funding:** Work supported by ISHTAR and Dipartimento di Eccellenza projects by Pisa University and MIUR.

**Acknowledgments:** Discussions with D. Monda, Pisa University, and G. Magazzù, INFN, are acknowledged.

**Conflicts of Interest:** The authors declare no conflict of interest.

## References

1. European Cooperation for Space Standard. *SpaceFibre—Very High-Speed Serial Link*; ECSS-E-ST-50-11C; ECSS: Noordwijk, The Netherlands, 2019.
2. Wang, F.; Agrawal, V.D. Single Event Upset: An Embedded Tutorial. In Proceedings of the 21st International Conference on VLSI Design (VLSID 2008), Hyderabad, India, 4–8 January 2008; pp. 429–434.
3. Schwank, J.R.; Shaneyfelt, M.R.; Fleetwood, D.M.; Felix, J.A.; Dodd, P.E.; Paillet, P.; Ferlet-Cavrois, V. Radiation Effects in MOS Oxides. *IEEE Trans. Nucl. Sci.* **2008**, *55*, 1833–1853. [CrossRef]

4. Suparta, W.; Zulkeple, S.K. An Assessment of the Space Radiation Environment in a Near-Equatorial Low Earth Orbit Based on the RazakSAT-1 Satellite. *arXiv* **2015**, arXiv:1511.03837.

5. Loveless, T.D.; Massengill, L.W.; Holman, W.T.; Bhuva, B.L.; McMorrow, D.; Warner, J.H. A Generalized Linear Model for Single Event Transient Propagation in Phase-Locked Loops. *IEEE Trans. Nucl. Sci.* **2010**, *57*, 2933–2947. [CrossRef]

6. Chen, Z.; Lin, M.; Ding, D.; Zheng, Y.; Sang, Z.; Zou, S. Analysis of Single-Event Effects in a Radiation-Hardened Low-Jitter PLL Under Heavy Ion and Pulsed Laser Irradiation. *IEEE Trans. Nucl. Sci.* **2017**, *64*, 106–112. [CrossRef]

7. Chen, L.; Wen, X.; You, Y.; Huang, D.; Li, C.; Chen, J. A radiation-tolerant ring oscillator phase-locked loop in 0.13 μm CMOS. In Proceedings of the IEEE 55th International Midwest Symposium on Circuits and Systems (MWSCAS) 2012, Boise, ID, USA, 5–8 August 2012; pp. 13–16.

8. Loveless, T.D.; Massengill, L.W.; Bhuva, B.L.; Holman, W.T.; Reed, R.A.; McMorrow, D.; Melinger, J.S.; Jenkins, P. A Single-Event-Hardened Phase-Locked Loop Fabricated in 130 nm CMOS. *IEEE Trans. Nucl. Sci.* **2007**, *54*, 2012–2020. [CrossRef]

9. Kumar, R.; Karkala, V.; Garg, R.; Jindal, T.; Khatri, S.P. A radiation tolerant Phase Locked Loop design for digital electronics. In Proceedings of the IEEE International Conference on Computer Design 2009, Lake Tahoe, CA, USA, 4–7 October 2009; pp. 505–510.

10. De Maria, N.; Barbero, M.B.; Fougeron, D.; Gensolen, F.; Godiot, S.; Menouni, M.; Pangaud, P.; Rozanov, A.; Wang, A.; Bomben, M.; et al. Recent progress of RD53 Collaboration towards next generation Pixel Read-Out Chip for HL-LHC. *J. Instrum.* **2016**, *11*, C12058. [CrossRef]

11. Prinzie, J.; Christiansen, J.; Moreira, P.; Steyaert, M.; Leroux, P. A 2.56-GHz SEU Radiation Hard *LC*-Tank VCO for High-Speed Communication Links in 65-nm CMOS Technology. *IEEE Trans. Nucl. Sci.* **2018**, *65*, 407–412. [CrossRef]

12. Prinzie, J.; Christiansen, J.; Moreira, P.; Steyaert, M.; Leroux, P. Comparison of a 65 nm CMOS Ring- and LC-Oscillator Based PLL in Terms of TID and SEU Sensitivity. *IEEE Trans. Nucl. Sci.* **2017**, *64*, 245–252. [CrossRef]

13. Prinzie, J.; Steyaert, M.; Leroux, P.; Christiansen, J.; Moreira, P. A single-event upset robust, 2.2 GHz to 3.2 GHz, 345 fs jitter PLL with triple-modular redundant phase detector in 65 nm CMOS. In Proceedings of the IEEE Asian Solid-State Circuits Conference (A-SSCC) 2016, Toyama, Japan, 7–9 Nov. 2016; pp. 285–288.

14. Loveless, T.D.; Massengill, L.W.; Bhuva, B.L.; Holman, W.T.; Witulski, A.F.; Boulghassoul, Y. A Hardened-by-Design Technique for RF Digital Phase-Locked Loops. *IEEE Trans. Nucl. Sci.* **2006**, *53*, 3432–3438. [CrossRef]

15. Mazza, G.; Panati, S. A Compact, Low Jitter, CMOS 65 nm 4.8–6 GHz Phase-Locked Loop for Applications in HEP Experiments Front-End Electronics. *IEEE Trans. Nucl. Sci.* **2018**, *65*, 1212–1217. [CrossRef]

16. Monda, D.; Ciarpi, G.; Mangraviti, G.; Berti, L.; Saponara, S. *Analysis and Comparison of Ring and LC-Tank Oscillators for 65 nm Integration of Rad-Hard VCO for SpaceFibre Applications, Lecture Notes in Electrical Engineering*; Springer: Cham, Switzerland, 2020; Volume 627.

17. Mestice, M.; Neri, B.; Saponara, S. *Analysis and Simulation of a PLL Architecture Towards a Fully Integrated 65 nm Solution for the New Spacefibre Standard, Lecture Notes in Electrical Engineering*; Springer: Cham, Switzerland, 2020; Volume 627.

18. Gardner, F.M. Charge-Pump Phase-Lock Loops. *IEEE Trans. Commun.* **1980**, *28*, 1849–1858. [CrossRef]

19. Razavi, B. *RF Microelectronics*, 2nd ed.; Pearson Education: Upper Saddle River, NJ, USA, 2011; pp. 646–647.

20. Rahman, L.; Ariffin, N.; Reaz, M.B.I.; Marufuzzaman, M. High Performance CMOS Charge Pumps for Phase-locked Loop. *Trans. Electr. Electron. Mater.* **2015**, *16*, 241–249. [CrossRef]

21. Zhang, C.; Au, T.; Syrzycki, M. A high performance NMOS-switch high swing cascode charge pump for phase-locked loops. In Proceedings of the IEEE 55th International MWSCAS, Boise, ID, USA, 5–8 August 2012; pp. 554–557.

22. Black, D.; Sternberg, A.L.; Alles, M.L.; Witulski, A.F.; Bhuva, B.L.; Massengill, L.W.; Benedetto, J.M.; Baze, M.P.; Wert, J.L.; Hubert, M.G. HBD layout isolation techniques for multiple node charge collection mitigation. *IEEE Trans. Nucl. Sci.* **2005**, *52*, 2536–2541. [CrossRef]

23. Messenger, G.C. Collection of Charge on Junction Nodes from Ion Tracks. *IEEE Trans. Nucl. Sci.* **1982**, *29*, 2024–2031. [CrossRef]

24. Ciarpi, G.; Magazzù, G.; Palla, F.; Saponara, S. Design, Implementation, and Experimental Verification of 5 Gbps, 800 Mrad TID and SEU-Tolerant Optical Modulators Drivers. *IEEE Trans. Circuits Syst. I* **2020**, *67*, 829–838. [CrossRef]

25. Ghosh, P.P.; Xiao, E. A 2.5 GHz radiation hard fully self-biased PLL using 0.25 μm SOS-CMOS technology. In Proceedings of the IEEE International Conference on IC Design and Technology 2009, Austin, TX, USA, 18–20 May 2009; pp. 121–124.

26. Herzel, F.; Osmany, S.A.; Schmalz, K.; Winkler, W.; Scheytt, J.C.; Podrebersek, T.; Follmann, R.; Heyer, H.V. An Integrated 18 GHz fractional-N PLL in SiGe BiCMOS technology for satellite communications. In Proceedings of the IEEE Radio Frequency Integrated Circuits Symposium 2009, Boston, MA, USA, 7–9 June 2009; pp. 329–332.

*Article*

# Machine Learning on Mainstream Microcontrollers [†]

**Fouad Sakr, Francesco Bellotti *, Riccardo Berta and Alessandro De Gloria**

Department of Electrical, Electronic and Telecommunication Engineering (DITEN)-University of Genoa, Via Opera Pia 11a, 16145 Genova, Italy; Fouad.Sakr@elios.unige.it (F.S.); riccardo.berta@unige.it (R.B.); alessandro.degloria@unige.it (A.D.G.)

* Correspondence: franz@elios.unige.it
† This paper is an extended version of the conference paper published in: Falbo, V.; Apicella, T.; Aurioso, D.; Danese, L.; Bellotti, F.; Berta, R.; De Gloria, A. Analyzing Machine Learning on Mainstream Microcontrollers. In Proceedings of the International Conference on Applications in Electronics Pervading Industry Environment and Society, ApplePies 2019, Pisa, Italy, 12–13 September 2019.

Received: 27 March 2020; Accepted: 29 April 2020; Published: 5 May 2020

**Abstract:** This paper presents the Edge Learning Machine (ELM), a machine learning framework for edge devices, which manages the training phase on a desktop computer and performs inferences on microcontrollers. The framework implements, in a platform-independent C language, three supervised machine learning algorithms (Support Vector Machine (SVM) with a linear kernel, k-Nearest Neighbors (K-NN), and Decision Tree (DT)), and exploits STM X-Cube-AI to implement Artificial Neural Networks (ANNs) on STM32 Nucleo boards. We investigated the performance of these algorithms on six embedded boards and six datasets (four classifications and two regression). Our analysis—which aims to plug a gap in the literature—shows that the target platforms allow us to achieve the same performance score as a desktop machine, with a similar time latency. ANN performs better than the other algorithms in most cases, with no difference among the target devices. We observed that increasing the depth of an NN improves performance, up to a saturation level. k-NN performs similarly to ANN and, in one case, even better, but requires all the training sets to be kept in the inference phase, posing a significant memory demand, which can be afforded only by high-end edge devices. DT performance has a larger variance across datasets. In general, several factors impact performance in different ways across datasets. This highlights the importance of a framework like ELM, which is able to train and compare different algorithms. To support the developer community, ELM is released on an open-source basis.

**Keywords:** machine learning; edge computing; embedded devices; edge analytics; ANN; k-NN; SVM; decision trees; ARM; X-Cube-AI; STM32 Nucleo

---

## 1. Introduction

The trend of moving computation towards the edge is becoming ever more relevant, leading to performance improvements and the development of new field data processing applications [1]. This computation shift from the cloud (e.g., [2]) to the edge has advantages in terms of response latency, bandwidth occupancy, energy consumption, security and expected privacy (e.g., [3]). The huge amount, relevance and overall sensitivity of the data now collected also raise clear concerns about their use, as is being increasingly acknowledged (e.g., [4]), meaning that this is a key issue to be addressed at the societal level.

The trend towards edge computing also concerns machine learning (ML) techniques, particularly for the inference task, which is much less computationally intensive than the previous training phase. ML systems "learn" to perform tasks by considering examples, in the training phase, generally without being programmed with task-specific rules. When running ML-trained models, Internet of Things

(IoT) devices can locally process their collected data, providing a prompter response and filtering the amount of bits exchanged with the cloud.

ML on the edge has attracted the interest of industry giants. Google has recently released the TensorFlow Lite platform, which provides a set of tools that enable the user to convert TensorFlow Neural Network (NN) models into a simplified and reduced version, then run this version on edge devices [5,6]. EdgeML is a Microsoft suite of ML algorithms designed to work off the grid in severely resource-constrained scenarios [7]. ARM has published an open-source library, namely Cortex Microcontroller Software Interface Standard Neural Network (CMSIS-NN), for Cortex-M processors, which maximizes NN performance [8]. Likewise, a new package, namely X-Cube-AI, has been released for implementing deep learning models on STM 32-bit microcontrollers [9].

While the literature is increasingly reporting on novel or adapted embedded machine learning algorithms, architectures and applications, there is a lack of quantitative analyses about the performance of common ML algorithms on state-of-the-art mainstream edge devices, such as ARM microcontrollers [10]. We argue that this has limited the development of new applications and the upgrading of existing ones through an edge computing extension.

In this context, we have developed the Edge Learning Machine (ELM), a framework that performs ML inference on edge devices using models created, trained, and optimized on a Desktop environment. The framework provides a platform-independent C language implementation of well-established ML algorithms, such as linear Support Vector Machine (SVM), k-Nearest Neighbors (k-NN) and Decision Tree. It also supports artificial neural networks by exploiting the X-Cube-AI package for STM 32 devices [9]. We validated the framework on a set of STM microcontrollers (families F0, F3, F4, F7, H7, and L4) using six different datasets, to answer a set of ten research questions exploring the performance of microcontrollers in typical ML Internet of Things (IoT) applications. The research questions concern a variety of aspects, ranging from inference performance comparisons (also with respect to a desktop implementation) to training time, and from pre-processing to hyperparameter tuning. The framework is released on an open-source basis (https://github.com/Edge-Learning-Machine), with a goal to support researchers in designing and deploying ML solutions on edge devices.

The remainder of this paper is organized as follows: Section 2 provides background information about the ML techniques that are discussed in the manuscript. Section 3 describes the related work in this field. Section 4 shows the implemented framework and the supported algorithms. Section 5 presents the extensive experimental analysis we conducted by exploiting the framework. Finally, Section 6 draws conclusions and briefly illustrates possible future research directions.

## 2. Background

The Edge Learning Machine framework aims to provide an extensible set of algorithms to perform inference on the edge. The current implementation features four well-established supervised learning algorithms, which we briefly introduce in the following subsections. They all support both classification and regression problems.

### 2.1. Artificial Neural Network (ANN)

An Artificial Neural Network is a model that mimics the structure of our brain's neural network. It consists of a number of computing neurons connected to each other in a three-layer system; one input layer, several hidden layers, and one output layer. Artificial Neural Networks (ANNs) can model complex and non-linear or hidden relationships between inputs and outputs [11]. This one of the most powerful and well-known ML algorithms, which is used in a variety of applications, such as image recognition, natural language processing, forecasting, etc.

### 2.2. Linear Kernel Support Vector Machine (SVM)

The SVM algorithm is a linear classifier that computes the hyperplane that maximizes the distance from it to the nearest samples of the two target classes. It is a memory-efficient inference algorithm and is able to capture complex relationships between data points. The downside is that the training time increases with huge and noisy datasets [12]. While the algorithm deals well with non-linear problems, thanks to the utilization of kernels that map the original data in higher dimension spaces, we implemented only the original, linear kernel [13] for simplicity of implementation into the edge device.

### 2.3. K-Nearest Neighbor (k-NN)

k-NN is a very simple algorithm based on feature similarity that assigns, to a sample point, the class of the nearest set of previously labeled points. k-NN's efficiency and performance depends on the number of neighbors K, the voting criterion (for K > 1) and the training data size. The training phase produces a very simple model (the K parameter), but the inference phase requires exploring the whole training set. Its performance is typically sensitive to noise and irrelevant features [12,14].

### 2.4. Decision Tree (DT)

This is a simple and useful algorithm, which has the advantage of clearly exposing the criteria of decisions that are made. In building the decision tree, at each step, the algorithm splits data so as to maximize the information gain, thus creating homogeneous subsets. The typical information gain criteria are Entropy and Gini. DT is able to deal with linearly inseparable data and can handle redundancy, missing values, and numerical and categorical types of data. It is negatively affected by high dimensionality and high numbers of classes, because of error propagation [12,15]. Typical hyperparameters that are tuned in the model selection phase concern regularization and typically include depth, the minimum number of samples for a leaf, and the minimum number of samples for a split, the maximum number of leaf nodes and the splitter strategy (the best one, which is the default, or a random one, which is typically used for random forests), etc. Several DTs can be randomly built for a problem, in order to create complex but high-performing random forests.

## 3. Related Work

A growing number of articles are being published on the implementation of ML on embedded systems, especially with a focus on the methodology of moving computation towards the edge. Zhang et al. [16] presented an object detector, namely MobileNet-Single Shot Detector (SSD), which was trained using a deep convolutional neural network with the popular Caffe framework. The pre-trained model was then deployed on NanoPi2, an ARM board developed by FriendlyARM, which uses Samsung Cortex-A9 Quad-Core S5P4418@1.4GHz SoC and 1 GB 32bit DDR3 RAM. MobileNet-SSD can run at 1.13FPS.

Yazici et al. [17] tested the ability of a Raspberry Pi to run ML algorithms. Three algorithms were tested, Support Vector Machine (SVM), Multi-Layer Perceptron, and Random Forests, with an accuracy above 80% and a low energy consumption. Fraunhofer Institute for Microelectronic Circuits and Systems have developed Artificial Intelligence for Embedded Systems (AIfES), a library that can run on 8-bit microcontrollers and recognize handwriting and gestures without requiring a connection to the cloud or servers [18]. Cerutti et al. [19] implemented a convolutional neural network on STM Nucleo-L476RG for people detection using CMSIS-NN, which is an optimized library that allows for the deployment of NNs on Cortex-M microcontrollers. In order to reduce the model size, weights are quantized to an 8-bit fixed point format, which slightly affects the performance. The network fits in 20 KB of flash and 6 KB of RAM with 77% accuracy.

Google has recently released Coral Dev Board, which includes a small low power Application-Specific Integrated Circuit (ASIC) called Edge TPU, and provides high-performance ML inferencing without running the ML model on any kind of server. Edge TPU can run TensorFlow Lite, with a low processing power and high performance [20]. There are a few application programming interfaces (APIs) in the Edge tencor processing unit (TPU) module that perform inference (ClassificationEngine) for image classification, for object detection (DetectionEngine) and others that perform on-device transfer learning [21].

Microsoft is developing EdgeML, a library of machine learning algorithms that are trained on the cloud/desktop and can run on severely resource-constrained edge and endpoint IoT devices (also with 2 KB RAM), ranging from the Arduino to the Raspberry Pi [7]. They are currently releasing tree- and k-NN-based algorithms, called Bonsai and ProtoNN, respectively, for classification, regression, ranking and other common IoT tasks. Their work also concerns recurrent neural networks [22]. A major achievement concerns the translation of floating-point ML models into fixed-point code [23], which is, however, not the case in state-of-the-art mainstream microcontrollers.

The Amazon Web Services (AWS) IoT Greengrass [24] supports machine learning inference locally on edge devices. The user could use his own pre-trained model or use models that are created, trained, and optimized in Amazon SageMaker (cloud), where massive computing resources are available. AWS IoT Greengrass features lambda runtime, a message manager, resource access, etc. The minimum hardware requirements are 1 GHz of computing speed and 128 MB of RAM.

Ghosh et al. [25] used autoencoders at the edge layer that are capable of dimensionality reduction to reduce the required processing time and storage space. The paper illustrates three scenarios. In the first one, data from sensors are sent to edge nodes, where data reduction is performed, and machine learning is then carried out in the cloud. In the second scenario, encoded data at the edge are decoded in the cloud to obtain the original amount of data and then perform machine learning tasks. Finally, pure cloud computing is performed, where data are sent from the sensors to the cloud. Results show that an autoencoder at the edge reduces the number of features and thus lowers the amount of data sent to the cloud.

Amiko's Respiro is a smart inhaler sensor featuring an ultra-low-power ARM Cortex-M processor [26]. This sensor uses machine learning to interpret vibration data from an inhaler. The processor allows for the running of ML algorithms where the sensor is trained to recognize breathing patterns and calculate important parameters. The collected data are processed in an application and feedback is provided.

Magno et al. [27] presented an open-source toolkit, namely FANNCortexM. It is built upon the Fast Artificial Neural Network (FANN) library and can run neural networks on the ARM Cortex-M series. This toolkit takes a neural network trained with FANN and generates code suitable for low-power microcontrollers. Another paper by Magno et al. [28] introduces a wearable multi-sensor bracelet for emotion detection that is able to run multilayer neural networks. In order to create, train, and test the neural network, the FANN library is used. To deploy the NN on the Cortex-M4F microcontroller, the above-mentioned library needs to be optimized using CMSIS and TI-Driverlib libraries.

FidoProject is a C++ machine learning library for embedded devices and robotics [29]. It implements a neural network for classification and other algorithms such as Reinforcement Learning. Alameh et al. [30] created a smart tactile sensing system by implementing a convolutional neural network on various hardware platforms like Raspberry Pi 4, NVidia Jetson TX2, and Movidius NCS2 for tactile data decoding.

As recent works used knowledge transfer (KT) techniques to transfer information from a large neural network to a small one in order to improve the performance of the latter, Sharma et al. [31] investigated the application of KT to edge devices, achieving good results by transferring knowledge from both the intermediate layers and the last layer of the teacher (original model) to a shallower student (target).

While most of the listed works use powerful edge devices (e.g., Cortex-A9, Raspberry PI) to test algorithms, especially NNs, there is a lack of performance analysis of common ML algorithms on mainstream microcontrollers. We intend to plug this gap by providing an open-source framework that we used for an extensive analysis.

## 4. Framework and Algorithm Understanding

The proposed Edge Learning Machine (EML) framework consists of two modules, one working on the desktop (namely DeskLM, for training and testing), and one on the edge (MicroLM, for inferencing and testing), as sketched in Figure 1.

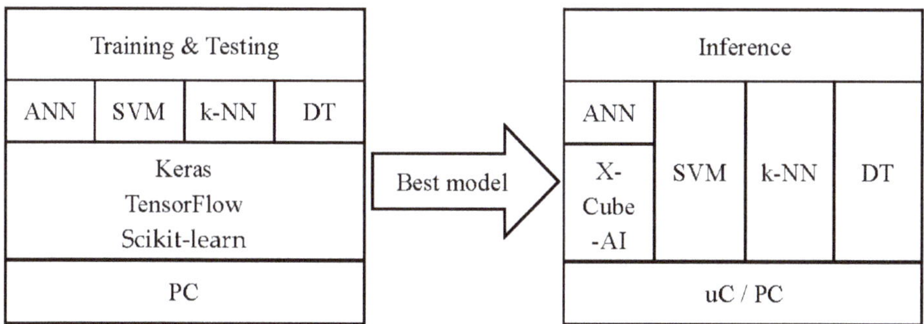


**Figure 1.** Block diagram of the Edge Learning Machine system architecture.

1. Desktop: the Desk-LM module is implemented in python and works on a PC to identify the best models for an input dataset. The current implementation involves four algorithms for both classification and regression: artificial neural networks (ANN), linear support vector machines (SVM), K-Nearest Neighbors (k-NN), and Decision Tree (DT) algorithms. For each algorithm, Desk-LM identifies the best model through hyperparameter tuning, as is described later in Table 2. Desk-LM relies on the scikit-learn python libraries [32] and exploits the TensorFlow [5] and Keras [33] packages for ANNs;

2. Edge: the MicroLM module reads and executes the models generated by Desk-LM. It is implemented in platform-independent C language (for linear kernel SVM, k-NN, DT) and can run on both microcontrollers and desktops, in order to perform inferences. ANNs are deployed using the X-Cube-AI expansion package for STM32 microcontrollers (TensorFlow and Keras on desktops).

The tool has been designed to support a four-step workflow, as shown in Figure 2.

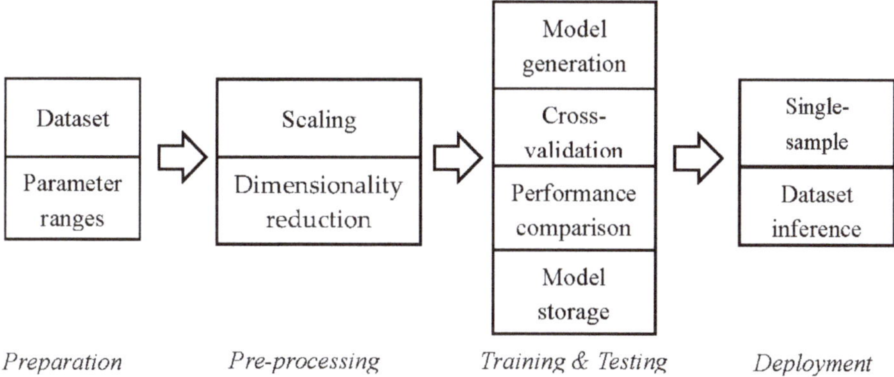


**Figure 2.** Supported workflow.

1. Preparation: in this first phase, the user provides the dataset and defines the range of the parameters to be investigated for each algorithm. The parameters are listed in Tables 1 and 2 (common and algorithm-specific, respectively—these common parameters are used by all the algorithms, even if they have different values);
2. Preprocessing: in this phase, data goes through the scaling and dimensionality reduction steps, which are important in order to allow optimal processing by the prediction algorithms [34]. The type of algorithm used for this step is one of the common parameters set by the user (Table 1);
3. Model generation: in this phase, all the configurations resulting from combining the values of the user-specified parameters (both common and algorithm-specific; see Tables 1 and 2, respectively) are evaluated through cross-validation, and their k values are, again, used as the parameters (Table 1). Most of the parameters (the algorithm's hyperparameters) can be assigned a list of values, each one of which is evaluated (scikit-learn exhaustive grid search), in order to allow for the selection of the best values. At the end of this step, the best model is saved in the disk, to be deployed on the edge. Desk-LM also saves the preprocessing parameters and, if needed for performance assessment purposes, the testing set (or a reduced version of it). All these files are then compiled in Micro-LM for the processing of data on the edge;
4. Deployment: in this final phase, the MicroLM module loads the model prepared on the desktop. The deployment process for our tests on microcontrollers is done using the STM32CubeIDE integrated development environment, which exploits the X-Cube-AI pack for ANNs. The software output by our framework supports both single-sample inference and whole dataset inference, for performance analysis purposes. In the latter case, Micro-LM exploits the testing set file produced by Desk-LM.

**Table 1.** Common configuration parameters.

| Common Parameters |
| :---: |
| Algorithm type (SVM, k-NN, DT, ANN) |
| Dataset |
| Content format (dataset start and end column, target column, etc.) |
| Number of classes (if classification) |
| Testing set size |
| Regression (True or False) |
| PCA (a specific number of features or MLE algorithm) |
| Normalization (standard or minmax) |
| K-fold cross-validation |
| Scoring metrics (accuracy, R2) |

**Table 2.** Algorithm-specific configuration parameters.

| Algorithm-Specific Configuration | | | |
| :---: | :---: | :---: | :---: |
| **ANN** | **Linear SVM** | **k-NN** | **DT** |
| Layer Shape | C | K (number of neighbors) | Splitting criterion |
| Activation Function | | Training set size for targets | max_depth |
| Dropout | | | min_samples_split |
| Loss metrics | | | min_samples_leaf |
| Number of epochs | | | max_leaf_nodes |
| Batch size | | | |
| Number of repeats | | | |

As anticipated, the current version of the EdgeLM framework features four well-established supervised learning algorithms, of which, in the following subsections, we briefly describe the implementation on both the desktop and edge side.

### 4.1. Artificial Neural Network (ANN)

In Desk-LM, ANNs are implemented through the TensorFlow [5] and its wrapper Keras [33] packages. As an optimizer, we use adaptive moment estimation ('adam') [35]. At each execution run, the DeskLM module performs the hyperparameter tuning by analyzing different ranges of parameters (Table 1 and first column of Table 2) specified by the user. The ANN model hyperparameters include layer shape (number and size of input, hidden, and output layers), activation function for the hidden layers (Rectified Linear Unit (ReLU), or Tangent Activation Function (Tanh)), number of epochs, batch size, number of repeats (in order to reduce result variance), and dropout rate. The best selected model is then saved in the high-efficiency Hierarchical Data Format 5 (HDF5) compressed format [36].

For the edge implementation, DeskLM relies on the STM X-Cube-AI expansion package, which is supported by STM32CubeIDE, and allows for its integration in the application of a trained Neural Network model. The package offers the possibility of compressing models up to eight times, with an accuracy loss which is estimated by the package. The tool also provides an estimation of the complexity, through the Multiply and Accumulate Operation (MACC) figure, and of the Flash and RAM memory footprint [37].

### 4.2. Linear Support Vector Machine (SVM)

As anticipated, for the simplicity of the implementation of the edge device, we implemented only the original, linear kernel SVM [13]. The linear model executes the y = w*x + b function, where w is the support vector and b is the bias. Model selection concerns the C regularization parameter [38] (Table 2). As an output model, Desk-LM generates a C source file containing the w and b values.

### 4.3. K-Nearest Neighbor (KNN)

For simplicity of implementation, we used a Euclidean distance criterion and majority voting (for K > 1). The training phase produces a very simple model (the K parameter), but deployment also requires the availability of the whole training set (Table 2).

### 4.4. Decision Tree (DT)

In order to cope with the limited resources of edge devices, our framework allows us to analyze different tree configurations in terms of depth, leaf size, and number of splits. Concerning the splitting criterion, for simplicity of implementation on the target microcontrollers, we implemented only the "Gini" method.

## 5. Experimental Analysis and Result

We conducted the experimental analysis using six ARM Cortex-M microcontrollers produced by STM, namely F091RC, F303RE, F401RE, F746ZG, H743ZI2, and L452RE. The F series represents a wide range of microcontroller families in terms of execution time, memory size, data processing and transfer capabilities [39], while the H series provides higher performance, security, and multimedia capabilities [40]. L microcontrollers are ultra-low-power devices used in energy-efficient embedded systems and applications [41]. All listed MCUs have been used in our experiments with their STM32CubeIDE default clock values, that could be increased for a faster response. Table 3 synthesizes the main features of these devices. In the analysis, we compare the performance of the embedded devices with that of a desktop PC hosting a 2.70 GHz Core i7 processor, with 16 GB RAM and 8 MB cache.

In order to characterize the performance of the selected edge devices, we have chosen six benchmark datasets to be representative of IoT applications (Table 4). These datasets represent different application scenarios: binary classification, multiclass classification, and regression. University of California Irvine (UCI) heart disease is a popular medical dataset [42]. Virus is a dataset developed by the University of Genova to deal with data traffic analysis [43–45]. Sonar represents the readings of a sonar system that analyses materials, distinguishing between rocks and metallic material [46,47]. Peugeot 207 contains various parameters collected from cars, which are used to predict either the road surface or the traffic (two labels were considered in our studies: label_14: road surface and label_15: traffic) [48]. The EnviroCar dataset records various vehicular signals through the onboard diagnostic (OBDII) interface to the Controller Area Network (CAN) bus [49–51]. The air quality index (AQI) dataset measures air quality in Australia during a period of one year [52]. Before processing, all data were converted to float32, according to the target execution platform.

**Table 3.** Microcontroller specifications.

| Microcontroller | Flash Memory | SRAM | Processor Speed Used (MHz) | Processor Cost ($) | Board Cost ($) |
| --- | --- | --- | --- | --- | --- |
| F091RC | 256 Kb | 32 Kb | 48 (max: 48) | 4.8 | 10.32 |
| F303RE | 512 Kb | 80 Kb | 72 (max: 72) | 7.72 | 10.32 |
| F401RE | 512 Kb | 96 Kb | 84 (max: 84) | 6.43 | 13 |
| F746ZG | 1 Mb | 340 Kb | 96 (max: 216) | 12.99 | 23 |
| H743ZI2 | 2 Mb | 1 Mb | 96 (max: 480) | 13.32 | 27 |
| L452RE | 512 Kb | 160 Kb | 80 (max: 80) | 7.03 | 14 |

**Table 4.** Dataset specifications.

| Dataset | Samples Features | Type |
| --- | --- | --- |
| Heart | $303 \times 13$ | Binary Classification |
| Virus | $24736 \times 13$ | Binary Classification |
| Sonar | $209 \times 60$ | Binary Classification |
| Peugeot 207 * | $8615 \times 14$ | Multiclass Classification |
| EnviroCar | $47077 \times 5$ | Regression |
| AQI | $367 \times 8$ | Regression |

* For Peugeot 207, we considered two different labels.

Our analysis was driven by a set of questions, synthesized in Table 5, aimed at investigating the performance of different microcontrollers in typical ML IoT contexts. We are also interested in comparing the inference performance of microcontrollers vs. desktops. The remainder of this section is devoted to the analysis of each research question. In a few cases, when the comparison is important, results are reported for every tested target platform. On the other hand, in most of the cases, when not differently stated, we chose the F401RE device as the reference for the embedded targets.

**Table 5.** Research questions.

| Research Questions (RQ) | Description |
| --- | --- |
| 1. Performance | Score (accuracy, R2) and inference time |
| 2. Scaling | Effect of scaling data |
| 3. PCA | Effect of dimensionality reduction on score |
| 4. ANN Layer Configuration | Different layer shapes (depth and thickness) |
| 5. ANN Activation Function | Effect of different neuron activation functions |
| 6. ANN Batch Size | Effect of batch size on score and time |
| 7. ANN Epochs | Effect of number of epochs in training |
| 8. ANN Dropout | Effect of a regularization technique to avoid overfitting |
| 9. SVM Regularization Training Time | SVM training time with different values of the "C" regularization parameter |
| 10. DT Parameters | Tuning the decision tree |

## 5.1. Performance

The first research question concerns the performance achieved both on desktop and on edge. For SVM, k-NN and DT on desktops, we report the performance of both our C implementation and the python scikit-learn implementation, while for ANN we have only the TensorFlow Keras implementation. The following set of tables show, for each algorithm, the obtained score, which is expressed in terms of accuracy (in percent, for classification problems), or coefficient of determination, R-Squared (R2, for regression problems). R2 is the proportion of the variance in the dependent variable that is predictable from the independent variable(s). The best possible score for R2 is 1.0. In scikit-learn, R2 can assume negative values, because the model can be arbitrarily worse. The second performance we consider is the inference time.

In the following (Tables 6–13), we report two tables for each algorithm. The first one provides the best performance (in terms of score) obtained in each dataset. The second shows the hyperparameter values of the best model.

**Table 6.** Artificial Neural Network (ANN) performance.

| | | | ANN | | | | | | |
| | | | Desktop | MCUs | | | | | |
| DatasetDataset | | Performance | | | | | | | |
| Type | Name | | Python | F0 | F3 | F4 | F7 | H7 | L4 |
|---|---|---|---|---|---|---|---|---|---|
| Binary | Heart | Accuracy | 84% | * | 84% | 84% | 84% | 84% | 84% |
| | | Inf. Time | <1 ms | * | 3 ms | 1 ms | <1 ms | <1 ms | 1 ms |
| | Virus | Accuracy | 99% | * | 99% | 99% | 99% | 99% | 99% |
| | | Inf. Time | <1 ms | * | 5 ms | 3 ms | 1 ms | 1 ms | 4 ms |
| | Sonar | Accuracy | 87% | * | 87% | 87% | 87% | 87% | 87% |
| | | Inf. Time | <1 ms | * | 16 ms | 8 ms | 3 ms | 3 ms | 10 ms |
| Multiclass | Peugeot_Target 14 | Accuracy | 99% | * | 99% | 99% | 99% | 99% | 99% |
| | | Inf. Time | <1 ms | * | 2 ms | 1 ms | <1 ms | <1 ms | 1 ms |
| | Peugeot_Target 15 | Accuracy | 99% | * | 99% | 99% | 99% | 99% | 99% |
| | | Inf. Time | <1 ms | * | 18 ms | 10 ms | 4 ms | 4 ms | 12 ms |
| Regression | Enviro Car | $R^2$ | 0.99 | * | 0.99 | 0.99 | 0.99 | 0.99 | 0.99 |
| | | Inf. Time | <1 ms | * | <1 ms | <1 ms | <1 ms | <1 ms | <1 ms |
| | AQI | $R^2$ | 0.86 | * | 0.86 | 0.86 | 0.86 | 0.86 | 0.86 |
| | | Inf. Time | <1 ms | * | 4 ms | 2 ms | 1 ms | 1 ms | 3 ms |

*: F0 not supported by the STM X-Cube-AI package.

**Table 7.** ANN corresponding configurations.

| Dataset | Best Configuration (Table 1, Table 2) | | | | |
|---|---|---|---|---|---|
| | AF | LC | PCA | Dropout | Scaling |
| Heart | Tanh | [500] | 30% | 0 | StandardScaler |
| Virus | Tanh | [100,100,100] | None | 0 | StandardScaler |
| Sonar | ReLU | [300,200,100,50] | 30% | 0 | MinMaxScaler |
| Peugeot_Target 14 | ReLU | [500] | None | 0 | StandardScaler |
| Peugeot_Target 15 | Tanh | [300,200,100,50] | None | 0 | StandardScaler |
| EnviroCar | Tanh | [50] | mle | 0 | MinMaxScaler |
| AQI | ReLU | [300,200,100,50] | None | 0 | MinMaxScaler |

Activation Function (AF), Layer Configuration (LC), Principal Component Analysis (PCA).

1.  ANN:

Remarkably, all the embedded platforms were able to achieve the same score (accuracy or R2) as the desktop python implementation. None of the chosen datasets required the compression of the models by the STM X-Cube-AI package. ANN performed well in general, except for the Heart and Virus datasets, where the accuracy is under 90%. The inference time is relatively low in both desktop and MCUs (with similar values, in the order of ms and sometimes less). However, there is an exception in some cases—especially for Peugeot_Target_15 and Sonar—when using the F3 microcontroller.

**Table 8.** Linear Support Vector Machine (SVM) performance.

| | | | Linear SVM | | | | | | | |
|---|---|---|---|---|---|---|---|---|---|---|
| Dataset | | Score | Desktop | | MCUs | | | | | |
| Type | Name | | Python | C | F0 | F3 | F4 | F7 | H7 | L4 |
| Binary | Heart | Acc. | 84% | 84% | 84% | 84% | 84% | 84% | 84% | 84% |
| | | Inf. Time | <1 ms | <1 ms | <1 ms | <1 ms | <1 ms | <1 ms | <1 ms | <1 ms |
| | Virus | Acc. | 94% | 94% | 94% | 94% | 94% | 94% | 94% | 94% |
| | | Inf. Time | <1 ms | <1 ms | 1 ms | <1 ms | <1 ms | <1 ms | <1 ms | <1 ms |
| | Sonar | Acc. | 78% | 78% | 78% | 78% | 78% | 78% | 78% | 78% |
| | | Inf. Time | <1 ms | <1 ms | 3 ms | <1 ms | <1 ms | <1 ms | <1 ms | <1 ms |
| Multiclass | Peugeot_Target 14 | Acc. | 91% | 91% | 91% | 91% | 91% | 91% | 91% | 91% |
| | | Inf. Time | <1 ms | <1 ms | 2 ms | <1 ms | <1 ms | <1 ms | <1 ms | <1 ms |
| | Peugeot_Target 15 | Acc. | 90% | 90% | 90% | 90% | 90% | 90% | 90% | 90% |
| | | Inf. Time | <1 ms | <1 ms | 2 ms | <1 ms | <1 ms | <1 ms | <1 ms | <1 ms |
| Regress | EnviroCar | $R^2$ | 0.99 | 0.99 | 0.99 | 0.99 | 0.99 | 0.99 | 0.99 | 0.99 |
| | | Inf. Time | <1 ms | <1 ms | 5 ms | 3 ms | <1 ms | <1 ms | <1 ms | <1 ms |
| | AQI | $R^2$ | 0.73 | 0.73 | 0.73 | 0.73 | 0.73 | 0.73 | 0.73 | 0.73 |
| | | Inf. Time | <1 ms | <1 ms | 5 ms | 3 ms | <1 ms | <1 ms | <1 ms | <1 ms |

**Table 9.** Linear SVM corresponding configuration.

| Dataset | Best Configuration (Table 1, Table 2) | | |
|---|---|---|---|
| | C | PCA | Scaling |
| Heart | 0.1 | 30% | StandardScaler |
| Virus | 1 | None | StandardScaler |
| Sonar | 0.01 | None | StandardScaler |
| Peugeot_Target 14 | 0.1 | mle | StandardScaler |
| Peugeot_Target 15 | 10 | mle | StandardScaler |
| EnviroCar | 0.1 | None | StandardScaler |
| AQI | 1 | mle | StandardScaler |

SVM regularization parameter (C), Principal Component Analysis (PCA).

**Table 10.** k-Nearest Neighbors (k-NN) performance.

| | | | k-NN | | | | | | | |
|---|---|---|---|---|---|---|---|---|---|---|
| Dataset | | Score | Desktop | | MCUs | | | | | |
| Type | Name | | Python | C | F0 | F3 | F4 | F7 | H7 | L4 |
| Bin. | Heart | Acc. | 83% | 83% | 83% | 83% | 83% | 83% | 83% | 83% |
| | | Inf Time | <1 ms | <1 ms | 366 ms | 71 ms | 7 ms | 4 ms | 4 ms | 8 ms |
| | Virus | Acc. | 99% | 99% | 95% | 95% | 95% | 95% | 95% | 95% |
| | | Inf Time | <1 ms | <1 ms | 199 ms | 38 ms | 4 ms | 3 ms | 3 ms | 4 ms |
| | Sonar | Acc. | 92% | 92% | 76% | 92% | 92% | 92% | 92% | 92% |
| | | Inf Time | <1 ms | <1 ms | 329 ms | 140 ms | 14 ms | 10 ms | 10 ms | 14 ms |
| Multi class | Peugeot_Target 14 | Acc. | 98% | 98% | 88% | 88% | 88% | 88% | 98% | 88% |
| | | Inf Time | <1 ms | <1 ms | 205 ms | 39 ms | 4 ms | 3 ms | 200 ms | 4 ms |
| | Peugeot_Target 15 | Acc. | 97% | 97% | 86% | 86% | 86% | 86% | 97% | 86% |
| | | Inf Time | <1 ms | <1 ms | 204 ms | 39 ms | 4 ms | 3 ms | 200 ms | 4 ms |
| Regr | Enviro | $R^2$ | 0.99 | 0.99 | 0.97 | 0.97 | 0.97 | 0.97 | 0.97 | 0.97 |
| | Car | Inf Time | <1 ms | <1 ms | 125 ms | 30 ms | 3 ms | 2 ms | 2 ms | 4 ms |
| | AQI | $R^2$ | 0.73 | 0.73 | 0.73 | 0.73 | 0.73 | 0.73 | 0.73 | 0.73 |
| | | Inf Time | <1 ms | <1 ms | 414 ms | 85 ms | 9 ms | 6 ms | 6 ms | 10 ms |

2.  Linear SVM:

As with ANN, for the linear SVM, we obtained the same score across all the target platforms, and relatively short inference times (again, with almost no difference between desktop and microcontroller implementations). However, we obtained significantly worse results than ANN for more than half of the investigated datasets. Table 9 stresses the importance of tuning the C regularization parameter, which implies the need for longer training times, particularly in the absence of normalization. We explore this in more depth when analyzing research question 9.

**Table 11.** k-NN corresponding configurations.

| Dataset | Best Configuration (Table 1, Table 2) | | | |
| --- | --- | --- | --- | --- |
| | K | PCA | Scaling | Notes |
| Heart | 10 | mle | StandardScaler | This configuration fits all targets |
| Virus | 1 | None | StandardScaler | In all MCUs K = 1, training set cap = 100 |
| Sonar | 1 | None | MinMaxScaler | In F0 K = 1, training set cap = 50 |
| Peugeot_Target 14 | 1 | mle | MinMaxScaler | In F0, F3, F4, F7, L4 K = 4, training set cap = 100 |
| Peugeot_Target 15 | 1 | mle | MinMaxScaler | In F0, F3, F4, F7, L4 K = 3, training set cap = 100 |
| EnviroCar | 3 | mle | MinMaxScaler | In all MCUs K = 2, training set cap = 100 |
| AQI | 1 | mle | StandardScaler | This configuration fits all targets |

Number of neighbors (K), Principal Component Analysis (PCA).

**Table 12.** Desktop (DT) performance.

| | | | DT | | | | | | | |
| --- | --- | --- | --- | --- | --- | --- | --- | --- | --- | --- |
| Dataset | | Score | Desktop | | MCUs | | | | | |
| Type | Name | | Python | C | F0 | F3 | F4 | F7 | H7 | L4 |
| Bin. | Heart | Accuracy | 78% | 78% | 78% | 78% | 78% | 78% | 78% | 78% |
| | | Inf. Time | <1 ms | <1 | <1 | <1 | <1 | <1 | <1 | <1 |
| | Virus | Accuracy | 99% | 99% | 99% | 99% | 99% | 99% | 99% | 99% |
| | | Inf. Time | <1 ms | <1 | <1 | <1 | <1 | <1 | <1 | <1 |
| | Sonar | Accuracy | 76% | 76% | 76% | 76% | 76% | 76% | 76% | 76% |
| | | Inf. Time | <1 ms | <1 | <1 | <1 | <1 | <1 | <1 | <1 |
| Multi class | Peugeot_Target 14 | Accuracy | 99% | 99% | 99% | 99% | 99% | 99% | 99% | 99% |
| | | Inf. Time | <1 ms | <1 | <1 | <1 | <1 | <1 | <1 | <1 |
| | Peugeot_Target 15 | Accuracy | 98% | 98% | 98% | 98% | 98% | 98% | 98% | 98% |
| | | Inf. Time | <1 ms | <1 | <1 | <1 | <1 | <1 | <1 | <1 |
| Regr | Enviro Car | $R^2$ | 0.99 | 0.99 | 0.99 | 0.99 | 0.99 | 0.99 | 0.99 | 0.99 |
| | | Inf. Time | <1 | <1 | 2 | 2 | <1 | <1 | <1 | <1 |
| | AQI | $R^2$ | 0.65 | 0.65 | 0.65 | 0.65 | 0.65 | 0.65 | 0.65 | 0.65 |
| | | Inf. Time | <1 ms | <1 | 2 | 2 | <1 | <1 | <1 | <1 |

**Table 13.** DT corresponding configurations. Time is in ms (omitted for reasons of space).

| Dataset | Best Configuration (Table 1, Table 2) | | | | | |
| --- | --- | --- | --- | --- | --- | --- |
| | Max Depth | Min Sample Split | Min Sample Leaf | Max Leaf Nodes | PCA | Notes |
| Heart | 7 | 2 | 10 | 80 | mle | * |
| Virus | None | 5 | 1 | 80 | None | * |
| Sonar | 7 | 2 | 10 | 80 | mle | * |
| Peugeot_Target 14 | None | 2 | 3 | 80 | None | * |
| Peugeot_Target 15 | None | 2 | 1 | 200 | None | * |
| EnviroCar | None | 2 | 1 | 5000 | None | Max leaf nodes = 1000 in F3, F4, L4; max leaf nodes = 200 for F0 |
| AQI | None | 10 | 3 | 80 | None | * |

Principal Component Analysis (PCA). * This configuration fits all targets.

3.  k-NN:

Notably, in some cases, the training set cap needed to be set to 100, because the Flash size was a limiting factor for some MCUs. Hence, for different training sets, we also had a different number of neighbors (K). Accordingly, the accuracy is also affected by the decrease in training set size, since the number of examples used for training is reduced. This effect is apparent for Sonar with an F0 device. This dataset has sixty features, much more than the others (typically 10–20 features). The inference time varies a lot among datasets, microcontrollers and in comparison with the desktop implementations.

This is because the k-NN inference algorithm always requires the exploration of the whole training set, and thus its size plays an important role in performance, especially for less powerful devices. In the multiclass problems, k-NN exploits the larger memory availability of H7 well, outperforming SVM, and reaching a performance level close to that of ANN. It is important to highlight that the Sonar labels were reasonably well predicted by k-NN compared to ANN and SVM (92% vs. 87% and 78%). In general, k-NN achieves performance levels similar to ANN, but requires a much larger memory footprint, which is possible only on the highest-end targets.

4.  DT:

When processing the EnviroCar dataset, the DT algorithm saturated the memory in most of the targets. We had to reduce the leaf size for all MCU families, apart from F7 and H7. However, this reduction did not significantly reduce the $R^2$ value. In addition, DT performs worse than the others in two binary classification datasets, Heart and Sonar, and in the AQI regression dataset as well, but performs at the same level as the ANNs for the multiclass datasets and in the EnviroCar regression problem. Notably, DT achieves the fastest inference time among all algorithms, with F0 and F3 performing worse than the others, particularly in the regression problems.

As a rough summary of the first research question, we can conclude that ANN and, surprisingly, k-NN, had the highest accuracy in most cases, and Decision Tree had the shortest response time, but accuracy results were quite dependent on the dataset. The main difference between ANN and k-NN results is represented by the fact that high performance in ANN is achieved by all the targets (but not F0, which is not supported by the STM X-Cube-AI package), while k-NN poses much higher memory requirements. Concerning the timing performance, microcontrollers perform similarly to desktop implementations on the studied datasets. The only exception is found in k-NN, for which each inference requires the exploration of the whole dataset, and the corresponding computational demand penalizes the performance, especially on low-end devices. When comparing the edge devices, the best time performance is achieved by F7 and H7 (and we used default clock speeds, that can be significantly increased). Unsurprisingly, given the available hardware, F0 performs worse than all the others. Considering the score, we managed to train all the edge devices to achieve the same level of performance as the desktop in each algorithm, with the exception of k-NN in the multiclass tests (Peugeot), where only H7 is able to perform like a desktop, but with a significant time performance penalty. On the other hand, F0 performs significantly worse than the other edge devices in the k-NN Sonar binary classification.

*5.2. Scaling*

Feature preprocessing is applied to the original features before the training phase, with the goal of increasing prediction accuracy and speeding up response times [34]. Since the range of values is typically different from one feature to another, the proper computation of the objective function requires normalized inputs. For instance, the computation of the Euclidean distance between points is governed by features with a broader value range. Moreover, gradient descent converges much faster on normalized values [53].

We considered three cases that we applied on ANN, SVM, and k-NN: no scaling, MinMax Scaler, and Standard Scaler (Std) [54]. The set of tables below (Tables 14–18) show the accuracy of $R^2$ for all datasets under various scaling conditions. Most common DT algorithms are invariant to monotonic transformations [55], so we did not consider DT in this analysis.

1. ANN:

**Table 14.** Performance and configuration of ANN with no scaling.

| | **ANN** | | | |
| | | **Configuration** | | |
| **Dataset** | **None** | **AF** | **LC** | **PCA** |
|---|---|---|---|---|
| Heart | 78% | ReLU | [500] | mle |
| Virus | 74% | Tanh | [100, 100, 100] | mle |
| Sonar | 85% | ReLU | [100, 100, 100] | mle |
| Peugeot_Target 14 | 95% | Tanh | [500] | mle |
| Peugeot_Target 15 | 92% | ReLU | [100, 100, 100] | mle |
| EnviroCar | 0.97 | Tanh | [50] | mle |
| AQI | 0.70 | ReLU | [300, 200, 100, 50] | None |

Activation Function (AF), Layer Configuration (LC), Principal Component Analysis (PCA).

**Table 15.** Performance and configuration of ANN with MinMax scaling.

| | **ANN** | | | |
| | | **Configuration** | | |
| **Dataset** | **MinMax** | **AF** | **LC** | **PCA** |
|---|---|---|---|---|
| Heart | 80% | Tanh | [300, 200, 100, 50] | 30% |
| Virus | 99% | ReLU | [100, 100, 100] | None |
| Sonar | 87% | ReLU | [300, 200, 100, 50] | 30% |
| Peugeot_Target 14 | 99% | ReLU | [100, 100, 100] | mle |
| Peugeot_Target 15 | 98% | Tanh | [300, 200, 100, 50] | mle |
| EnviroCar | 0.99 | ReLU | [50] | mle |
| AQI | 0.86 | ReLU | [300, 200, 100, 50] | None |

Activation Function (AF), Layer Configuration (LC), Principal Component Analysis (PCA).

**Table 16.** Performance and configuration of ANN with StandardScaler normalization.

| | **ANN** | | | |
| | | **Configuration** | | |
| **Dataset** | **Std** | **AF** | **LC** | **PCA** |
|---|---|---|---|---|
| Heart | 84% | Tanh | [500] | 30% |
| Virus | 99% | Tanh | [100, 100, 100] | None |
| Sonar | 86% | ReLU | [100, 100, 100] | mle |
| Peugeot_Target 14 | 99% | ReLU | [500] | None |
| Peugeot_Target 15 | 99% | Tanh | [300, 200, 100, 50] | None |
| EnviroCar | 0.99 | Tanh | [50] | mle |
| AQI | 0.84 | ReLU | [300, 200, 100, 50] | None |

Activation Function (AF), Layer Configuration (LC), Principal Component Analysis (PCA).

2. SVM:

**Table 17.** Performance and configuration of SVM for different scaling techniques.

| | **SVM** | | | | | | | | |
| | | **Configuration** | | | **Configuration** | | | **Configuration** | |
| **Dataset** | **None** | **C** | **PCA** | **MinMax** | **C** | **PCA** | **Std** | **C** | **PCA** |
|---|---|---|---|---|---|---|---|---|---|
| Heart | 78% | 0.01 | mle | 79% | 1 | mle | 84% | 0.1 | 30% |
| Virus | 71% | 0.01 | mle | 94% | 100 | None | 94% | 1 | None |
| Sonar | 77% | 0.1 | mle | 73% | 0.1 | None | 78% | 0.01 | None |
| Peugeot_Target 14 | 50% | 0.1 | mle | 91% | 10 | mle | 91% | 0.1 | mle |
| Peugeot_Target 15 | 76% | 0.01 | mle | 90% | 10 | mle | 90% | 10 | mle |
| EnviroCar | 0.97 | Slow | mle | 0.98 | 0.1 | None | 0.99 | 0.1 | None |
| AQI | 0.7 | 100 | mle | 0.62 | 100 | 30% | 0.73 | 1 | mle |

SVM regularization parameter (C), Principal Component Analysis (PCA).

3. k-NN:

**Table 18.** Performance and configuration of k-NN for different scaling techniques.

| | | k-NN | | | | | | | | |
|---|---|---|---|---|---|---|---|---|---|---|
| | | | Configuration | | | Configuration | | | Configuration | |
| Dataset | None | | K | PCA | MinMax | K | PCA | Std | K | PCA |
| Heart | 63% | | 3 | mle | 75% | 3 | None | 83% | 10 | mle |
| Virus | 95% | | 1 | mle | 99% | 1 | None | 99% | 1 | None |
| Sonar | 77% | | 1 | mle | 92% | 1 | None | 87% | 1 | None |
| Peugeot_Target 14 | 91% | | 1 | mle | 98% | 1 | mle | 97% | 1 | mle |
| Peugeot_Target 15 | 89% | | 2 | mle | 97% | 1 | mle | 97% | 1 | None |
| EnviroCar | 0.98 | | 5 | mle | 0.99 | 3 | mle | 0.99 | 3 | None |
| AQI | 0.73 | | 4 | mle | 0.7 | 3 | mle | 0.73 | 6 | mle |

Number of neighbors (K), Principal Component Analysis (PCA).

These results clearly show the importance across all the datasets and algorithms of scaling the inputs. For instance, MinMax scaling allowed ANNs to reach 99% accuracy in Virus (from a 74% baseline), and Peugeot 14 (from 95%) and 0.86 R2 (from 0.70) in AQI. The application of MinMax allowed SVM to achieve 94% accuracy in Virus (form 71%) and 91% accuracy in Peugeot 14 (from 50%). Standard input scaling improved the k-NN accuracy of Heart from 63% to 83%. For large regression datasets, especially with SVM (see also research question 9), input scaling avoids large training times.

*5.3. Principal Component Analysis (PCA)*

Dimensionality reduction allows us to reduce the effects of noise, space and processing requirements. One well-known method is Principal Component Analysis (PCA), which performs an orthogonal transformation to convert a set of observations of possibly correlated variables into a set of values of linearly independent variables, which are called principal components [24]. We tried different values of PCA dimension reduction: none, 30% (i.e., the algorithm selects a number of components such that the amount of variance that needs to be explained is greater than 30%), and automatic maximum likelihood estimation (mle) [56], whose results are shown in Tables 19–26.

1. SVM:

**Table 19.** SVM performance and configuration for various PCA values.

| | SVM | | | | | | | | | |
|---|---|---|---|---|---|---|---|---|---|---|
| Dataset | PCA = None | | | PCA = 30% | | | PCA = mle | | | |
| | Score | | Configuration | | Score | Configuration | | Score | Configuration | |
| | | C | Scaling | | | C | Scaling | | C | Scaling |
| Heart | 78% | 0.01 | Std | | 84% | 0.1 | Std | 79% | 0.1 | Std |
| Virus | 99% | 1 | Std | | 86% | 100 | MinMax | 94% | 0.1 | Std |
| Sonar | 78% | 0.01 | Std | | 75% | 100 | Std | 77% | 0.01 | Std |
| Peugeot_Target 14 | 91% | 10 | MinMax | | 83% | 0.1 | MinMax | 91% | 0.1 | Std |
| Peugeot_Target 15 | 90% | 10 | MinMax | | 86% | 10 | MinMax | 90% | 10 | Std |
| EnviroCar | 0.99 | 0.1 | Std | | 0.94 | 0.1 | MinMax | 0.98 | 0.1 | MinMax |
| AQI | 0.73 | 10 | Std | | 0.71 | 100 | Std | 0.73 | 1 | Std |

SVM regularization parameter (C).

2. k-NN:

**Table 20.** k-NN performance and configuration for various PCA techniques.

| | k-NN | | | | | | | | |
|---|---|---|---|---|---|---|---|---|---|
| Dataset | PCA = None | | | PCA = 30% | | | PCA = mle | | |
| | Score | Configuration | | Score | Configuration | | Score | Configuration | |
| | | K | Scaling | | K | Scaling | | K | Scaling |
| Heart | 77% | 3 | Std | 76% | 13 | Std | 83% | 10 | Std |
| Virus | 99% | 1 | Std | 99% | 1 | Std | 99% | 1 | Std |
| Sonar | 92% | 1 | MinMax | 84% | 1 | MinMax | 97% | 1 | Std |
| Peugeot Targ 14 | 98% | 1 | MinMax | 92% | 3 | MinMax | 98% | 1 | Std |
| Peugeot Targ 15 | 97% | 1 | MinMax | 90% | 6 | MinMax | 97% | 1 | Std |
| EnviroCar | 0.99 | 3 | MinMax | 0.99 | 9 | Std | 0.99 | 3 | MinMax |
| AQI | 0.73 | 6 | Std | 0.57 | 3 | MinMax | 0.73 | 6 | Std |

Number of neighbors (K).

3. ANN:

**Table 21.** ANN performance and configuration for PCA = None.

| | ANN | | | |
|---|---|---|---|---|
| Dataset | Score | Configuration | | |
| | | AF | LC | Scaling |
| Heart | 81% | Tanh | [100, 100, 100] | Std |
| Virus | 99% | Tanh | [100, 100, 100] | Std |
| Sonar | 84% | ReLU | [50] | Std |
| Peugeot_Target 14 | 99% | ReLU | [500] | Std |
| Peugeot_Target 15 | 99% | Tanh | [300, 200, 100, 50] | Std |
| EnviroCar | 0.99 | Tanh | [50] | MinMax |
| AQI | 0.86 | ReLU | [300,200,100,50] | MinMax |

Activation Function (AF), Layer Configuration (LC).

**Table 22.** ANN performance and configuration for PCA = 30%.

| | ANN | | | |
|---|---|---|---|---|
| Dataset | Score | Configuration | | |
| | | AF | LC | Scaling |
| Heart | 84% | Tanh | [500] | Std |
| Virus | 98% | ReLU | [100, 100, 100] | Std |
| Sonar | 87% | ReLU | [300, 200, 100, 50] | MinMax |
| Peugeot_Target 14 | 92% | ReLU | [50] | MinMax |
| Peugeot_Target 15 | 90% | ReLU | [50] | MinMax |
| EnviroCar | 0.98 | ReLU | [50] | MinMax |
| AQI | 0.71 | ReLU | [100,100,100] | MinMax |

Activation Function (AF), Layer Configuration (LC).

**Table 23.** ANN performance and configuration for PCA = mle.

| | ANN | | | |
|---|---|---|---|---|
| Dataset | Score | Configuration | | |
| | | AF | LC | Scaling |
| Heart | 83% | Tanh | [300,200,100,50] | Std |
| Virus | 99% | Tanh | [100,100,100] | Std |
| Sonar | 86% | ReLU | [100,100,100] | Std |
| Peugeot_Target 14 | 99% | Tanh | [100,100,100] | Std |
| Peugeot_Target 15 | 99% | Tanh | [300,200,100,50] | Std |
| EnviroCar | 0.99 | Tanh | [50] | MinMax |
| AQI | 0.82 | Tanh | [300,200,100,50] | MinMax |

Activation Function (AF), Layer Configuration (LC).

4.  DT:

**Table 24.** DT performance and configuration for PCA = None.

| | | DT | | | |
| --- | --- | --- | --- | --- | --- |
| | | Configuration | | | |
| Dataset | Score | Max-Depth | Min_Sample _Split | Min_Sample _Leaf | Max_Leaf _Nodes |
| Heart | 67% | 7 | 2 | 10 | 80 |
| Virus | 99% | None | 5 | 1 | 80 |
| Sonar | 65% | 7 | 2 | 10 | 80 |
| Peugeot_Target 14 | 99% | None | 2 | 3 | 80 |
| Peugeot_Target 15 | 98% | None | 2 | 1 | 200 |
| EnviroCar | 0.99 | None | 2 | 1 | 5000 |
| AQI | 0.65 | None | 10 | 3 | 80 |

**Table 25.** DT performance and configuration for PCA = 30%.

| | | DT | | | |
| --- | --- | --- | --- | --- | --- |
| | | Configuration | | | |
| Dataset | Score | Max-Depth | Min_Sample _Split | Min_Sample _Leaf | Max_Leaf _Nodes |
| Heart | 62% | 3 | 2 | 1 | 80 |
| Virus | 96% | None | 2 | 1 | 1000 |
| Sonar | 75% | 7 | 2 | 3 | 80 |
| Peugeot_Target 14 | 84% | None | 2 | 1 | 200 |
| Peugeot_Target 15 | 87% | 7 | 2 | 10 | 80 |
| EnviroCar | 0.98 | None | 2 | 10 | 1000 |
| AQI | 0.62 | 7 | 10 | 1 | 80 |

The results reported in the above tables are quite varied. The 30% PCA value is frequently too low, except for the Sonar dataset, which has 60 features, much more than the others, and thus looks less sensitive to such a coarse reduction. For SVM, PCA does not perform better than or equal to mle, while the opposite is true for k-NN. Moreover, in ANNs, mle tends to provide better results, except AQI. In DT, there is a variance of outcomes. Mle does not perform any better than the other algorithms in Heart (78% vs. 67% accuracy) and Sonar (76% vs. 65%), while performance decreases for Peugeot 14 (93% vs. 99%) and AQI (0.49 vs. 0.65). For AQI, PCA never improves performance. The opposite is true for Heart and (except SVM) Sonar.

**Table 26.** DT performance and configuration for PCA = mle.

| | | DT | | | |
| --- | --- | --- | --- | --- | --- |
| | | Configuration | | | |
| Dataset | Score | Max-Depth | Min_Sample _Split | Min_Sample _Leaf | Max_Leaf _Nodes |
| Heart | 78% | 7 | 2 | 10 | 80 |
| Virus | 99% | None | 2 | 1 | 200 |
| Sonar | 76% | 7 | 2 | 10 | 80 |
| Peugeot_Target 14 | 93% | None | 5 | 1 | 80 |
| Peugeot_Target 15 | 92% | None | 10 | 1 | 80 |
| EnviroCar | 0.98 | None | 2 | 10 | 5000 |
| AQI | 0.49 | 7 | 5 | 1 | 80 |

### 5.4. ANN Layer Configuration

To answer this question, we investigated performance among four ANN hidden-layer configurations, as follows:

1. One hidden layer of 50 neurons;
2. One hidden layer of 500 neurons;
3. Three hidden layers of 100 neurons each;
4. Four hidden layers with 300, 200, 100, 50 neurons, respectively.

Tables 27–29 indicate the highest performance for each layer shape.

**Table 27.** Results for layer configuration LC = [50] and LC = [500]. We have omitted columns with all zero Dropout values.

| Dataset | LC = [50] | Configuration | | | LC = [500] | Configuration | | |
|---|---|---|---|---|---|---|---|---|
| | | AF | PCA | Scaling | | AF | PCA | Scaling |
| Heart | 83% | Tanh | 30% | Std | 84% | Tanh | 30% | Std |
| Virus | 98% | ReLU | None | Std | 98% | ReLU | None | Std |
| Sonar | 84% | ReLU | None | Std | 81% | ReLU | mle | Std |
| Peugeot_Target 14 | 98% | ReLU | mle | Std | 99% | ReLU | None | Std |
| Peugeot_Target 15 | 98% | ReLU | mle | Std | 98% | ReLU | None | Std |
| EnviroCar | 0.99 | Tanh | mle | MinMax | 0.99 | ReLU | mle | MinMax |
| AQI | 0.76 | Tanh | mle | MinMax | 0.78 | ReLU | mle | MinMax |

Activation Function (AF), Principal Component Analysis (PCA).

**Table 28.** Layer configuration LC = [100,100,100] results.

| Dataset | LC = [100,100,100] | Configuration | | | |
|---|---|---|---|---|---|
| | | AF | PCA | Scaling | Dropout |
| Heart | 84% | Tanh | 30% | Std | 0 |
| Virus | 99% | Tanh | None | Std | 0 |
| Sonar | 87% | ReLU | 30% | Std | 0 |
| Peugeot Target 14 | 99% | ReLU | mle | Std | 0 |
| Peugeot Target 15 | 98% | Tanh | mle | Std | 0.1 |
| EnviroCar | 0.99 | Tanh | mle | MinMax | 0 |
| AQI | 0.80 | ReLU | mle | MinMax | 0 |

Activation Function (AF), Principal Component Analysis (PCA).

**Table 29.** Layer Configuration (LC) = [300,200,100,50] results.

| Dataset | LC = [300,200,100,50] | Configuration | | |
|---|---|---|---|---|
| | | AF | PCA | Scaling |
| Heart | 84% | Tanh | 30% | Std |
| Virus | 99% | ReLU | None | Std |
| Sonar | 87% | ReLU | 30% | MinMax |
| Peugeot Target 14 | 99% | ReLU | mle | MinMax |
| Peugeot Target 15 | 99% | Tanh | None | Std |
| EnviroCar | 0.99 | Tanh | mle | MinMax |
| AQI | 0.86 | ReLU | None | MinMax |

Activation Function (AF), Principal Component Analysis (PCA).

By observing the results, we can see that deepening the network tends to improve the results, but only up to a certain threshold. For the Heart dataset, which has the lowest overall accuracy, we tried additional, deeper shapes beyond those reported in Tables 27–29, but with no better results. On the other hand, widening the first layer provides only slightly better results (and in one case worsens them).

## 5.5. ANN Activation Function

Another relevant design choice concerns the activation function in the hidden layers. Activation functions are attached to each neuron in the network and define its output. They introduce a non-linear factor in the processing of a neural network. Two activation functions are typically used: Rectified Linear Unit (ReLU) and Tangent Activation Function (Tanh). On the other hand, for the output layer, we used a sigmoid for binary classification models as an activation function, and a softmax for multiclassification tasks. For regression problems, we created an output layer without any activation function (i.e., we use the default "linear" activation), as we are interested in predicting numerical values directly, without transformation. Tables 30 and 31 show the highest accuracy achieved in hidden layers for each function, alongside its corresponding configuration.

**Table 30.** ReLU activation function results.

| Dataset | Score | Best Configuration | | |
|---|---|---|---|---|
| | | LC | PCA | Scaling |
| Heart | 83% | [500] | 30% | Std |
| Virus | 99% | [100,100,100] | mle | Std |
| Sonar | 87% | [300,200,100,50] | 30% | MinMax |
| Peugeot_Target 14 | 99% | [500] | None | Std |
| Peugeot_Target 15 | 99% | [300,200,100,50] | None | Std |
| EnviroCar | 0.99 | [50] | mle | MinMax |
| AQI | 0.86 | [300,200,100,50] | None | MinMax |

Layer Configuration (LC), Principal Component Analysis (PCA).

**Table 31.** Tanh activation function results.

| Dataset | Score | Best Configuration | | | |
|---|---|---|---|---|---|
| | | LC | PCA | Scaling | Dropout |
| Heart | 84% | [500] | 30% | Std | 0 |
| Virus | 99% | [100,100,100] | None | Std | 0 |
| Sonar | 81% | [50] | 30% | MinMax | 0 |
| Peugeot_Target 14 | 99% | [100,100,100] | mle | Std | 0 |
| Peugeot_Target 15 | 99% | [300,200,100,50] | None | Std | 0 |
| EnviroCar | 0.99 | [50] | mle | MinMax | 0 |
| AQI | 0.82 | [300,200,100,50] | mle | MinMax | 0.1 |

Layer Configuration (LC), Principal Component Analysis (PCA).

The results are similar, with a slight prevalence of ReLU, with a valuable difference for Sonar (+7% accuracy) and AQI (+5% R2).

## 5.6. ANN Batch Size

The batch size is the number of training examples processed in one iteration before the model being trained is updated. To test the effect of this parameter, we considered three values, one, 10, and 20, keeping the number of epochs fixed to 20. Table 32 shows the accuracy of each dataset for various batch sizes.

**Table 32.** Performance on difference batch size.

| Dataset | Epoch = 20 | | |
| --- | --- | --- | --- |
| | Batch Size = 1 | Batch Size = 10 | Batch Size = 20 |
| Heart | 84% | 84% | 84% |
| Virus | 99% | 99% | 99% |
| Sonar | 87% | 87% | 84% |
| Peugeot_Target 14 | 98% | 99% | 98% |
| Peugeot_Target 15 | 97% | 99% | 99% |
| EnviroCar | 0.99 | 0.99 | 0.99 |
| AQI | 0.63 | 0.86 | 0.76 |

The results show that the value of 10 provides optimal results in terms of accuracy. Actually, the difference becomes relevant only for the case of AQI. A batch size equal to one poses an excessive time overhead (approximately 30% slower than the batch size of 10), while a batch size of 20 achieves a speedup of about 40%.

*5.7. ANN Accuracy vs. Epochs*

ANN training goes through several epochs, where an epoch is a learning cycle in which the learner model sees the whole training data set. Figures 3 and 4 show that the training of ANN on all datasets converges quickly within 10 epochs.

*5.8. ANN Dropout*

Dropout is a simple method to prevent overfitting in ANNs. It consists of randomly ignoring a certain number of neuron outputs in a layer during the training phase.

The results in Table 33 show that this regularization step provides no improvement in the considered cases, but has a slight negative effect in a couple of datasets (Sonar and AQI).

**Table 33.** Dropout effect on ANN.

| Dataset | Dropout | |
| --- | --- | --- |
| | 0 | 0.1 |
| Heart | 84% | 84% |
| Virus | 99% | 99% |
| Sonar | 87% | 83% |
| Peugeot_Target 14 | 99% | 99% |
| Peugeot_Target 15 | 99% | 99% |
| EnviroCar | 0.99 | 0.99 |
| AQI | 0.86 | 0.82 |

*5.9. SVM Regularization Training Time*

In SVM, *C* is a key regularization parameter, that controls the tradeoff between errors of the SVM on training data and margin maximization [13,57]. The classification rate is highly dependent on this coefficient, as confirmed by Tables 8 and 9. Desk-LM uses the grid search method to explore the C values presented by the user, which require long waiting times in some cases. To quantify this, we measured the training latency time in a set of typical values ($C = 0.01, 0.1, 1, 10,$ and $100$), with the results provided in Table 34.

Different values of the *C* parameter have an impact on the training time. The table shows that higher *C* values require higher training time. We must stress that the above results represent the training time for the best models. In particular, when no normalization procedure was applied, the training time using large values of C became huge (also up to one hour), especially for regression datasets.

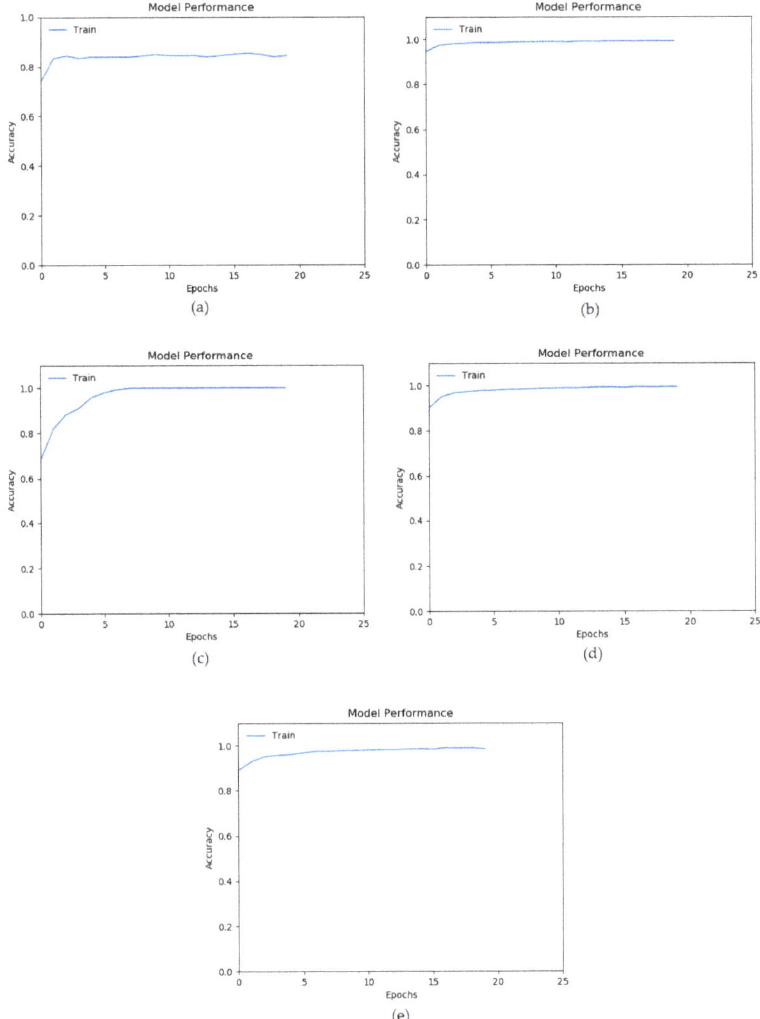

**Figure 3.** Accuracy vs. Epochs for (**a**) Heart, (**b**) Virus, (**c**) Sonar, (**d**) Peugeot target 14, and (**e**) Peugeot target 15.

**Table 34.** Training time for different values of the C parameter.

| Dataset | C Parameter | | | | |
|---|---|---|---|---|---|
| | 0.01 | 0.1 | 1 | 10 | 100 |
| Heart | <1 ms | <1 ms | <1 ms | <1 ms | <1 ms |
| Virus | 100 ms | 300 ms | 500 ms | 600 ms | 600 ms |
| Sonar | <1 ms | <1 ms | <1 ms | <1 ms | <1 ms |
| Peugeot_Target 14 | <1 ms | 100 ms | 200 ms | 300 ms | 400 ms |
| Peugeot_Target 15 | <1 ms | 100 ms | 300 ms | 300 ms | 300 ms |
| EnviroCar | 100 ms | 100 ms | 100 ms | 100 ms | 100 ms |
| AQI | <1 ms | <1 ms | <1 ms | <1 ms | 300 ms |

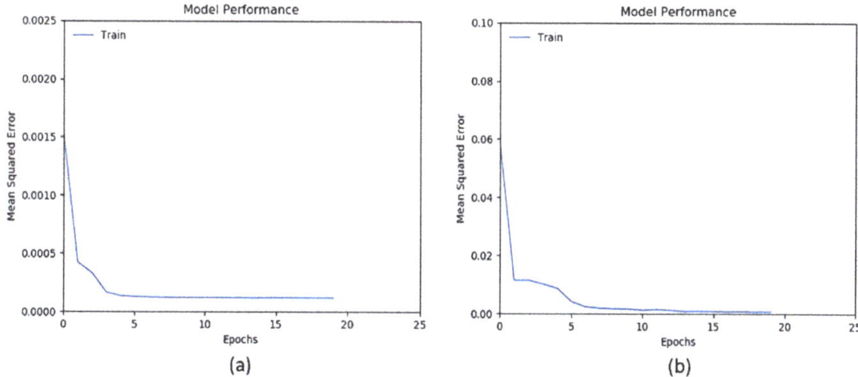

**Figure 4.** Mean Squared Error vs. Epochs for (**a**) EnviroCar, and (**b**) air quality index (AQI).

## 5.10. DT Parameters

Tuning a decision tree requires us to test the effect of various hyperparameters, such as *max_depth*, *min_simple_split*. Figure 5 shows the distribution of the tested parameter values for the best models in the different datasets (see also Table 13 to see the best results).

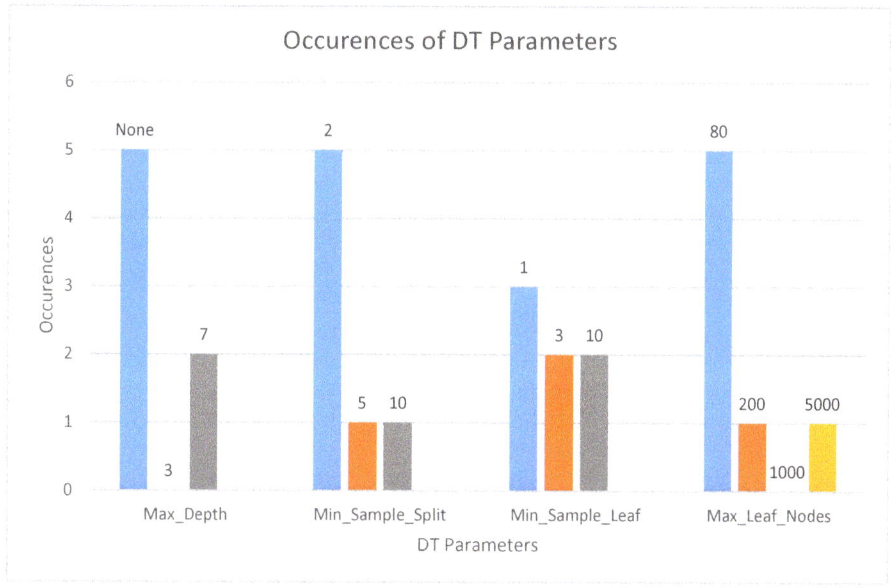

**Figure 5.** Number of occurrences of each DT parameter.

In most cases, the whole tree depth is needed, and this does not exceed the memory available in the microcontrollers. However, *Max_Leaf_Nodes* values usually need a low threshold (80). EnviroCar required a high value of 5000, which had to be reduced down to 1000 for F3, F4, L4 and to 200 for F0 because of the limited RAM availability.

## 6. Conclusions and Future Work

This paper presented the Edge Learning Machine (ELM), a machine learning platform for edge devices. ELM performs training on desktop computers, exploiting TensorFlow, Keras, and scikit-learn, and makes inferences on microcontrollers. It implements, in platform-independent C language, three supervised machine learning algorithms (Linear SVM, k-NN, and DT), and exploits the STM X-Cube-AI package for implementing ANNs on STM32 Nucleo boards. The training phase on Desk-LM searches for the best configuration across a variety of user-defined parameter values. In order to investigate the performance of these algorithms on the targeted devices, we posed ten research questions (RQ 1–10, in the following) and analyzed a set of six datasets (four classifications and two regressions). To the best of our knowledge, this is the first paper presenting such an extensive performance analysis of edge machine learning in terms of datasets, algorithms, configurations, and types of devices.

Our analysis shows that, on a set of available IoT data, we managed to train all the targeted devices to achieve, with at least one algorithm, the best score (classification accuracy or regression R2) obtained through a desktop machine (RQ1). ANN performs better than the other algorithms in most of the cases, without differences among the target devices (apart from F0, that is not supported by STM X-Cube-AI). k-NN performs similarly to ANN, and in one case even better, but requires that all the training sets are kept in the inference phase, posing a significant memory demand, which penalizes time performance, particularly on low-end devices. The performance of Decision Tree performance varied widely across datasets. When comparing edge devices, the best time performance is achieved by F7 and H7. Unsurprisingly, given the available hardware, F0 performs worse than all the others.

The preprocessing phase is extremely important. Results across all the datasets and algorithms show the importance of scaling the inputs, which lead to improvements of up to 82% in accuracy (SVM Virus) and 23% in R2 (k-NN Heart) (RQ2). The applications of PCA have various effects across algorithms and datasets (RQ3).

In terms of the ANN hyperparameters, we observed that increasing the depth of a NN typically improves its performance, up to a saturation level (RQ4). When comparing the neuron activation functions, we observed a slight prevalence of ReLU over Tanh (RQ5). The batch size has little influence on score, but it does have an influence on training time. We established that 10 was the optimal value for all the examined datasets (RQ6). In all datasets, the ANN training quickly converges within 10 epochs (RQ7). The dropout regularization parameter only led to some slight worsening in a couple of datasets (RQ8).

In SVM, the C hyperparameter value selection has an impact on training times, but only when inputs are not scaled (RQ9). In most datasets, the whole tree depth is needed for DT models, and this does not exceed the memory available in the microcontrollers. However, the values of *Max_Leaf_Nodes* usually require a low threshold value (80) (RQ10).

As synthesized above, in general, several factors impact performance in different ways across datasets. This highlights the importance of a framework like ELM, which is able to test different algorithms, each one with different configurations. To support the developer community, ELM is released on an open-source basis.

As a possible direction for future work, we consider that the analysis should be extended to include different types of NNs (Convolutional Neural Networks, Recurrent Neural Networks) with more complex datasets (e.g., also including images and audio streams). An extensive analysis should also be performed on unsupervised algorithms that look particularly suited for immediate field deployment, especially in low-accessibility areas. As the complexity of IoT applications is likely to increase, we also expect that distributed ML at the edge will probably be a significant challenge in the coming years.

**Author Contributions:** F.S.: conceptualization, data curation, investigation, software, validation; F.B.: conceptualization, data curation, investigation, methodology, software, validation; R.B.: conceptualization, methodology, software, validation; A.D.G.: conceptualization, methodology, supervision, validation. All authors have read and agreed to the published version of the manuscript.

**Funding:** This research received no external funding.

**Conflicts of Interest:** The authors declare no conflict of interest.

## References

1. Lin, L.; Liao, X.; Jin, H.; Li, P. Computation offloading toward edge computing. *Proc. IEEE* **2019**, *107*, 1584–1607. [CrossRef]
2. Gubbi, J.; Buyya, R.; Marusic, S.; Palaniswami, M. Internet of Things (IoT): A vision, architectural elements, and future directions. *Future Gener. Comput. Syst.* **2013**, *29*, 1645–1660. [CrossRef]
3. Shi, W.; Cao, J.; Zhang, Q.; Li, Y.; Xu, L. Edge computing: Vision and challenges. *IEEE Internet Things J.* **2016**, *3*, 637–646. [CrossRef]
4. Zuboff, S. *The Age of Surveillance Capitalism: The Fight for a Human Future at the New Frontier of Power*; PublicAffairs: New York, NY, USA, 2019.
5. TensorFlow Lite. Available online: http://www.tensorflow.org/lite (accessed on 10 February 2020).
6. Louis, M.; Azad, Z.; Delhadtehrani, L.; Gupta, S.L.; Warden, P.; Reddi, V.; Joshi, A. Towards deep learning using tensorFlow lite on RISC-V. *Workshop Comput. Archit. Res. RISC-V* **2019**. [CrossRef]
7. Dennis, D.K.; Gopinath, S.; Gupta, C.; Kumar, A.; Kusupati, A.; Patil, S.G.; Simhadri, H.V. EdgeML Machine LEARNING for Resource-Constrained Edge Devices. Available online: https://github.com/Microsoft/EdgeML (accessed on 24 April 2020).
8. Suda, N.; Loh, D. *Machine Learning on ARM Cortex-M Microcontrollers*; Arm Ltd.: Cambridge, UK, 2019.
9. X-CUBE-AI—AI Expansion Pack for STM32CubeMX—STMicroelectronics. Available online: http://www.st.com/en/embedded-software/x-cube-ai.html (accessed on 10 February 2020).
10. Bai, N. *Practical Microcontroller Engineering with ARM Technology*; Wiley-IEEE Press: Hoboken, NJ, USA, 2016.
11. Goodfellow, I.; Bengio, Y.; Courville, A. *Deep Learning*; The MIT Press: Cambridge, UK, 2016.
12. Shalev-Shwartz, S.; Ben-David, S. *Understanding Machine Learning: From Theory to Algorithms*; Cambridge University Press: Cambridge, UK, 2014.
13. Vapnik, V.N.; Lerner, A.Y. Recognition of Patterns with help of Generalized Portraits. *Avtomat. Telemekh.* **1963**, *24*, 774–780.
14. Shakhnarovich, G.; Darrell, T.; Indyk, P. *Nearest-Neighbor Methods in Learning and Vision*; The MIT Press: Cambridge, UK, 2005.
15. Breiman, L.; Friedman, J.; Stone, C.J.; Olshen, R. *Classification and Regression Trees*; Chapman and Hall/CRC: Boca Raton, FL, USA, 1984.
16. Zhang, Y.; Bi, S.; Dong, M.; Liu, Y. The Implementation of CNN-Based Object Detector on ARM Embedded Platforms. In Proceedings of the 2018 IEEE 16th Intl Conf on Dependable, Autonomic and Secure Computing, Athens, Greece, 12–15 Auguest 2018.
17. Yazici, M.T.; Basurra, S.; Gaber, M.M. Edge Machine learning: Enabling smart Internet of Things applications. *Big Data Cogn. Comput.* **2018**, *2*, 26. [CrossRef]
18. Embedded Machine Learning on 8-Bit Microcontrollers, Including Arduino—Hackster.io. Available online: http://www.hackster.io/news/embedded-machine-learning-on-8-bit-microcontrollers-includingarduino-783155e7a135 (accessed on 10 February 2020).
19. Cerutti, G.; Prasad, R.; Farella, E. Convolutional Neural Network on Embedded Platform for People Presence Detection in Low Resolution Thermal Images. In Proceedings of the ICASSP, IEEE International Conference on Acoustics, Speech and Signal Processing, Brighton, UK, 12–17 May 2019.
20. Frequently Asked Questions | Coral. Available online: http://www.coral.ai/docs/edgetpu/faq/ (accessed on 11 February 2020).
21. Edge TPU Python API Overview | Coral. Available online: http://www.coral.ai/docs/edgetpu/api-intro (accessed on 11 February 2020).

22. Kusupati, A.; Singh, M.; Bhatia, K.; Kumar, A.; Jain, P.; Varma, M. FastGRNN: A Fast, Accurate, Stable and Tiny Kilobyte Sized Gated Recurrent Neural Network. In Proceedings of the 32nd Conference on Neural Information Processing Systems (NeurIPS 2018), Montréal, QC, Canada, 3–8 December 2018.

23. Gopinath, S.; Ghanathe, N.; Seshadri, V.; Sharma, R. Compiling KB-Sized Machine Learning Models to Tiny IoT Devices. In Proceedings of the 40th ACM SIGPLAN Conference on Programming Language Design and Implementation (PLDI 2019), Phoenix, AZ, USA, 22–26 June 2019.

24. AWS Greengrass Machine Learning Inference—Amazon Web Services. Available online: http://www.aws.amazon.com/greengrass/ml/ (accessed on 10 February 2020).

25. Ghosh, A.M.; Grolinger, K. Deep Learning: Edge-Cloud Data Analytics for IoT. In Proceedings of the IEEE Canadian Conference of Electrical and Computer Engineering (CCECE), Edmonton, AB, Canada, 5–8 May 2019.

26. AI Technology Helping Asthma Sufferers Breathe Easier—Hackster.io. Available online: http://www.hackster.io/news/ai-technology-helping-asthma-sufferers-breathe-easier-50775aa7b89f (accessed on 12 February 2020).

27. Magno, M.; Cavigelli, L.; Mayer, P.; von Hagen, F.; Luca Benini, F. Fanncortexm: An open source toolkit for deployment of multi-layer neural networks on arm cortex-m family microcontrollers: Performance analysis with stress detection. In Proceedings of the IEEE 5th World Forum on Internet of Things, Limerick, Ireland, 15–18 April 2019.

28. Magno, M.; Pritz, M.; Mayer, P.; Benini, L. DeepEmote: Towards Multi-Layer Neural Networks in a Low Power Wearable Multi-Sensors Bracelet. In Proceedings of the 7th IEEE International Workshop on Advances in Sensors and Interfaces (IWASI), Vieste, Italy, 15–16 June 2017.

29. FidoProject/Fido: A lightweight C++ Machine Learning Library for Embedded Electronics and Robotics. Available online: http://www.github.com/FidoProject/Fido (accessed on 12 February 2020).

30. Alameh, M.; Abbass, Y.; Ibrahim, A.; Valle, M. Smart tactile sensing systems based on embedded CNN implementations. *Micromachines* **2019**, *11*, 103. [CrossRef] [PubMed]

31. Sharma, R.; Blookaghazadeh, S.; Li, B.; Zhao, M. Are Existing Knowledge Transfer Techniques Effective for Deep Learning with Edge Devices? In Proceedings of the 2018 IEEE International Conference on Edge Computing (EDGE), San Francisco, CA, USA, 2–7 July 2018.

32. Scikit-Learn: Machine Learning in Python—Scikit-Learn 0.22.1 Documentation. Available online: http://www.scikit-learn.org/stable/ (accessed on 2 March 2020).

33. Home—Keras Documentation. Available online: http://www.keras.io/ (accessed on 2 March 2020).

34. Bengio, Y.; LeCun, Y. Scaling learning algorithms towards AI. In *Large Scale Kernel Machines*; Bottou, L., Chapelle, O., DeCoste, D., Weston, J., Eds.; MIT Press: Cambridge, UK, 2007.

35. Kingma, D.P.; Ba, L.J. Adam: A Method for Stochastic Optimization. Available online: https://arxiv.org/abs/1412.6980 (accessed on 24 April 2020).

36. The HDF5 Group. Available online: https://www.hdfgroup.org/solutions/hdf5 (accessed on 2 March 2020).

37. STMicroelectronics. Getting Started with X-CUBE-AI Expansion Package for Artificial Intelligence (AI) User Manual | Enhanced Reader. Technical Report. Available online: https://www.st.com/en/embedded-software/x-cube-ai.html (accessed on 24 April 2020).

38. Parameters | SVMS.org. Available online: http://www.svms.org/parameters/ (accessed on 4 March 2020).

39. STM32 High Performance Microcontrollers (MCUs)—STMicroelectronics. Available online: http://www.st.com/en/microcontrollers-microprocessors/stm32-high-performance-mcus.html (accessed on 13 February 2020).

40. STM32H7—Arm Cortex-M7 and Cortex-M4 MCUs (480 MHz)—STMicroelectronics. Available online: http://www.st.com/en/microcontrollers-microprocessors/stm32h7-series.html (accessed on 13 February 2020).

41. STM32L4—ARM Cortex-M4 ultra-low-power MCUs—STMicroelectronics. Available online: http://www.st.com/en/microcontrollers-microprocessors/stm32l4-series.html (accessed on 13 February 2020).

42. Heart Disease UCI | Kaggle. Available online: http://www.kaggle.com/ronitf/heart-disease-uci (accessed on 13 February 2020).

43. Boero, L.; Cello, M.; Marchese, M.; Mariconti, E.; Naqash, T.; Zappatore, S. Statistical fingerprint—Based intrusion detection system (SF-IDS). *Int. J. Commun. Syst.* **2017**, *30*, e3225 [CrossRef]

44. Fausto, A.; Marchese, M. Implementation Details to Reduce the Latency of an SDN Statistical Fingerprint-Based IDS. In Proceedings of the IEEE International Symposium on Advanced Electrical and Communication Technologies (ISAECT), Rome, Italy, 27–29 November 2019.

45. Falbo, V.; Apicella, T.; Aurioso, D.; Danese, L.; Bellotti, F.; Berta, R.; Gloria, A.D. Analyzing Machine Learning on Mainstream Microcontrollers. In Proceedings of the International Conference on Applications in Electronics Pervading Industry Environment and Society (ApplePies 2019), Pisa, Italy, 26–27 September 2019.

46. Benchmark Datasets Used for Classification: Comparison of Results. Available online: http://www.fizyka.umk.pl/kis-old/projects/datasets.html#Sonar (accessed on 13 February 2020).

47. Parodi, A.; Bellotti, F.; Berta, R.; Gloria, A.D. Developing a machine learning library for microcontrollers. In Proceedings of the International Conference on Applications in Electronics Pervading Industry, Environment and Society, Pisa, Italy, 26–27 September 2018.

48. Traffic, Driving Style and Road Surface Condition | Kaggle. Available online: http://www.kaggle.com/gloseto/traffic-driving-style-road-surface-condition (accessed on 13 February 2020).

49. EnviroCar—Datasets—The Datahub. Available online: http://www.old.datahub.io/dataset/envirocar (accessed on 13 February 2020).

50. Massoud, R.; Poslad, S.; Bellotti, F.; Berta, R.; Mehran, K.; Gloria, A.D. A fuzzy logic module to estimate a driver's fuel consumption for reality-enhanced serious games. *Int. J. Serious Games* **2018**, *5*, 45–62. [CrossRef]

51. Massoud, R.; Bellotti, F.; Poslad, S.; Berta, R.; De Gloria, A. Towards a reality-enhanced serious game to promote eco-driving in the wild. In *Games and Learning Alliance. GALA 2019*; Liapis, A., Yannakakis, G., Gentile, M., Ninaus, M., Eds.; Springer: Berlin, Germany, 2019.

52. Search for and download air quality data | NSW Dept of Planning, Industry and Environment. Available online: http://www.dpie.nsw.gov.au/air-quality/search-for-and-download-air-quality-data (accessed on 13 February 2020).

53. Ioffe, S.; Szegedy, C. Batch Normalization: Accelerating Deep Network Training by Reducing Internal474Covariate Shift; 2015. Available online: https://dblp.uni-trier.de/rec/html/conf/icml/IoffeS15 (accessed on 17 February 2020).

54. Sklearn.Preprocessing Data—Scikit-Learn 0.22.2 Documentation. Available online: https://scikit-learn.org/stable/modules/preprocessing.html (accessed on 17 February 2020).

55. Decision Trees: How to Optimize My Decision-Making Process. Available online: http://www.medium.com/cracking-the-data-science-interview/decision-trees-how-to-optimize-my-decision-making-process-e1f327999c7a (accessed on 17 February 2020).

56. Sklearn. Decomposition.PCA—Scikit-Learn 0.22.1 Documentation. Available online: http://www.scikit-learn.org/stable/modules/generated/sklearn.decomposition.PCA.html (accessed on 17 February 2020).

57. Sentelle, C.G.; Anagnostopoulos, G.C.; Georgiopoulos, M. A simple method for solving the SVM regularization path for semidefinite kernels. *IEEE Trans. Neural Netw. Learn. Syst.* **2015**, *27*, 709–722. [CrossRef] [PubMed]

*Article*

# Cryptographically Secure Pseudo-Random Number Generator IP-Core Based on SHA2 Algorithm [†]

**Luca Baldanzi, Luca Crocetti, Francesco Falaschi *, Matteo Bertolucci, Jacopo Belli, Luca Fanucci and Sergio Saponara**

Department of Information Engineering, University of Pisa, Via G. Caruso n. 16, 56122 Pisa, Italy;
luca.baldanzi@ing.unipi.it (L.B.); luca.crocetti@phd.unipi.it (L.C.); francesco.falaschi@phd.unipi.it (F.F.);
matteo.bertolucci@phd.unipi.it (M.B.); jacopo.belli23@gmail.com (J.B.); luca.fanucci@unipi.it (L.F.);
sergio.saponara@unipi.it (S.S.)
* Correspondence: francesco.falaschi@phd.unipi.it
† This paper is an extended version of our paper published in Springer LNEE vol. 627, 2020, Luca Baldanzi,
  Luca Crocetti, Francesco Falaschi, Jacopo Belli, Luca Fanucci and Sergio Saponara, Digital Random Number
  Generator Hardware Accelerator IP-Core for Security Applications, in Proceedings of the Applications in
  Electronics Pervading Industry, Environment and Society, Pisa, Italy, 11–13 September 2019.

Received: 7 February 2020 ; Accepted: 17 March 2020; Published: 27 March 2020

**Abstract:** In the context of growing the adoption of advanced sensors and systems for active vehicle safety and driver assistance, an increasingly important issue is the security of the information exchanged between the different sub-systems of the vehicle. Random number generation is crucial in modern encryption and security applications as it is a critical task from the point of view of the robustness of the security chain. Random numbers are in fact used to generate the encryption keys to be used for ciphers. Consequently, any weakness in the key generation process can potentially leak information that can be used to breach even the strongest cipher. This paper presents the architecture of a high performance Random Number Generator (RNG) IP-core, in particular a Cryptographically Secure Pseudo-Random Number Generator (CSPRNG) IP-core, a digital hardware accelerator for random numbers generation which can be employed for cryptographically secure applications. The specifications used to develop the proposed project were derived from dedicated literature and standards. Subsequently, specific architecture optimizations were studied to achieve better timing performance and very high throughput values. The IP-core has been validated thanks to the official NIST Statistical Test Suite, in order to evaluate the degree of randomness of the numbers generated in output. Finally the CSPRNG IP-core has been characterized on relevant Field Programmable Gate Array (FPGA) and ASIC standard-cell technologies.

**Keywords:** intelligent sensors; autonomous driving; cyber security; HW accelerator; on-chip random number generator (RNG); SHA2; FPGA; ASIC standard-cell

---

## 1. Introduction

The rapid technology development of intelligent sensors in the automotive field in recent years, driven by institutions and supported by manufacturers to integrate advanced systems for active safety and hazard prevention, has generated additional collateral technological needs. All the equipment distributed on board the vehicle are, in fact, interconnected with real communication networks and exchange critical data and sensitive information that must be protected from potential attacks and violations. Cybersecurity thus becomes a central issue and all mechanisms ensuring authentication, confidentiality and integrity of messages become an enabling technology for the further development of Advanced Driver Assistance Systems (ADAS) and Autonomous Driving Systems (AD).

In particular, projects like the one carried on in the framework of the European Processor Initiative (EPI) programme [1] anticipate requirements from future scenarios, with very high throughput hardware accelerators for cybersecurity applications. Highly automated vehicles will be equipped with advanced sensors (e.g., camera arrays, radar, lidar), capable of generating significant data flows. Moreover, over-the-air update (OTA) of the software of control units will be implemented, for which it will be very important to keep the update time as short as possible. In this context, it is essential to increase information security, and it is therefore necessary to implement encrypted data transfers (i.e., encryption and decryption) and verified content (i.e., digital signature) with throughput compatible with the use of buses such as automotive ethernet (i.e., 1–10 Gbps). For this reason, hardware accelerators with throughput requirements of the order of tens Gbps are necessary, also to take into considerations the expected lifespan of vehicles.

Random numbers are widely used in encryption and security applications, usually to generate encryption keys or secret data to be shared between communication entities. Therefore a Random Number Generator (RNG) is a very important primitive for cryptographically secure applications [2]. In particular, it is used as a fundamental part for the development of several globally distributed applications related to the field of cyber security such as digital payments, online authentication and instant messaging [3]. Cryptographic keys are based on random numbers and must be characterized by a high degree of unpredictability to be considered secure: this is necessary to prevent an attacker from violating the security chain based on this cryptographic key. Random numbers are also the starting point for generating nonces within authentication protocols as a countermeasure against replay attacks, so the higher the degree of randomness, the more robust the countermeasure will be. Moreover, strong random number generation helps digital signature procedures to prevent private keys to be disclosed, thus violating the signature itself [4]. Several development techniques for RNG engines are reported in literature, and many of them exploit physical sources (e.g., analog noise) as random processes to obtain the randomness characteristic of the generated bit sequences [3–9].

These circuits are identified as True Random Number Generators (TRNG) and their output sequences are considered high quality random numbers. However, TRNGs have non-negligible disadvantages that must be considered: the use of physical sources leads to high energy consumption and insufficient throughput for fast and advanced integrated systems. TRNGs are also sensitive to changing operating conditions, which means that post-processing must be implemented to ensure reliable random output data, further reducing the throughput under non-ideal condition.

To overcome these limitations, a powerful Deterministic Random Bit Generator (DRBG) circuit can be used in addition to a very low-area, low-power and low-throughput TRNG implementation. This means that the RNG engine would be mainly based on a deterministic algorithm that generates pseudo-random output sequences. In this case, the required degree of randomness is obtained through additional mechanisms to increase the level of entropy of the generated sequences, which would otherwise be deterministic. This operation is called *reseed* and it consists in providing a trigger to restart the circuit of the deterministic algorithm from a new high entropy starting point (i.e., the new seed). DRBG-based solutions use periodic *reseed* to allow the RNG to generate pseudo-random binary output sequences that are equivalent and indistinguishable from true random ones. The limitations of TRNGs for high-speed devices are thus overcome by restricting its use to the periodic seed generation operation only, which has the characteristic of being a very light task. For this reason the new seed is usually provided by a very low complexity and target specific TRNG module [10], which also may be powered down when not necessary.

For the proposed architecture, the DRBG mechanism was chosen from those approved by National Institute of Standards and Technology (NIST) [6]. The standard provides general information for PRNGs based on cryptographic primitives, some of which are incontrovertible and proven (e.g., Hash DRBG and HMAC DRBG). For the proposed Cryptographically Secure Pseudo-Random Number Generator (CSPRNG) IP-core the algorithm selection was made based on a compromise between performance, area and security strength. The Hash DRBG with SHA-256 as cryptographic core (i.e., based on the SHA2 algorithm) proved to be the most efficient solution between logical complexity and expected throughput during random bit generation, offering 256 bits of security strength.

The reminder of this paper is organized as it follows: Section 2 presents the trade-off analysis among the different suitable DRBG algorithms, Section 3 details the implementation of the SHA2 algorithm being chosen to develop the DRBG core, Section 4 describes the Hash DRBG design architecture as CSPRNG IP-core, Section 5 collects the characterization results. Finally, conclusions are discussed in Section 6.

## 2. DRBG Algorithms Trade-Off Analysis

As mentioned in Section 1, the different deterministic algorithms suitable for implementing DRBG circuits have been evaluated by the NIST and those approved-recommended are collected in the NIST SP 800-90A Rev.1 pubblication [6]. Such mechanisms present common features and functionalities:

- they rely on a one-way cryptographic function, thus providing backtracking resistance;
- the internal status memories are secret and inaccessible to the user;
- the following essential operations are allowed:

  - *instance*, to acquire a random seed (i.e., concatenation of input entropy content, possibly input or internal random nonce, and personalization string) and to initialize the internal state to a random value derived from the seed;
  - *reseed*, to acquire a random seed (i.e., concatenation of internal state, input entropy content, and personalization string) and update the internal state to a random value derived from the seed;
  - *generate*, to generate an output bits sequence based on current state and then to update the state to a random value derived from previous state;
  - *uninstantiate*, to delete the internal state;

- they support a maximum security strength (112, 128, 192 or 256) and all of the lower ones;
- a *reseed counter* counter, and the corresponding threshold called *seed lifetime*, is present to signal the user that the mechanism needs a new *seed*;
- user is always able to run a command with an associated personalization string, which needs not to be secret but it contributes to the internal state randomization.

Most of the DRBG engines implementations rely on hash functions and counter mode (CTR) of symmetric-key encryption processes. Hash functions family includes SHA1 and SHA2 algorithms, but the former is going to be deprecated because of its low security strength and high vulnerability [11], therefore only SHA2 cryptographic primitives are taken into exam for Hash DRBG mechanisms. The main parameters related to DRBG cores based on SHA2 primitive are reported in Table 1.

**Table 1.** Hash DRBG mechanisms parameters (SHA2 only).

| | SHA2 Algorithm | | | |
| --- | --- | --- | --- | --- |
| | **SHA-224** | **SHA-256** | **SHA-384** | **SHA-512** |
| Highest Security Strength | 192 bits | 256 bits | 256 bits | 256 bits |
| Output Block Length (*outlen*) | 224 bits | 256 bits | 384 bits | 512 bits |
| Min. Entropy for *instance* and *reseed* | 192 bits | 256 bits | 256 bits | 256 bits |
| Seed Length (*seedlen*) | 440 bits | 440 bits | 888 bits | 888 bits |
| Max. Num. of Bit per Request | $2^{19}$ | $2^{19}$ | $2^{19}$ | $2^{19}$ |
| Max. Num. of Request between *reseeds* | $2^{48}$ | $2^{48}$ | $2^{48}$ | $2^{48}$ |

CTR (CTR is abbreviation for *Counter*) DRBG mechanisms are based onto block cipher cores used in *counter mode*. Different block cipher cores are suitable to develop a DRBG circuit and the main parameters related to the different implementations are collected in Table 2.

**Table 2.** CTR DRBG mechanisms parameters. $B = (2^{ctrl\_len} - 4)\, blocklen$.

| | AES Algorithm | | | |
| --- | --- | --- | --- | --- |
| | **3 Key TDEA** | **AES-128** | **AES-192** | **AES-256** |
| Highest Security Strength | 112 bits | 128 bits | 192 bits | 256 bits |
| Input/Output Block Length (*blocklen*) | 64 bits | 128 bits | 128 bits | 128 bits |
| Key Length Length (*keylen*) | 168 bits | 128 bits | 192 bits | 256 bits |
| Counter Field Length (*ctr_len*) | | 4 <*ctr_len* <*blocklen* | | |
| Min. Enctropy for *instance* and *reseed* | 112 bits | 128 bits | 192 bits | 256 bits |
| Seed Length (*seedlen*) | 232 bits | 256 bits | 320 bits | 384 bits |
| Max. Num. of Bit per Request | $\min(B, 2^{13})$ | $\min(B, 2^{19})$ | $\min(B, 2^{19})$ | $\min(B, 2^{19})$ |
| Max. Num. of Request between *reseeds* | $2^{48}$ | $2^{48}$ | $2^{48}$ | $2^{48}$ |

Table 3 summarizes area and latency values obtained for our version of SHA2 IP-core, where complexity values are related to the characterization on 45nm ASIC standard-cell technology. In the perspective to implement a Hash DRBG circuit, solutions for SHA-224 and SHA-384 are discarded in favor of SHA-256 and SHA-512 because area and latency values are the same, but the former couple offers a shorter output block. Concerning SHA-256 and SHA-512 comparison, the following considerations can be done in order to select the best candidate for Hash DRBG:

- SHA-256 has lower latency per block than SHA-512;
- SHA-512 offers a higher throughput with respect to SHA-256, since it provides 512 bits every 83 clock cycles instead of 256 bits every 67 clock cycles;
- SHA-256 is more compact in terms of area, which reflects also on internal state registers area footprint. As shown in Table 1, *seedlen* is 440 for SHA-256 and 888 for SHA-512, meaning that the internal state requires around 900 registers for the former and 1800 for the latter.

**Table 3.** SHA2 IP-core specifications.

| SHA2 Algorithm | Area | Latency per Block | Output Block Size |
| --- | --- | --- | --- |
| SHA-224 | 15 kGE | 67 clock cycles | 224 bits |
| SHA-256 | 15 kGE | 67 clock cycles | 256 bits |
| SHA-384 | 30 kGE | 83 clock cycles | 384 bits |
| SHA-512 | 30 kGE | 83 clock cycles | 512 bits |

The throughput values of the SHA-256 core and SHA-512 core, when operating in generation phase for a Hash DRBG implementation, can be obtained through Equations (1) and (2), respectively.

$$T_{SHA-256} = 256/67 \cdot f_{clk} \cdot n_{parallel\_core} = 3.82 \cdot f_{clk} \cdot n_{parallel\_core} \quad bit/s \tag{1}$$

$$T_{SHA-512} = 512/83 \cdot f_{clk} \cdot n_{parallel\_core} = 6.17 \cdot f_{clk} \cdot n_{parallel\_core} \quad bit/s \tag{2}$$

Concerning the CTR DRBG circuit, the AES IP-core is proved to be best in class for both area and throughput. Table 4 collects the area and latency values for our versions of AES-128 and AES-256 IP-cores characterized on 45nm ASIC standard-cell technology.

**Table 4.** AES IP-core specifications.

| AES Algorithm | Area | Latency per Block | Output Block Size |
| --- | --- | --- | --- |
| AES-128 | 11 kGE | 11 clock cycles | 128 bits |
| AES-256 | 12.5 kGE | 15 clock cycles | 128 bits |

Given that the target is to identify the most suitable core for implementation of DRBG circuit with highest level of security strength possible, the AES-256 algorithm is the only block cipher core being considered for the trade-off. As shown in Table 4, its area is lower than the one for SHA-256, while the throughput value is higher than that reported for SHA-512. In particular, the throughput of AES-256 to be considered for a CTR DRBG implementation is calculated as below:

$$T_{AES-256} = 128/15 \cdot f_{clk} \cdot n_{parallel\_core} = 8.53 \cdot f_{clk} \cdot n_{parallel\_core} \qquad bit/s \qquad (3)$$

Figure 1 shows a comparison of DRBG mechanisms implemented as in [5] and based on Hash algorithms (i.e., SHA-256 or SHA-512) and block ciphers (i.e., AES-256) used in CTR mode. The logic throughput and complexity values are obtained with synthesis on 45 nm ASIC standard-cell technology.

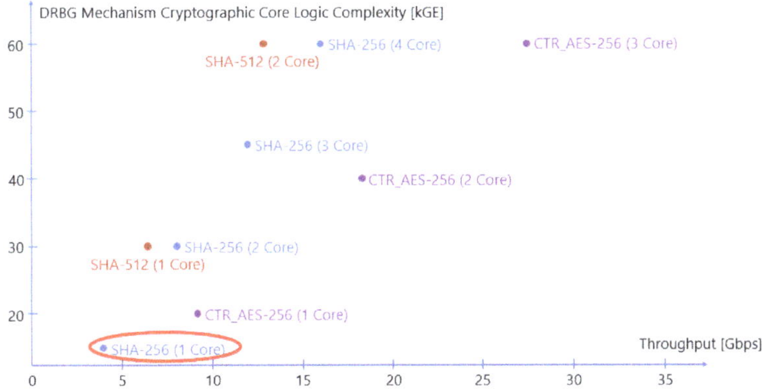


**Figure 1.** Comparison between NIST approved DRBG mechanisms based on logic complexity in kGE and throughput.

The algorithm chosen for the development of the DRBG circuit inside the CSPRNG IP-core was SHA-256 (i.e., the primitive SHA2), therefore we decided to use an Hash DRBG despite the better area and latency values collected for AES cores (i.e., for CTR DRBG circuit). This is because M.Schmid [12] explained how block cipher-based DRBGs should not be used as they are indeed not able to reach maximum security strength. The author declares that the pseudo-random permutation inside each AES round, coupled with counter mode of operation, generates a binary sequence which results to be distinguishable with respect to what a random source could give, thus being unable to satisfy the security requirements. This is not the case with Hash-based DBRGs, so the use of SHA-256 cores offers better robustness to the entire security chain where the CSPRNG is based on this algorithm. The SHA-256 core ensures a compact implementation for the mechanism and the possibility to extend the design for supporting multiple cores to increase the throughput. In a context with multiple cryptographic cores, 2 SHA-256 perform better than a single SHA-512, having a higher throughput and requiring lower internal state.

## 3. SHA-256 Core Implementation

As explained in Section 2, the fundamental element of the proposed Hash DRBG circuit is the SHA-256 core, based on SHA2 cryptographic primitive. In order to achieve high throughput for the whole CSPRNG IP-core, it is then essential to optimize the SHA-256 implementation performances to the maximum. To do so, the canonical logic implementation derived from the standard [13] has been improved through the use of *Carry-Save Adder* (CSA) units for consecutive additions and by application of retiming-pipelining to perform delay balancing. To better understand the implemented optimizations, a brief description of the standard is given.

The SHA-256 standard may be defined by two separate, ideally consecutive, procedures:

1. the *message schedule*;
2. the *compression function*.

The *message schedule* is in charge of creating a *key* schedule starting from the 512-bits *input message* to be then provided to the *compression function*. The operation is performed through the $\sigma_0$, $\sigma_1$ and modulo $2^{32}$ adder operations defined in Equations (4)–(6) respectively.

$$\sigma_0(x) = RotateRight_7(x) \oplus RotateRight_{18}(x) \oplus ShiftRight_3(x) \tag{4}$$

$$\sigma_1(x) = RotateRight_{17}(x) \oplus RotateRight_{19}(x) \oplus ShiftRight_{10}(x) \tag{5}$$

$$x \boxplus y = x + y \quad (\text{mod } 2^{32}) \tag{6}$$

Usually the message schedule operation is also called *expansion* due to the fact that the 512-bits input message is expanded to $32 \cdot 64 = 2048 bits$. The canonical serial architecture of the *message schedule* block, derived from Equation (7), is depicted in Figure 2.

$$W_t = \begin{cases} M_t & 0 < t < 15 \\ \sigma_1(W_{t-2}) \boxplus W_{t-7} \boxplus \sigma_0(W_{t-15}) \boxplus W_{t-16} & 16 < t < 63 \end{cases} \tag{7}$$

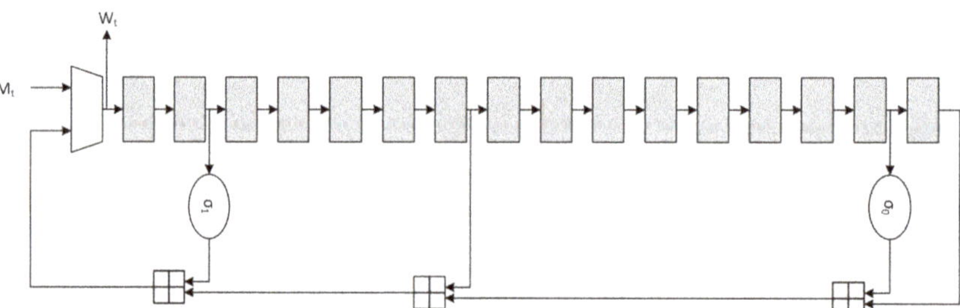

**Figure 2.** SHA2 standard *message schedule* architecture.

The optimization for the message schedule is performed on the adder chain through the use of CSA, which are essentially *Full-Adder Arrays* (FAA), producing partial sums (*ps*) and shift-carries (*sc*).

$$\begin{aligned} ps_i &= a_i \oplus b_i \oplus b_i \\ sc_i &= (a_i \wedge b_i) \vee (a_i \wedge c_i) \vee (b_i \wedge c_i) \end{aligned} \tag{8}$$

CSAs have advantages on both area and critical path. Implementation on 45nm and 7nm ASIC standard-cell technologies demonstrated that, when compared to *Carry-Lookahead Adder* (CLA) units, the delay relationship is $T_{CLA-32} = 1.78 \cdot T_{CSA-32}$ and $T_{CLA-32} = 1.87 \cdot T_{CSA-32}$ respectively for the two technologies. The optimized serial implementation is shown in Figure 3, where the high-level timing block analysis shows that the critical path is reduced from $T_{\sigma_0} + 3 \cdot T_{\boxplus}$ to $T_{\sigma_0} + 2 \cdot T_{CSA} + T_{\boxplus}$. Further optimizations are possible through the use of retiming, but they are not considered due to the critical path being mainly located in the *compression function* architecture.

The SHA-256 *compression function* is composed of three consecutive steps: initialization, one-way compression and termination. In the first step, the variables A-H are initialized with the intermediate Hash value $H^{(t-1)}$ (the first 512-bits message block at t=1 uses a constant $H^{(0)}$ provided by the standard). The one-way compression then performs 64 loops according to:

$$T_1 \leftarrow H \boxplus \Sigma_1(E) \boxplus Ch(E,F,G) \boxplus K_j \boxplus W_j$$
$$T_2 \leftarrow \Sigma_0(A) \boxplus Maj(A,B,C)$$
$$H \leftarrow G$$
$$F \leftarrow E$$
$$E \leftarrow D \boxplus T_1$$
$$D \leftarrow C$$
$$C \leftarrow B$$
$$B \leftarrow A$$
$$A \leftarrow T_1 \boxplus T_2$$

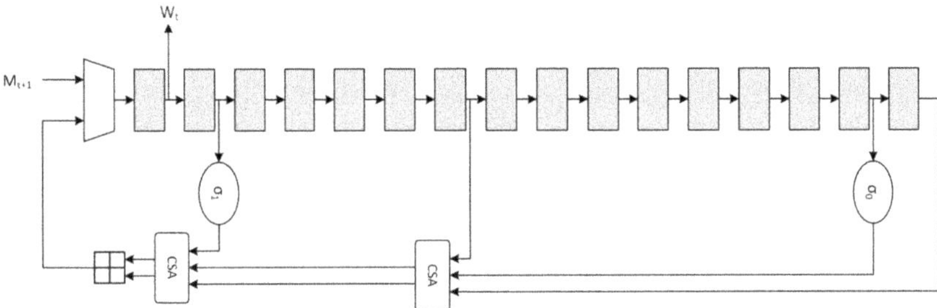


**Figure 3.** SHA2 optimized *message schedule* architecture.

Finally the intermediate Hash value at time $t$ is calculated by a $2^{32}$ modulo addition between the variables A-H at initialization time and the variables A-H after the one way compression. The functions $Maj$, $Ch$, $\Sigma_0$ and $\Sigma_1$ are defined as:

$$Maj(x,y,z) = (x \wedge y) \oplus (x \wedge z) \oplus (y \wedge z)$$
$$Ch(x,y,z) = (x \wedge y) \oplus (\neg x \wedge z)$$
$$\Sigma_0(x) = RotateRight_2(x) \oplus RotateRight_{13}(x) \oplus RotateRight_{22}(x) \tag{9}$$
$$\Sigma_1(x) = RotateRight_6(x) \oplus RotateRight_{11}(x) \oplus RotateRight_{25}(x)$$

The canonical scheme corresponding to the described procedure is represented in Figure 4, where the output stage performing the termination phase is not represented.

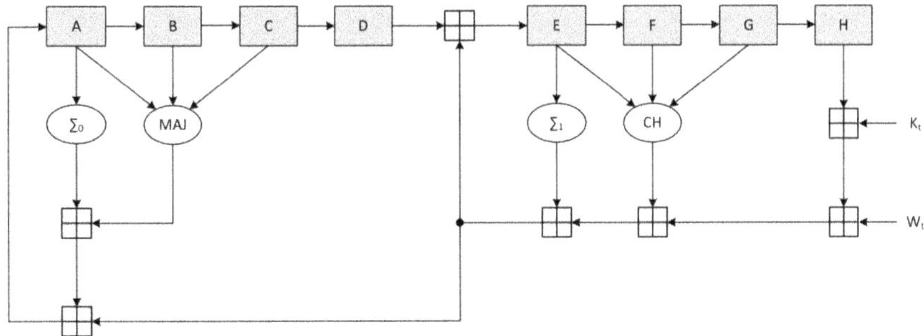


**Figure 4.** SHA2 standard *compression function* architecture

The high-level block timing analysis showed that the critical path on the non-optimized architecture is located between register H and register E, involving 5 $\boxplus$ operations. Optimization of the *compression function* was achieved through the use of CSA, retiming and delay balancing. In particular, all the adder chains were converted to CSA with the exception of register B and E inputs.

Moreover, the path going from $K_t$-$W_t$ was duplicated to allow the value of register D to be added immediately to $H + K_t + W_t$. Finally a pipeline stage $L_1, L_2$ was added (with the associated C-D multiplexer to ensure the functionality) and the a register was split to move the CLA position.

Finally, both SHA-2 implementations have been synthesized on 45 nm and 7 nm ASIC technologies, whose results are represented on Table 5 for canonical and optimized architectures.

**Table 5.** Canonical and Optimized SHA-256 implementation results.

| Canonical | Area | Max. Frequency | Optimized | Area | Max. Frequency |
| --- | --- | --- | --- | --- | --- |
| 45 nm ASIC | 16.85 kGE | 640 MHz | 45 nm ASIC | 15.38 kGE | 1.00 GHz |
| 7 nm ASIC | 17.18 kGE | 3.15 GHz | 7 nm ASIC | 15.45 kGE | 5.15 GHz |

A detail analysis of the critical path on the implemented design shows that the real critical path on 45 nm ASIC technology is located between register L1 and register E (Figure 5), while for the 7 nm ASIC technology it is located in the message schedule between the second right register and the left one (Figure 3). This behavior can easily be attributed to the synthesizer, which is able to use complex ASIC cells and better merge the 4·CSAs than the 2·SCA+1·CLA. a separate synthesis, to emulate the main paths of the canonical and optimized architectures on the 45 nm technology, shows the critical path of these extracted sub-elements. As visible from Table 6 the 2·SCA+1·CLA sub-clock is the slowest path w.r.t. The optimized implementation, with an equivalent frequency $f_{2 \cdot CSA+1 \cdot CLA} = 1200$ MHz.

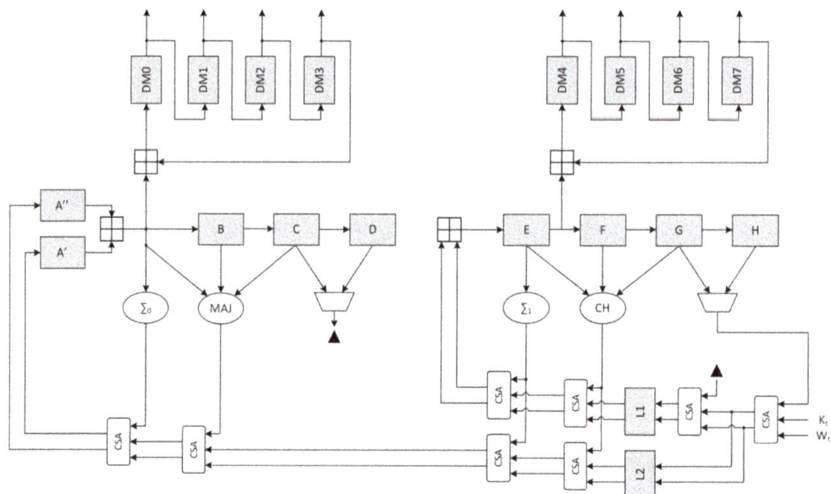

**Figure 5.** SHA2 optimized *compression function* architecture.

**Table 6.** SHA-256 sub-block implementation results.

| Sub-Block | $T_{4 \cdot CSA}$ | $T_{2 \cdot CSA+1 \cdot CLA}$ | $T_{2 \cdot CLA}$ | $T_{5 \cdot CLA}$ |
| --- | --- | --- | --- | --- |
| Max. Latency | 606.06 ps | 833.33 ps | 714.29 ps | 1111.11 ps |

Looking for a comparison between the canonical ASIC implementation and the ASIC optimized one, the latter results to be about 9% smaller, while providing about a 56% maximum frequency increase.

## 4. CSPRNG Design Architecture

The architecture of the proposed CSPRNG IP-core, meaning of the Hash DRBG with optimized SHA-256 core, is shown in Figure 6. The proposed design is based on the following building blocks:

- state registers for 440-bits $V$, 440-bits $C$ and 20-bits *reseed counter*; in addition to them, a 128-bits register is available to store optional personalization string (i.e., to randomize the internal state), while a 512-bits entropy register is used to store the input entropy content for a size larger than necessary: to ensure the minimum number of entropy bits per bit string, *instance* requires 394 bits, while *reseed* requires 256 bits;
- a SHA-256 core, with 512-bits input and 256-bits output; the core is designed to execute operations on a 512-bits block of a message in 67 clock cycles; if the message is composed by a single block (*length* $\leq$ 443 bits), the complete Hash is performed in 67 clock cycles, while if the message is longer, it must be divided in blocks to be processed sequentially;
- a serial adder with 440-bits inputs and modulo 440-bits output; since the adder always runs in parallel with the SHA-256 core, which at least requires 67 cycles to output data, there were no need to implement a low latency adder; intermediate results are stored to one of the input registers to minimize the area occupation;
- multiplexer network to address all data in internal state and from the previous operation to the inputs of the SHA-256 core and adder; as an example, consider the following configuration, where the adder calculates at every Hash cycle the incremented value of $V$ and provides this data to the input of the SHA-256 for it to be hashed:

```
sha_256_in = adder_out || 1'd1 || 7'd0 || 64'd440
adder_x = adder_out
adder_y = 440'd1
```

- a Finite State Machine (FSM), which controls the flow of operations: it is articulated in three main branches (i.e., *instance*, *reseed* and *generate*), and after the completion of a command, it remains stable in idle mode until another command is issued;
- a DRBG self test module (i.e., not shown in Figure 6), which provides built-in self-test functionalities to diagnose possible failures; multiplexers are used to feed the procedures with known values, and compare logic check if outputs match with the expected values.

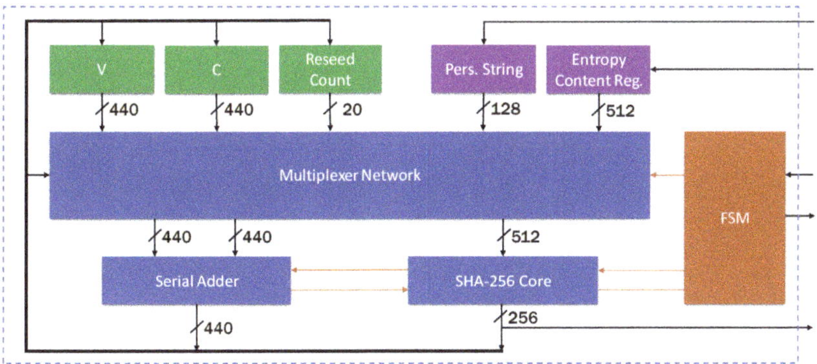


**Figure 6.** CSPRNG (Hash DRBG) design architecture developed.

The *instance* procedure acquires 512 entropy bit from the entropy content input and then hashes eight blocks to create the internal state. With $\tau$ equal to the number of clock cycles necessary to acquire 8 bitt from the entropy content input, total execution time is approximately:

$$t_{instance} = (64 \cdot \tau + 8 \cdot 67) \cdot T_{clk} = (64 \cdot \tau + 536) \cdot T_{clk} \tag{10}$$

The *reseed* procedure acquires 384 entropy bit from the entropy content input and then hashes 8 blocks to update the internal state. Execution time is approximately:

$$t_{reseed} = (48 \cdot \tau + 8 \cdot 67) \cdot T_{clk} = (48 \cdot \tau + 536) \cdot T_{clk} \tag{11}$$

In the *generation* phase, if a personalization string is inserted by the user, a new value of $V$ is immediately calculated before generating output bits. This operations requires a sum and two hash cycles. Since the serial adder latency is 14 clock cycles:

$$t_{gen\_pers\_string} = (14 + 2 \cdot 67) \cdot T_{clk} \tag{12}$$

After generation a new state is derived within the same time:

$$t_{gen\_new\_state} = t_{gen\_pers\_string} = (14 + 2 \cdot 67) \cdot T_{clk} \tag{13}$$

For a clock frequency of 100 MHz and $\tau = 1$, these values result to be:

$$t_{instance} = 6.000 \ \mu s$$
$$t_{reseed} = 5.840 \ \mu s$$
$$t_{gen\_pers\_string} = 1.480 \ \mu s$$
$$t_{gen\_new\_state} = 1.480 \ \mu s$$

For the same clock frequency and $\tau = 8$:

$$t_{instance} = 10.480 \ \mu s$$
$$t_{reseed} = 9.200 \ \mu s$$
$$t_{gen\_pers\_string} = 1.480 \ \mu s$$
$$t_{gen\_new\_state} = 1.480 \ \mu s$$

## 5. Results

In order to validate the CSPRNG IP-core, evaluation of the randmoness degree of the sequences generated was obtained by using the NIST Statistical Test Suite [14]. The test suite ran on a sequence of 128 MB acquired from the CSPRNG with the following strategy:

- every *generation* command requests four blocks, i.e., 1024 bits;
- during a seed lifetime, 1000 *generation* commands are issued;
- the *reseed* operation is commanded 1000 times to reseed the generator.

In this way the total number of acquired bits is $n_{bit} = 1024 \cdot 1000 \cdot 1000 = 1,024,000,000$. This sequence has been then converted to a binary file, which is subsequently given as input to the test suite. The NIST Statistical Test Suite parameters and the corresponding results (using $\alpha = 0.01$) on the 128 MB data block are collected on Table 7.

Three technologies were identified as potential targets for characterization of the CSPRNG hardware accelerator IP-core, one FPGA and two ASIC standard-cell: Intel Stratix IV FPGA, 45 nm Silvaco [15], and 7 nm Artisan TSMC [16]. In all of these cases different implementation effort corners were tested, in order to evaluate the trade-off between throughput and area. The synthesis performed on Intel Stratix IV (EP4SGX230KF40C2) FPGA technology with high performance constraints, configuring a single instance of the SHA-256 core, provides a maximum operating frequency of 180 MHz; a throughput of 690 Mbps; and an overall resource utilization of 4713 ALMs. The 45 nm Silvaco ASIC standard-cell implementation increases its throughput to 3.82 Gbps, since the maximum frequency is 1 GHz (being the critical path in the SHA-2 sub-block), with a logical complexity of 49.19 kGE. Finally, the proposed design, with a single SHA-256 core, brought on the 7 nm Artisan ASIC standard-cell reaches a throughput value of 19.67 Gbps, given a maximum clock frequency of 5.15 GHz, requiring an overall complexity of 46.56 kGE.

**Table 7.** NIST Statistical Test Suite parameters and results.

| Test | Block/Template Length | Pass Rate |
| --- | --- | --- |
| Frequency (Monobit) | - | 0.9924 |
| Frequency Within a Block | 256 | 0.9876 |
| Runs | - | 0.9901 |
| Longest-Run-of-Ones in a Block | - | 0.9878 |
| Binary Matrix Rank | - | 0.9901 |
| Discrete Fourier Transform (Spectral) | - | 0.9874 |
| Non-overlapping Template Matching | 10 | [0.9801–0.9974] |
| Overlapping Template Matching | 10 | 0.9848 |
| Maurer's Universal Statistical | - | 0.9901 |
| Linear Complexity | 1024 | 0.9900 |
| Serial | 16 | 0.9825, 0.9876 |
| Approximate Entropy | 10 | 0.9901 |
| Cumulative Sums (Cusums) | - | 0.9901 |
| Random Excursions | - | [0.9826–0.9947] |
| Random Excursions Variant | - | [0.9875–0.9975] |

The diagrams in Figures 7 and 8 show the occupation percentage for the different parts of the architecture proposed respectively for 45nm and 7nm ASIC target technologies.

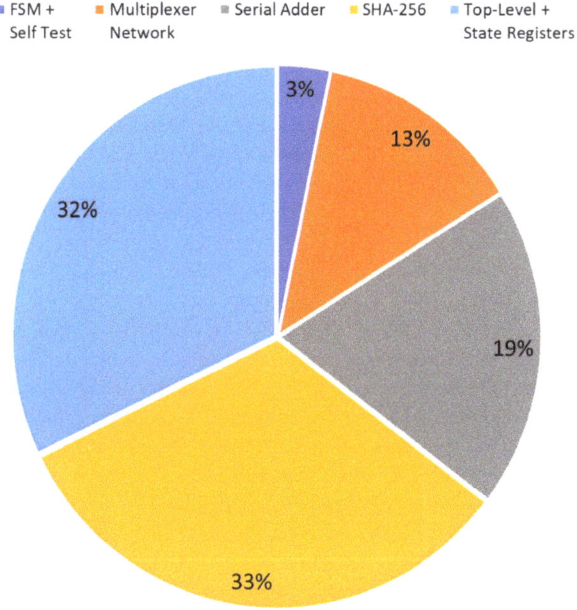


**Figure 7.** CSPRNG (Hash DRBG) IP-core occupation diagram on 45 nm ASIC technology (kGE based).

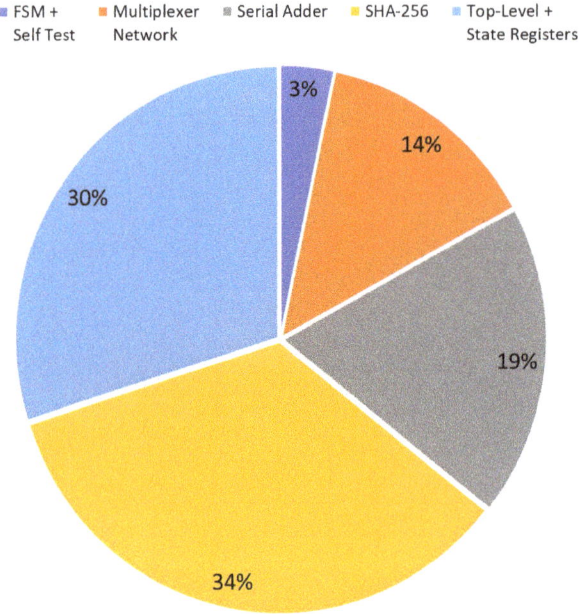


**Figure 8.** CSPRNG (Hash DRBG) IP-core occupation diagram on 7 nm ASIC technology (kGE based).

## 6. Conclusions

This paper presented the architecture and implementation of a high performance digital Cryptographically Secure Pseudo-Random Number Generator (CSPRNG). Specifically, a Hash-based Deterministic Random Bit Generator (DRBG) circuit was presented, following recommendation given by NIST in [6], using SHA256 cryptographic primitive. CSPRNG is a key component to implement efficient cybersecurity applications for authentication, confidentiality and message integrity. In addition, the security of critical information exchanged between the different subsystems in modern vehicles is proving to be a key issue in the automotive sector: more advanced and complex devices and sensors provide the platform for active assistance and security on which passengers rely, so it is crucial to ensure that these systems are adequately protected from cyber attacks. Hash algorithm selection was done according to a trade-off analysis on throughput, area and security strength: among the solutions able to satisfy the security requirements, the SHA-256 core was proved to be the most efficient solution in terms of throughput-complexity ratio. The detailed description of the optimized SHA-256 core architecture being developed for DRBG circuit implementation is also given. The proposed CSPRNG IP-core was tested by means of NIST Statistical Test Suite, thus stating that the sequences of bits generated cannot be distinguished from a true random sequence of numbers, and therefore validating its use for cryptographic applications. It has been also implemented on FPGA and ASIC standard-cell technologies for characterization: on Intel Stratix IV FPGA it is reported a throughput of 690 Mbps at 180 MHz with a maximum occupation of 4713 ALMs, on 45 nm ASIC standard-cell [15] the throughput is equal to 3.82 Gbps at 1 GHz with a logic complexity of 49.19 kGE, and finally on 7 nm ASIC standard-cell [16] the throughput reaches a value of 19.67 Gbps at 5.15 GHz with the logic complexity of 46.56 kGE.

**Author Contributions:** Conceptualization, L.F. and S.S.; methodology, L.B., L.C., F.F., L.F. and S.S.; software, L.B. and L.C.; validation, M.B. and J.B.; formal analysis, L.B., L.C., F.F., M.B. and J.B.; investigation, M.B. and J.B.; resources, L.F. and S.S.; data curation, L.B., L.C. and F.F.; writing—original draft preparation, L.B., L.C., F.F. and M.B.; writing—review and editing, F.F., L.F. and S.S.; visualization, F.F., L.F. and S.S.; supervision, L.F. and S.S.;

project administration, L.F. and S.S.; funding acquisition, L.F. and S.S. All authors have read and agreed to the published version of the manuscript.

**Funding:** This research was partially funded by the European Union's Horizon 2020 research and innovation programme "European Processor Initiative" under grant agreement No. 826647.

**Conflicts of Interest:** The authors declare no conflict of interest.

## References

1. Baldanzi, L.; Crocetti, L.; Di Matteo, S.; Fanucci, L.; Saponara, S.; Hameau, P. Crypto Accelerators for Power-Efficient and Real-Time on-Chip Implementation of Secure Algorithms. In Proceedings of the 2019 26th IEEE International Conference on Electronics, Circuits and Systems (ICECS), Genova, Italy, 27–29 November 2019; pp. 775–778.
2. Baldanzi, L.; Crocetti, L.; Falaschi, F.; Belli, J.; Fanucci, L.; Saponara, S. Digital Random Number Generator Hardware Accelerator IP-Core for Security Applications. In Proceedings of the Applications in Electronics Pervading Industry, Environment and Society, Pisa, Italy, 11–13 September 2019; Volume 627, pp. 117–123.
3. Paar, C.; Pelzl, J. *Understanding Cryptography*; Springer: Berlin/Heidelberg, Germany, 2011.
4. Lo Bello, L.; Mariani, R.; Mubeen, S.; Saponara, S. Recent Advances and Trends in On-Board Embedded and Networked Automotive Systems. *IEEE Trans. Ind. Inf.* **2019**, *15*, 1038–1051. [CrossRef]
5. Barker, E.; Kelsey, J. *Recommendation for Random Bit Generator (RBG) Constructions, 2nd Draft*; US Department of Commerce, National Institute of Standards and Technology: Gaithersburg, MD, USA, 2016.
6. Barker, E.; Kelsey, J. *Recommendation for Random Number Generation Using Deterministic Random Bit Generators*; US Department of Commerce, National Institute of Standards and Technology: Gaithersburg, MD, USA, 2015.
7. Dang, Q.H. *Secure Hash Standard—Technical Report*; US Department of Commerce, National Institute of Standards and Technology: Gaithersburg, MD, USA, 2015.
8. Dichtl, M.; Golić, J.D. High Speed True Random Number Generation with Logic Gates Only. Cryptographic Hardware and Embedded Systems. *Lect. Notes Comput. Sci.* **2007**, *4727*, 45–62.
9. Vasyltsov, I.; Hambardzumyan, E.; KimBohdan, Y.S.; Karpinskyy, B. Fast Digital TRNG Based on Metastable Ring Oscillator. Cryptographic Hardware and Embedded Systems. *Lect. Notes Comput. Sci.* **2008**, *5154*, 164–180.
10. Yang, B.; Mentens, N. ES-TRNG: A high-throughput, low-area true random number generator based on edge sampling. *IACR Trans. Cryptogr. Hardw. Embedded Syst.* **2018**, *2018*, 267–292.
11. Stevens, M.; Karpman, P.; Peyrin, T. Freestart collision for full SHA-1. In *Annual International Conference on the Theory and Applications of Cryptographic Techniques*; Springer: Berlin, Germany, 2016; pp. 459–483.
12. ECDSA—Application and Implementation Failures. Available online: https://www.semanticscholar.org/paper/ECDSA-Application-and-Implementation-Failures-Schmid/f528fb05f0709923e22c04239503b4289a24cb4f (accessed on 25 March 2020).
13. U.S. Department of Commerce and National Institute of Standards and Technology. *Secure Hash Standard—SHS: Federal Information Processing Standards Publication 180-4*; CreateSpace Independent Publishing Platform: North Charleston, SC, USA, 2012.
14. Bassham, L.E.; Rukhin, A.; Soto, J.; Nechvatal, J.; Smid, M.; Barker, E.; Leigh, S.; Levenson, M.; Vangel, M.; Banks, D.; et al. *A Statistical Test Suite for Random and Pseudorandom Number Generators for Cryptographic Applications—Technical Report*; US Department of Commerce, National Institute of Standards and Technology: Gaithersburg, MD, USA, 2010.
15. Silvaco PDK 45 nm Open Cell Library. Available online: https://www.silvaco.com/products/nangate/FreePDK45_Open_Cell_Library/index.html (accessed on 25 March 2020).
16. Artisan TSMC 7 nm Cell Library. Available online: https://www.tsmc.com/english/dedicatedFoundry/technology/7nm.htm (accessed on 25 March 2020).

Article

# Data Processing and Information Classification—An In-Memory Approach

**Milena Andrighetti** [1], **Giovanna Turvani** [1,*], **Giulia Santoro** [1], **Marco Vacca** [1], **Andrea Marchesin** [1], **Fabrizio Ottati** [1], **Massimo Ruo Roch** [1], **Mariagrazia Graziano** [2] and **Maurizio Zamboni** [1]

[1] Department of Electronics and Telecommunication (DET), Politecnico di Torino, Corso Castelfidardo 39, 10129 Torino, Italy; milena.andrighetti@studenti.polito.it (M.A.); giulia.santoro@polito.it (G.S.); marco.vacca@polito.it (M.V.); andrea.marchesin@studenti.polito.it (A.M.); fabrizio.ottati@studenti.polito.it (F.O.); massimo.ruoroch@polito.it (M.R.R.); maurizio.zamboni@polito.it (M.Z.)

[2] Department of Applied Science and Technology (DISAT), Politecnico di Torino, Corso Castelfidardo 39, 10129 Torino, Italy; mariagrazia.graziano@polito.it

[*] Correspondence: giovanna.turvani@polito.it

Received: 31 January 2020; Accepted: 13 March 2020; Published: 18 March 2020

**Abstract:** To live in the information society means to be surrounded by billions of electronic devices full of sensors that constantly acquire data. This enormous amount of data must be processed and classified. A solution commonly adopted is to send these data to server farms to be remotely elaborated. The drawback is a huge battery drain due to high amount of information that must be exchanged. To compensate this problem data must be processed locally, near the sensor itself. But this solution requires huge computational capabilities. While microprocessors, even mobile ones, nowadays have enough computational power, their performance are severely limited by the Memory Wall problem. Memories are too slow, so microprocessors cannot fetch enough data from them, greatly limiting their performance. A solution is the Processing-In-Memory (PIM) approach. New memories are designed that can elaborate data inside them eliminating the Memory Wall problem. In this work we present an example of such a system, using as a case of study the Bitmap Indexing algorithm. Such algorithm is used to classify data coming from many sources in parallel. We propose a hardware accelerator designed around the Processing-In-Memory approach, that is capable of implementing this algorithm and that can also be reconfigured to do other tasks or to work as standard memory. The architecture has been synthesized using CMOS technology. The results that we have obtained highlights that, not only it is possible to process and classify huge amount of data locally, but also that it is possible to obtain this result with a very low power consumption.

**Keywords:** bitmap indexing; processing in memory; memory wall; big data; internet of things

---

## 1. Introduction

Nowadays many applications used everyday, defined as data-intensive, require a lot of data to process. Examples are the databases manipulation and image processing. This requirement is the effect of the fast improvement of CMOS technology, that has lead to the creation of very powerful and flexible portable devices. These devices are full of sensors that continuously acquire data. Data can be elaborated remotely by powerful servers, but sending a lot of information through electromagnetic waves requires a huge amount of energy, severely impacting the battery life of mobile devices. The only solution is to elaborate data locally, on the mobile device itself.

Thanks to the scaling of transistors size, mobile microprocessors are now theoretically capable of such computation. Unfortunately, memory scaling has been following a different path, resulting still in slow accesses compared to processors computing speed. This discrepancy in performance harms the computing abilities of the CPU, since the memory cannot provide data as quickly as required by the CPU. This problem is called *Von Neumann bottleneck* or *Memory Wall*. The idea that took form to solve this problem is to null the distance between processor and memory, removing the cost of data transfer and create a unit which is capable of storing information and of performing operation on them. This idea takes the name of Processing-in-Memory.

Many in literature have approached the "in-memory" idea. Some narrowing the physical distance between memory and computation unit by creating and stacking different layers together. But even if the two units are moved very close to each other, they are still distinct components. Others exploited intrinsic functionality of the memory array or slightly modified peripheral circuitry to perform computation.

Among the many example provided by literature, one of the best fitting representative of the PIM concept is presented in Reference [1]. In this work the proposed architecture is a memory array in which the cell itself is capable of performing logical operations aimed at solving Convolutional Neural Networks (CNN). In this paper, our main goal is to introduce a proper example of Processing-in-Memory, choosing Bitmap Indexing as an application around which the architecture is shaped. In the design, it was not used a specific memory technology because the idea is to provide a worst-case estimation and it was also meant to leave space for future exploration to implement the cell with a custom model of the memory cell. The Bitmap Indexig algorithm has been chosen because it is used for data classification. This is one of the most important task that must be performed by such mobile devices. Being able to classify data allows to understand which data must be sent to remote servers and which not, greatly reducing the overall power consumption. The presented architecture is a memory array in which each cell is both capable of storing information and to perform simple logical operation on them. A characteristic of our architecture is its modularity. The architecture is divided in independent memory banks. A memory bank can work both on its own or interacting with other banks. Moreover it is possible to build the array with as many banks as needed. This feature lead to great flexibility and high degree of parallelism. The structure was eventually synthesized for analysis purposes, in a 8.5 KB square array, using CMOS 45 nm and 28 nm. The storage segment of the proposed PIM cell was synthesized as a latch. The evaluation showed great results, achieving a maximum throughput of 2.45 Gop/s and 9.2 Gop/s respectively for the two technologies used. This paper is the extended version of our prior work [2]. In the conference paper the general idea was introduced. Here we greatly expand the architecture, moving from the idea to the real implementation. The novelty of this work, in comparison with other works presented in the literature, consists in an enhanced architecture characterized by a high level of granularity and flexibility.

## 2. Background

The Processing-in-Memory paradigm was born to solve the *Von Neumann bottleneck*, which is characterized by the gap in performance between memory and processor. Processing-in-Memory thus tries to reduce the disparity by merging together storage and processing units. Processing-in-Memory (PIM) can be approached in different ways, depending on the architecture or the technologies to use. A lot of examples can be found in literature, some of them will be depicted in the following, grouped in categories.

### 2.1. Magnet-Based

*Magnetic Random Access Memory* (MRAM) is a non-volatile memory that uses Magneto-Tunnel Junctions as its basic storage element. Thanks to their dual storage-logic properties, MTJs are suitable to implement hybrid logic circuits with CMOS technology suited to implement the PIM principle. In Reference [3] is presented a MTJ-CMOS Full Adder, which compared to a standard only-CMOS solution showed better results. In Reference [4] the authors proposed an MTJ-based TCAM, in which the logic part and the storage element are merged together, and an MTJ-based Non-Volatile FPGA exploiting MTJs and combinatorial blocks. Both structures resulted in a more compact solution with respect to conventional ones.

In Reference [5] it is proposed a different way to implement Nano Magnetic Logic (NML) exploiting the MRAM structure. Since the basic concept of the NML technology is the transmission of information through magnetodynamic interaction between neighbouring magnets, the MRAM structure has been modified so that MTJs could interact with each other. Another example is represented by PISOTM [6], an architecture based on SOT-RAM. It is a reconfigurable architecture in which the main advantage is that the storage and logic element result identical and for this reason technology conflict is avoided.

### 2.2. 3D-Stacking

According to the 3D-Stacking approach multiple layers of DRAM memory are stacked together with a logic layer that can be application-specific ([7,8]) or general purpose [9]. In Reference [7] the XNOR-POP architecture was designed to accelerate CNNs for mobile devices. It is composed of Wide-IO2 DRAM memory with the logic layer modified according to the XNOR-Net requirements. In Reference [8] it is proposed an architecture for data intensive applications, where a PIM layer made of memory and application-specific logic is sandwiched between DRAM dies connected together using TSVs. An example of general purpose 3D-stacking is 3D-MAPS in Reference [9]. A multi-core structure is used, and every core is composed of a memory layer and a computing layer.

### 2.3. ReRAM-Based

*Resistive RAM* is a non-volatile memory that uses a metal-insulator-metal element as storage component. The information is represented by the resistance of the device that can be either high (HRS) or low (LRS). To switch between states the appropriate voltage has to be applied to the cell. The common structure of a ReRAM array is a crossbar, a structure used in matrix-vector multiplication, commonly found in neural networks applications. PRIME [10], an architecture aimed at accelerating Artificial Neural Networks is an example of this kind of implementations. PRIME is compliant with the in-memory principle, since the computation is performed directly into the memory array with few modifications to the peripheral circuitry. Memory banks are divided intro three sub-arrays each with a specific role in the architecture. In Reference [11] is proposed a 3D-ReCAM based architecture to accelerate the BLAST algorithm for DNA sequence alignment. The architecture, named RADAR, aims to move the operations in memory, this way there is no need to transfer the DNA database. In Reference [12] is presented a non-volatile intelligent processor built on a 150 nm CMOS process with HfO RRAM. The structure is capable of both general computing and the acceleration of neural networks, in fact it is provided with a FCNN Turbo Unit, enhanced with low-power MVM engines to perform FCNN tasks.

Another application that is limited by the Memory Wall problem is Graph Processing. In Reference [13] is proposed a ReRAM-based in-memory architecture as a possible solution. The structure is composed of multiple ReRAM banks, divided into 2 types: graph banks that are used to map the graph and to store its adjacency list and a master bank which stores metadata of the graph banks. This allows to process the graphs that are stored inside the memory. In Reference [14] is presented PLiM, a programmable system composed of a PIM controller and a multi-bank ReRAM which can work both as a standard memory and as a computational unit, according to the controller signals. PLiM implemented only serial operation to keep the controller as simple as possible. In Reference [15] the authors presented ReVAMP, an architecture composed of two ReRAM crossbars, supporting parallel computations and VLIW-like instructions. To perform logic operations ReVAMP exploits the native properties of ReRAM cells that implement a majority voting logic function.

### 2.4. PIM

In Reference [16] the authors presented TOP-PIM, a system composed of an host processor surrounded by several units characterized by 3D-stacked memories with an in-memory processor embedded on the logic die. In Reference [17] is proposed DIVA, a system in which multiple PIM chips serve as smart-memory co-processors to a standard microprocessor aimed at improving bandwidth performance for data intensive applications executing computation directly in memory and enabling a dedicated communication line between the PIM chips. In Reference [18] is presented Terasys, a massively parallel PIM array. The goal of Terasys was to embed an SIMD PIM array very close to an host processor in order for it to be seen both as a processor array and conventional memory. As solution for large-scale graph processing performance bottleneck, in Reference [19] the authors proposed Tesseract, a PIM architecture used as an accelerator for an host processor. Each element of Tesseract has a single-issue in-order core to execute operations, moreover, the host processor has access to the entire Tesseract's memory whilst each core of Tesseract can interact only with its own. Tesseract does not depend on a particular memory organization, but it was analyzed exploiting Hybrid Memory Cube (HMC) as baseline. Such a structure proved to perform better than traditional approaches thanks to the fact that Tesseract was able to use more of the available bandwidth. In Reference [20] is presented Prometheus, a PIM-based framework, which proposes the approach of distributing data across different vaults in HMC-based systems with the purpose of reducing energy consumption, improving performance and exploiting the high intra-vault memory bandwidth.

In Reference [21] is proposed a solution to accelerate Bulk Bitwise Operations. PINATUBO is an architecture based on resistive cell memories, such as ReRAMs. The structure is composed of multiple banks which are also subdivided into mats. Pinatubo is able to eliminate the movement of data, since computation is performed directly inside memory, executing operations between banks, mats and subarrays. This way PINATUBO interacts with CPU only for row addresses and control commands. Another example of PIM architecture to accelerate bulk bitwise operations was conceived by the authors of Reference [22], who presented Ambit, an in-memory accelerator which exploits DRAM technology to achieve total usage of the available bandwidth. The DRAM array is slightly modified to perform AND, OR and NOT operations. Moreover, the CPU can access Ambit directly, this way it is not necessary to transfer data between CPU memory and the accelerator. In Reference [23] is proposed APIM, an Approximate Processing-in-Memory architecture which aims to achieve better performance despite a decrease in accuracy. It is based on emerging non-volatile memories, such as ReRAM and it is composed of a cross-bar structure grouped in blocks. All the blocks are structurally identical but divided into data and processing blocks. They are linked together through configurable interconnections. Furthermore APIM is able to configure computation precision dynamically, so that it is possible to tune the accuracy runtime.

In Reference [24] is presented ApproxPIM, an HMC-based system in which each vault is independent from one another and communication with the host processor is based on a parcel transmission protocol. This results in energy and speedup improvements with respect to the used baselines. In Reference [25] the authors presented MISK, a proposal to reduce the gap between memory and processor. Since data movement imply a great energy cost, MISK is intended to reduce it by implementing a monolithic structure, avoiding physical separation between memory and CPU. In fact, MISK is to be integrated into the cache and it is not conceived to work on its own, but embedded in the CPU. This way it is possible to achieve great results in terms of energy-per-cycle and execution time. In Reference [26] is introduced Gilgamesh, a system based on distributed and shared memory. It is characterized by a multitude of chips, called MIND chips, which are connected together through a global interconnection network. Each chip is a general purpose unit equipped with multiple DRAM bank and processing logic. In Reference [27] Smart Memory Cube is presented, a PIM processor built near the memory, in particular HMC, which is connected to an host processor. HMC vault controls are modified to perform atomic operations. The PIM processor interacts with the host processor so that smaller tasks are executed directly side by side the memory.

In References [28,29], the authors presented in-memory architectures on which the Advanced Encryption Standard (AES) algorithm was mapped, showing great result in speed and energy saving compared to other solutions. In Reference [1], the authors presented an architecture based on the in-memory paradigm aimed at Convolutional Neural Networks (CNN). The structure is a memory array in which each cell is provided with both storage and computation properties and with the support of an additional weight memory which is designed to support CNN data flow and computation inside the array. This structure showed great result compared with a conventional CNN accelerator in terms of memory accesses and clock cycles.

## 3. The Algorithm

The Processing-in-Memory principle requires that the storage and logic components are merged together. In order to implement an architecture compliant with such a requirement it was necessary to firstly shape it according to a suitable application. For this purpose Bitmap indexing was selected. Bitmap indexes are often used in database management systems.

Taking as an example the simple database in Figure 1A, each column of the database represents a particular characteristic of the profile of the entry described in one row. Suppose a search on the database is to be performed to create a statistic on how many men possess a sport car or a motorbike. Such a query would imply looking for all the men and then excluding the ones that do not own the specified vehicles. If the database is big this operation would require a long response time. Bitmap indexing was introduced to solve this issue. Bitmap indexing transforms each column of a table in as many indexes as the number of distinct key-values that particular column can have.

A bitmap index is a bit array in which the $i$-th bit is set to 1 if the value in the $i$-th row of the column is equal to the value represented by the index, otherwise it is set to 0 (Figure 1A). Thus, bitmap indexing allows to fragment search queries in simple logic bitwise operations (Figure 1B). This way it is not necessary to analyze the whole database discarding unwanted data, but only to operate on selected indexes. Bitmap indexing can provide great results in response time and in storage requirements since it can be compressed. Bitmap indexing is suited for entries with a number of possible values smaller than the depth of the whole table. This technique is mostly functional for queries regarding the identification of the position of specific features, for this reason to answer an "how many" query it is necessary to insert a component that counts the hits obtained. Summing up, a query can be decomposed in simple logic operations which are performed between indexes, processing bits belonging to the same position in the array (Figure 1C).

Clearly, Bitmap indexing results compatible with the Processing-in-Memory paradigm, since it is characterized by simple logic bitwise operations and its data format make it easy to embed in memory. However, bitmap indexing involves operations between columns of a table. If we consider memory organization and imagine to maintain the column-row distribution of the table in memory, this would imply to access multiple rows and then discard all the data that do not belong to the desired indexes. This approach would be too costly. For this reason for our implementation a column-oriented was preferred, which means that the entire table is stored transposed, so that now, applying bitmap indexing, indexes lie on rows (Figure 2).

Thanks to this method, to access an index it is only necessary to access a row and consequently operations between indexes result in operations between memory rows. In this implementation we thus consider the indexes distributed on rows in a memory array. We also take into account two types of query, *simple* and *composed*. A simple query is composed of only one operation (e.g., "Who is female and married?") whilst a composed one is characterized by intertwined operations (e.g., Figure 1B). Considering the composed query depicted in Figure 1B the operations to perform would be:

1.  Access the first operand;
2.  Access the second operand;
3.  Execute bitwise operation between the two operands;
4.  Read result;

5.  Execute bitwise operation between computed result and third index;
6.  Count the hits obtained;
7.  Read final result;

While to answer a simple query only steps 1–4 are needed. The goal is then to implement the just introduced algorithm directly inside a memory array.

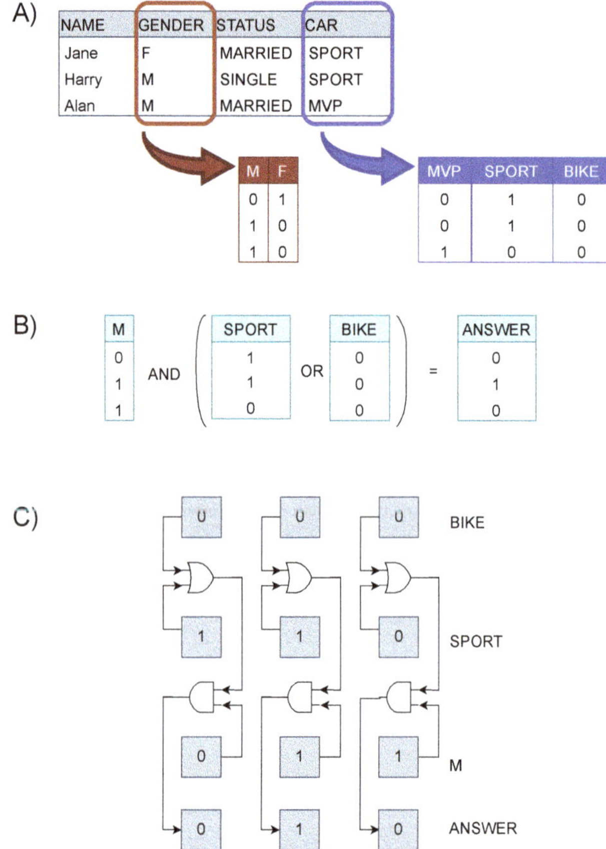

Figure 1. (**A**) Given a table, bitmap indexing transforms each column in as many bitmap as the number of possible key-values for that column (**B**) In order to answer a query logic bitwise operations are to be performed (**C**) Practical scheme of the execution of the query.

| NAME | Jane | Harry | Alan |
|---|---|---|---|
| GENDER | F | M | M |
| STATUS | MARRIED | SINGLE | MARRIED |
| CAR | SPORT | SPORT | MPV |

| F | 1 | 0 | 0 |
|---|---|---|---|
| M | 0 | 1 | 1 |

Figure 2. Column-oriented memory organization.

## 4. The Architecture

The architecture proposed in this paper present a possible solution for the Von Neumann bottleneck implementing a proper *in-memory* architecture, where logic functions are implemented directly inside each memory cell, in contrast with the *near-memory* approach seen in some state-of-the-art implementations, where logic operations are performed with logic circuits located on the border of the memory array. Moreover, this architecture was intended to overcome the limits provided by specific technologies by keeping the development of the architecture technology-independent, in order to implement a configurable architecture with the highest degree of parallelism achievable.

A memory array is composed of many storage units, each of which is made of multiple memory cells. Cells are the basic element of the memory itself. Therefore, in order to implement an entire memory array aimed at executing the Bitmap indexing algorithm, firstly it is necessary to define the structure of the memory cell.

According to the specifications required by the Bitmap indexing, the cell has to be able to perform simple logic operations interacting with other cells in the array. This means that our cell should have both storage and logic properties. Indeed, the basic cell of the PIM array is provided with an element that store information and a configurable logic element which performs AND, OR, XOR operations with all the combinations of input (e.g., $A, \overline{A}$), between the stored information and the one coming from another cell (Figure 3). The system has indeed the granularity of a single bit, meaning that every memory cell executes a logic operation.

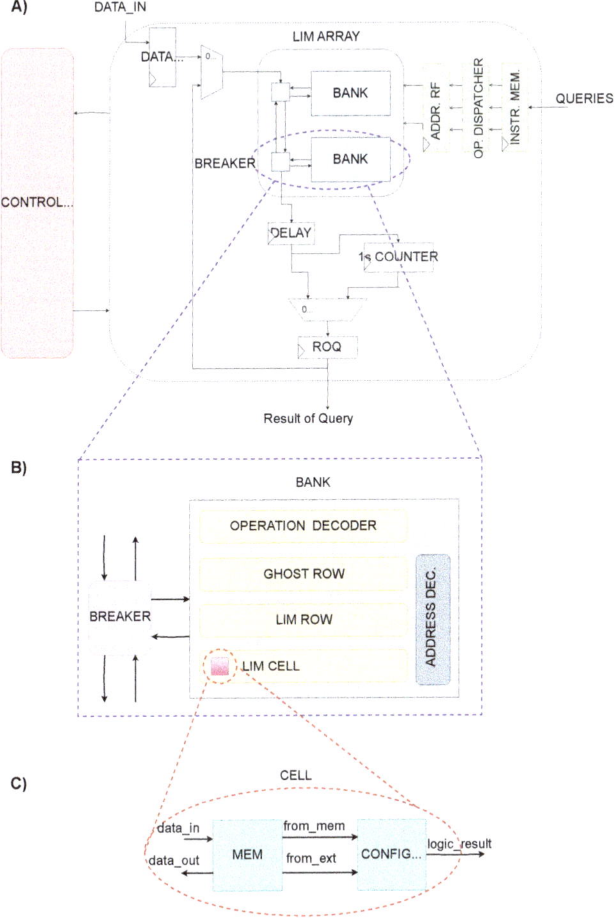


**Figure 3.** (**A**) Overview of the complete architecture. (**B**) Structure of the duo Bank-Breaker. (**C**) Insight of the Processing-In-Memory (PIM) cell.

Other than standard memory features the PIM cell can interact with other cells, according to its control input. As every single cell in the array has the ability to perform computation, it is necessary to choose which cell will be executing the operation and which will be read. In order to implement it, the designated passive cell is read and the stored data travels to the operative cell. To avoid interference between inactive cells, the output lines of cells that are not used are interrupted. To implement the bitwise feature each cell of a row has its input and output line common to any other cell belonging to the same column of different rows.

In Figure 3, the whole structure is depicted. Noticeably, other than the array, the architecture is composed of a control unit and some additional components, such as the counter (for counting ones) and register files. Focusing on the array, like any standard memory, it was divided into multiple banks. Each bank is associated with a *breaker* that manages data flow from and to the bank. A bank represents the smallest degree of parallelism of the architecture. This means that in a bank it is possible to execute one operation at a time. The system has also a second level of granularity because thanks to the breakers every bank can work independently. This solution provides at the same time a high level of granularity and flexibility. Banks can execute operations between its rows or can work with other banks, making interact rows belonging to different banks, while other banks work on different operations in parallel. As a consequence, supposing each bank in the array works on a different operation by itself, the maximum degree of parallelism achievable is equal to the number of banks in the array. The *Bidirectional Breaker* is in charge of managing relations between its bank and the rest of the array. According to the control input, the breaker can be passive, that is, letting data pass through without disturbing its bank so that the bank can work on its own or be silent. The breaker can also be active and diverting data to or from its bank.

A bank is composed of multiple PIM rows and one *Ghost row* which is provided only with memory properties used to store temporary operation results. The Ghost row has the input line connected to the logic result output line of the PIM rows, whilst its output line is common with the PIM rows. This way it is possible to read the Ghost row or use its content for further computation. As in standard memories, each row is fragmented in multiple words. This means that operations are actually performed between words belonging to different rows. The result is then temporary saved in the Ghost word corresponding to the same word address of the word which executed the operation. This was implemented to avoid the need to manage a third address. To handle all the configuration signals needed to manage the correct execution, two decoders were needed inside each bank. One that sets the configuration for the logic operation to execute, sending it to the right row. The second was implemented to control addresses, data flow inside the bank and to distinguish between standard memory mode and PIM operation mode. Since a simple AND operation can be performed in one bank in a single clock cycle, imaging of having multiple banks definitely increase the number of operations that can be executed in one clock cycle in parallel. The same reasoning goes for a composed operation which takes two clock cycles. The throughput is directly proportional to the number of banks in the memory block. So, the larger the number of banks, the larger the memory block and also the larger the throughput.

In Figure 3, it is highlighted that, other than the array, there are some additional components which are used to guarantee the correct functioning of the entire structure.

The *Instruction Memory* is used to collect the queries to execute. It consists in a register file, having as many registers as the number of banks, with an input parallelism equal to the length of a complete query (i.e., two complete addresses and a logic operation configuration string). A composed query is treated as the combination of two distinct queries, which means that a composed query will occupy two consecutive registers of the Instruction Memory. Clearly, even if the architecture was configured to exploit its maximum potential by implementing the bitmap indexing algorithm, it can be configured to perform additional algorithms. For reconfigurability purposes the instruction memory had to be implemented as wide as possible, but most likely it will not be updated fully each time. In order to avoid conflicts the *Operation Dispatcher* is in charge of blocking any old query. Since a query can take place between any couple of addresses in the array, it is necessary to sent the addresses to their respective bank. The Operation Dispatcher thus reorders addresses and sends them to their own bank. After the correct reordering, to ensure synchronization the addresses are sampled by the *Address Register File* which loads the addresses and sends them to the array.

As illustrated previously, results of bitwise logic operations answer to queries in where clause. To count the number of ones ("1") in the "how many" clause it was inserted a ones counter of logic "1" connected with the output of a delay register. The register was added to ensure timing constraints given by the counter. A simple counter that processes the data input bit-by-bit and increments by one for each "1" found was too slow. Therefore, a tree-structured counter was implemented. Firstly, the data array is fragmented into D segments, each of $\frac{N}{D}$-bits. All segments are then analyzed at the same time and the ones contained in each segment are counted. Finally, all the factors are added together to obtain the final sum. Also, all the adders that form the tree-structure are of the same dimension computed to avoid overflow.

The architecture was conceived to incorporate as many features as possible and at the same time trying to keep the control circuits as simple as possible. The implemented structure is versatile and can work in 8 different operation modes, discerned among traditional memory operations and PIM operations based on the position of the two operands and the desired parallelism: (1) Write; (2) Read; (3) Save result; (4) PIM simple single bank; (5) PIM simple different banks; (6) PIM multiple banks; (7) PIM composed; (8) PIM multiple composed. Each operation mode is the starting point of a query, which is composed as shown in Figure 4A. The FSM chart of all operation modes are reported in Figure 4B.

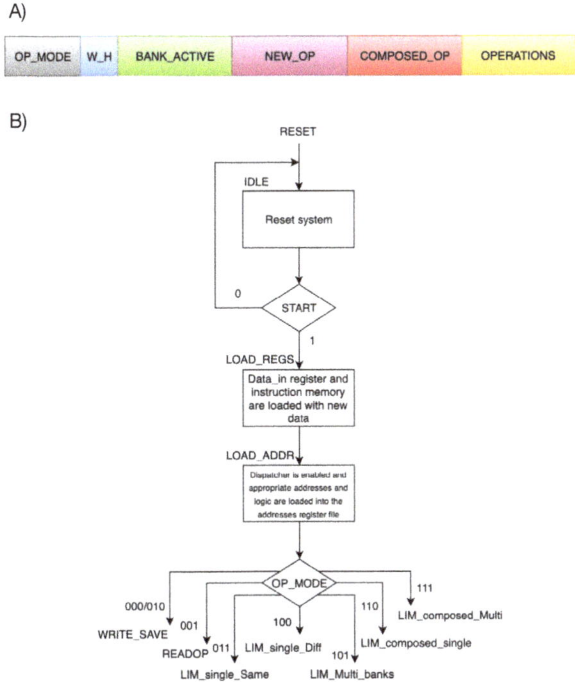


**Figure 4.** (**A**) Composition of a complete query. (**B**) Preliminary stages.

The developed architecture is a modular configurable parallel architecture that implements the concept of Processing-in-Memory to perform bitwise logic operations directly inside the memory, making it suitable for other applications other than Bitmap indexing, as long as they are based on bitwise.

## 5. Results and Conclusions

The architecture was fully developed in VHDL (VHSIC Hardware Description Language). In order to evaluate its performance a 8.704 KB square memory array was analysed. The array distribution consisted in 16 banks with 16 bit data size. All the internal structures have been kept para- metric to give the possibility to implement the architecture composed of how many banks, rows and words needed according to the target database. From a MATLAB script (or from an external source in the case of the bitmap) were extracted both the bitmap and the queries to execute. The files were then set as input for the VHDL Testbench and finally it was run a simulation of the queries to feed the PIM architecture. When started, the script enters a loop that terminates only when the user decides not to create any more queries and a file generated as output. The completion of the query is assisted by two pop-up windows: one shows the internal composition of the memory and the other shows the available logic operations and their correspondent code.

All eight operation modes were tested with Modelsim to ensure the correct functioning. Two examples of operation mode are reported in Figure 5, it shows two examples of logic behavior (expected and simulated) of the proposed architecture.

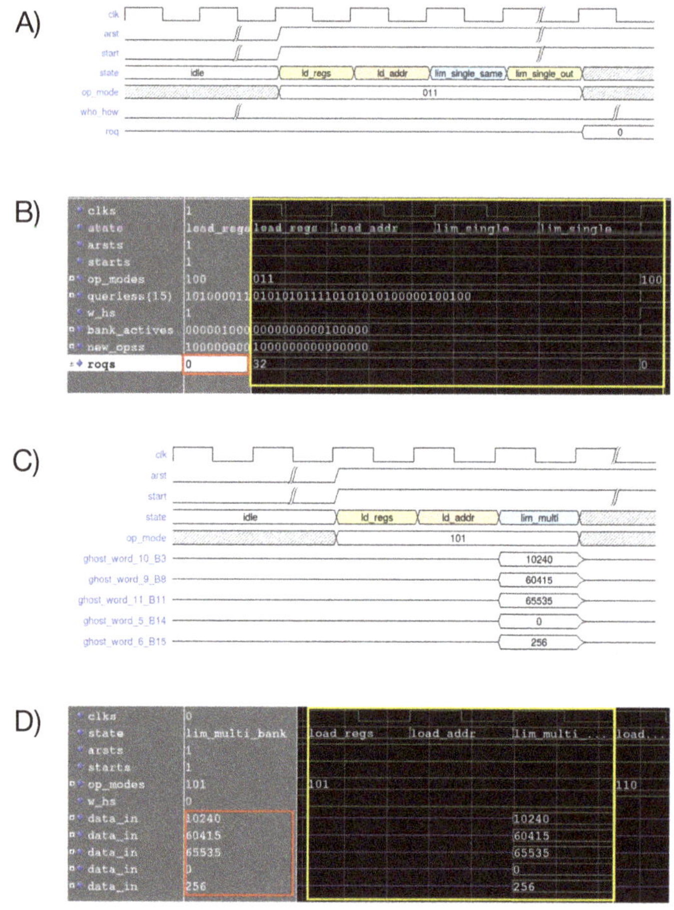

**Figure 5. (A)** Expected waveform of a LIM single same bank AND operation. **(B)** Waveform of a LIM single same bank AND operation. **(C)** Expected waveform of a PIM multiple operations. **(D)** Simulated waveform of a PIM multiple-bank operation.

The architecture was later synthesized with Synopsys Design Compiler using 45 nm BULK and 28 nm FDSOI CMOS technologies (Table 1). By using Synopsys Design Compiler latches and logic gates are used to implement the memory cell, so the results are not optimized as they will be if a custom transistor layout was created for the memory cell.

As the fundamental element of the whole structure, the Cell was analyzed and optimized. The obtained results are reported in Tables 1 and 2.

From, Table 1 it is possible to evince the the area overhead is 55%. The overhead in terms of power dissipation is similar.

**Table 1.** Synthesis of the fundamental element.

|  | Memory | Logic | Cell |
|---|---|---|---|
| Non-Combinational Area [mm$^2$] | 9.31 | 2.12 | 11.43 |
| Combinational Area [mm$^2$] | 5.32 | 15.43 | 20.75 |
| Total Area [mm$^2$] |  |  | 32.18 |
| Delay [ns] |  |  | 0.45 |

**Table 2.** Synthesis results for 45 nm and 28 nm CMOS technologies.

| Parameter | Value (45 nm) | Value (28 nm) |
|---|---|---|
| Total area [mm$^2$] | 2.33 | 1.058 |
| $f_{CLK}$ [MHz] | 153.4 | 574.7 |
| Total Power [mW] | 49.7 | 14.07 |

An interesting point is the relation between the number of the segments and the resulting delay. An analysis was carried out with 8 bit and 16 bit input data size (Figure 6). As it shows the delay reduces considerably with a bigger amount of segments. Indeed, the architecture under consideration was synthesized with a value D of 8 to achieve best speed.

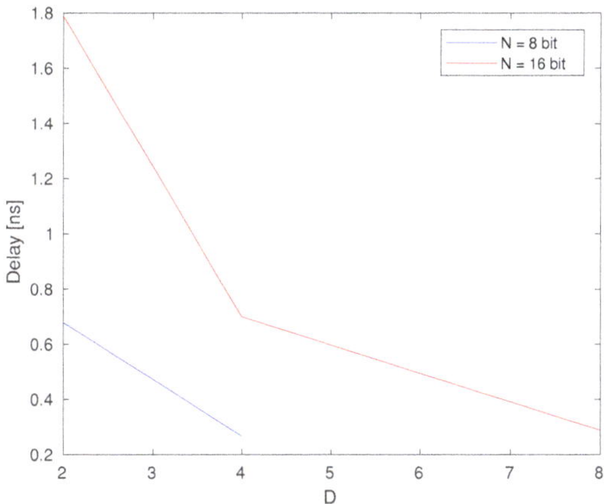


**Figure 6.** Relation between number of segments in the counter and resulting delay.

One of the main goal this paper aimed to fulfill is the high level of concurrency. This was accomplished thanks to the internal structure of the array, distributed on banks which are capable of working both independently and with each other, providing flexibility in the position of the operands that are called to act in the query. To execute a simple query only one cycle is required. Thanks to the modular structure of the array, the maximum throughput achievable working in parallel in PIM multiple banks mode is:

$$throughput_{max_{simple}} = f_{CLK} \cdot N_{ops}.$$

As for composed query two cycles are required to complete the operations. The resulting maximum throughput operating in PIM multiple composed mode is:

$$throughput_{max_{composed}} = \frac{f_{CLK}}{2} \cdot N_{ops}.$$

So, assuming to execute a different query in each of the 16 available banks, we will reach a maximum throughput of 2.45 Gop/s and 9.2 Gop/s for 45 nm and 28 nm respectively. The performance of the proposed PIM architecture was compared with results of other in-memory proposals found in Reference [29] (Table 3).

**Table 3.** Clock cycles comparison for a single query execution.

|  | $f = A \cdot B$ | $f = A \cdot (\overline{B} \cdot C)$ |
|---|---|---|
| **Pinatubo** [21] | 5 | 9 |
| **RIMPA** [28] | 3 | 5 |
| **PIMA-Logic** [29] | 1 | 3 |
| **PIM** | 1 | 2 |

Noticeably, operations in the proposed PIM array take less clock time compared to other solutions. Moreover, it should be taken into consideration that executing multiple parallel operations would not change the number of clock cycles required. This shows how the throughput mentioned above is obtained. Thus, the maximum degree of parallelism achievable is correspondent to the number of the available banks. Moreover, it is possible to scale the architecture to bigger dimensions as it was conceived as modular, meaning it can be composed with as many banks as wanted. Another possibility is to develop a 3D structure in order to enhance performance. Nonetheless, it would be easy to modify the architecture to make it fit for other types of operations. These results, coupled with the flexibility of the architecture, highlight the potential of the proposed architecture.

**Author Contributions:** Conceptualization, G.T., G.S., M.G., M.V., M.Z., M.R.R.; methodology, G.T., M.V.; software, M.A.; validation, M.A., G.T.; investigation, M.A., G.T., M.G., M.V., M.Z., M.R.R.; resources, X.X.; data curation, X.X.; writing–original draft preparation, M.A., G.T.; writing–review and editing, M.A., G.T., M.V., A.M., F.O.; supervision, M.G., M.Z. All authors have read and agreed to the published version of the manuscript.

**Funding:** This research received no external funding.

**Conflicts of Interest:** The authors declare no conflict of interest.

## References

1. Santoro, G.; Turvani, G.; Graziano, M. New Logic-In-Memory Paradigms: An Architectural and Technological Perspective. *Micromachines* **2019**, *10*, 368. [CrossRef]
2. Andrighetti, M.; Turvani, G.; Santoro, G.; Vacca, M.; Ruo Roch, M.; Graziano, M.; Zamboni, M. Bitmap Index: A Processing-in-Memory reconfigurable implementation. In Proceedings of the Applications in Electronics Pervading Industry, Environment and Society (ApplePies), Pisa, Italy, 12–13 September 2019.
3. Matsunaga, S.; Hayakawa, J.; Ikeda, S.; Miura, K.; Endoh, T.; Ohno, H.; Hanyu, T. MTJ-based nonvolatile logic-in-memory circuit, future prospects and issues. In Proceedings of the 2009 Design, Automation Test in Europe Conference Exhibition, Nice, France, 20–24 April 2009; pp. 433–435. [CrossRef]
4. Hanyu, T. Challenge of MTJ-Based Nonvolatile Logic-in-Memory Architecture for Dark-Silicon Logic LSI. *SPIN* **2013**, *3*, 1340014. [CrossRef]
5. Turvani, G.; Bollo, M.; Vacca, M.; Cairo, F.; Zamboni, M.; Graziano, M. Design of MRAM-Based Magnetic Logic Circuits. *IEEE Trans. Nanotechnol.* **2017**, *16*, 851–859. [CrossRef]
6. Chang, L.; Wang, Z.; Zhang, Y.; Zhao, W. Reconfigurable processing in memory architecture based on spin orbit torque. In Proceedings of the 2017 IEEE/ACM International Symposium on Nanoscale Architectures (NANOARCH), Newport, RI, USA, 25–26 July 2017; pp. 95–96. [CrossRef]

7.	Jiang, L.; Kim, M.; Wen, W.; Wang, D. XNOR-POP: A processing-in-memory architecture for binary Convolutional Neural Networks in Wide-IO2 DRAMs. In Proceedings of the 2017 IEEE/ACM International Symposium on Low Power Electronics and Design (ISLPED), Taipei, Taiwan, 24–26 July 2017; pp. 1–6. [CrossRef]

8.	Zhu, Q.; Akin, B.; Sumbul, H.E.; Sadi, F.; Hoe, J.C.; Pileggi, L.; Franchetti, F. A 3D-stacked logic-in-memory accelerator for application-specific data intensive computing. In Proceedings of the 2013 IEEE International 3D Systems Integration Conference (3DIC), San Francisco, CA, USA, 2–4 October 2013; pp. 1–7. [CrossRef]

9.	Kim, D.H.; Athikulwongse, K.; Healy, M.B.; Hossain, M.M.; Jung, M.; Khorosh, I.; Kumar, G.; Lee, Y.J.; Lewis, D.L.; Lin, T.W.; et al. Design and Analysis of 3D-MAPS (3D Massively Parallel Processor with Stacked Memory). *IEEE Trans. Comput.* **2015**, *64*, 112–125. [CrossRef]

10.	Chi, P.; Li, S.; Xu, C.; Zhang, T.; Zhao, J.; Liu, Y.; Wang, Y.; Xie, Y. PRIME: A Novel Processing-in-Memory Architecture for Neural Network Computation in ReRAM-Based Main Memory. In Proceedings of the 2016 ACM/IEEE 43rd Annual International Symposium on Computer Architecture (ISCA), Seoul, Korea, 18–22 June 2016; pp. 27–39. [CrossRef]

11.	Huangfu, W.; Li, S.; Hu, X.; Xie, Y. RADAR: A 3D-ReRAM based DNA Alignment Accelerator Architecture. In Proceedings of the 2018 55th ACM/ESDA/IEEE Design Automation Conference (DAC), San Francisco, CA, USA, 24–28 June 2018; pp. 1–6. [CrossRef]

12.	Su, F.; Chen, W.H.; Xia, L.; Lo, C.P.; Tang, T.; Wang, Z.; Hsu, K.H.; Cheng, M.; Li, J.Y.; Xie, Y.; et al. A 462GOPs/J RRAM-based nonvolatile intelligent processor for energy harvesting IoE system featuring nonvolatile logics and processing-in-memory. In Proceedings of the 2017 Symposium on VLSI Technology, Kyoto, Japan, 5–8 June 2017; pp. T260–T261. [CrossRef]

13.	Han, L.; Shen, Z.; Shao, Z.; Huang, H.H.; Li, T. A novel ReRAM-based processing-in-memory architecture for graph computing. In Proceedings of the 2017 IEEE 6th Non-Volatile Memory Systems and Applications Symposium (NVMSA), Hsinchu, Taiwan, 16–18 August 2017; pp. 1–6. [CrossRef]

14.	Gaillardon, P.E.; Amarú, L.; Siemon, A.; Linn, E.; Waser, R.; Chattopadhyay, A.; Micheli, G.D. The Programmable Logic-in-Memory (PLiM) computer. In Proceedings of the 2016 Design, Automation Test in Europe Conference Exhibition (DATE), Dresden, Germany, 14–18 March 2016; pp. 427–432.

15.	Bhattacharjee, D.; Devadoss, R.; Chattopadhyay, A. ReVAMP: ReRAM based VLIW architecture for in-memory computing. In Proceedings of the Design, Automation Test in Europe Conference Exhibition (DATE), Lausanne, Switzerland, 27–31 March 2017; pp. 782–787. [CrossRef]

16.	Zhang, D.; Jayasena, N.; Lyashevsky, A.; Greathouse, J.L.; Xu, L.; Ignatowski, M. TOP-PIM: Throughput-oriented Programmable Processing in Memory. In Proceedings of the 23rd International Symposium on High-performance Parallel and Distributed Computing, Vancouver, BC, Canada, 23–27 June 2014; pp. 85–98. [CrossRef]

17.	Draper, J.; Chame, J.; Hall, M.; Steele, C.; Barrett, T.; LaCoss, J.; Granacki, J.; Shin, J.; Chen, C.; Kang, C.W.; et al. The Architecture of the DIVA Processing-in-memory Chip. In Proceedings of the 16th International Conference on Supercomputing, ICS '02, New York, NY, USA, 22–26 June 2002; pp. 14–25. [CrossRef]

18.	Gokhale, M.; Holmes, B.; Iobst, K. Processing in memory: The Terasys massively parallel PIM array. *Computer* **1995**, *28*, 23–31. [CrossRef]

19.	Ahn, J.; Hong, S.; Yoo, S.; Mutlu, O.; Choi, K. A scalable processing-in-memory accelerator for parallel graph processing. In Proceedings of the 2015 ACM/IEEE 42nd Annual International Symposium on Computer Architecture (ISCA), Portland, OR, USA, 13–17 June 2015; pp. 105–117. [CrossRef]

20.	Xiao, Y.; Nazarian, S.; Bogdan, P. Prometheus: Processing-in-memory heterogeneous architecture design from a multi-layer network theoretic strategy. In Proceedings of the 2018 Design, Automation Test in Europe Conference Exhibition (DATE), Dresden, Germany, 19–23 March 2018; pp. 1387–1392. [CrossRef]

21.	Li, S.; Xu, C.; Zou, Q.; Zhao, J.; Lu, Y.; Xie, Y. Pinatubo: A processing-in-memory architecture for bulk bitwise operations in emerging non-volatile memories. In Proceedings of the 2016 53nd ACM/EDAC/IEEE Design Automation Conference (DAC), Austin, TX, USA, 5–9 June 2016; pp. 1–6. [CrossRef]

22.	Seshadri, V.; Lee, D.; Mullins, T.; Hassan, H.; Boroumand, A.; Kim, J.; Kozuch, M.A.; Mutlu, O.; Gibbons, P.B.; Mowry, T.C. Ambit: In-memory Accelerator for Bulk Bitwise Operations Using Commodity DRAM Technology. In Proceedings of the 50th Annual IEEE/ACM International Symposium on Microarchitecture, MICRO-50 '17, Boston, MA, USA, 14–18 October 2017; pp. 273–287. [CrossRef]

23. Imani, M.; Gupta, S.; Rosing, T. Ultra-Efficient Processing In-Memory for Data Intensive Applications. In Proceedings of the 54th Annual Design Automation Conference 2017, DAC '17, Austin, TX, USA, 14–22 June 2017; pp. 6:1–6:6. [CrossRef]
24. Tang, Y.; Wang, Y.; Li, H.; Li, X. ApproxPIM: Exploiting realistic 3D-stacked DRAM for energy-efficient processing in-memory. In Proceedings of the 2017 22nd Asia and South Pacific Design Automation Conference (ASP-DAC), Chiba, Japan, 16–19 January 2017; pp. 396–401. [CrossRef]
25. Yang, K.; Karam, R.; Bhunia, S. Interleaved logic-in-memory architecture for energy-efficient fine-grained data processing. In Proceedings of the 2017 IEEE 60th International Midwest Symposium on Circuits and Systems (MWSCAS), Boston, MA, USA, 6–9 August 2017; pp. 409–412. [CrossRef]
26. Sterling, T.L.; Zima, H.P. Gilgamesh: A Multithreaded Processor-In-Memory Architecture for Petaflops Computing. In Proceedings of the Supercomputing, ACM/IEEE 2002 Conference, Baltimore, MD, USA, 16–22 November 2002; p. 48. [CrossRef]
27. Azarkhish, E.; Rossi, D.; Loi, I.; Benini, L. Design and Evaluation of a Processing-in-Memory Architecture for the Smart Memory Cube. In Proceedings of the 29th International Conference on Architecture of Computing Systems–ARCS 2016, Nuremberg, Germany, 4–7 April 2016; Volume 9637, pp. 19–31. [CrossRef]
28. Angizi, S.; He, Z.; Parveen, F.; Fan, D. RIMPA: A New Reconfigurable Dual-Mode In-Memory Processing Architecture with Spin Hall Effect-Driven Domain Wall Motion Device. In Proceedings of the 2017 IEEE Computer Society Annual Symposium on VLSI (ISVLSI), Bochum, Germany, 3–5 July 2017; pp. 45–50. [CrossRef]
29. Angizi, S.; He, Z.; Fan, D. PIMA-Logic: A Novel Processing-in-Memory Architecture for Highly Flexible and Energy-Efficient Logic Computation. In Proceedings of the 2018 55th ACM/ESDA/IEEE Design Automation Conference (DAC), San Francisco, CA, USA, 24–28 June 2018; pp. 1–6. [CrossRef]

*Article*

# Digital Circuit for Seamless Resampling ADC Output Streams[†]

**Mauro D'Arco *, Ettore Napoli and Efstratios Zacharelos**

Department Electrical and Information Technologies Engineering (DIETI), University of Naples Federico II, via Claudio 21, 80125 Naples, Italy; ettore.napoli@unina.it (E.N.); efstratios.zacharelos@unina.it (E.Z.)
*   Correspondence: mauro.darco@unina.it; Tel.: +39-081-768-3237
†   This paper is an extended version of our paper published in Applications in Electronics Pervading Industry, Environment and Society (ApplePies 2019).

Received: 6 February 2020; Accepted: 11 March 2020; Published: 14 March 2020

**Abstract:** Fine resolution selection of the sample rate is not available in digital storage oscilloscopes (DSOs), so the user has to rely on offline processing to cope with such need. The paper first discusses digital signal processing based methods that allow changing the sampling rate by means of digital resampling approaches. Then, it proposes a digital circuit that, if included in the acquisition channel of a digital storage oscilloscope, between the internal analog-to-digital converter (ADC) and the acquisition memory, allows the user to select any sampling rate lower than the maximum one with fine resolution. The circuit relies both on the use of a short digital filter with dynamically generated coefficients and on a suitable memory management strategy. The output samples produced by the digital circuit are characterized by a sampling rate that can be incoherent with the clock frequency regulating the memory access. Both a field programmable gate array (FPGA) implementation and an application specific integrated circuit (ASIC) design of the proposed circuit are evaluated.

**Keywords:** resampling; interpolating polynomial; polyphase filter; digital circuit design; FPGA; ASIC

---

## 1. Introduction

In the majority of digital storage scopes (DSOs) the analog-to-digital converter (ADC) always works at its maximum sampling rate, imposed by an internal fixed frequency clock [1,2]. The user can also select lower sampling rates, which are achieved by seamlessly resampling the ADC output stream. Resampling is performed by means of a digital circuit that interfaces ADC and acquisition memory, and merely consists in decimating the input stream, which involves grouping the samples at the maximum sampling rate into consecutive sets, and acquiring, that is, storing in the acquisition memory, only the first sample of each set. All sets have the same size, which is equal to the required decimation factor—for instance, grouping samples into sets with size equal to 2 means acquiring one every other sample, thus halving the input sampling rate [3,4].

Resampling based on decimation is characterized by the following drawbacks: (i) the selection of the sampling rate is limited to the values that can be obtained dividing the maximum sampling rate by integer values; (ii) if the selected sampling rate is less than the Nyquist rate of the analog input, the acquired signal is corrupted by aliasing [5,6].

In general, fine selection of the sample rate improves the performance of the DSO, allowing more efficient usage of memory resources. In fact, a limited set of sample rates implies a limited set of time windows for signal observation. Due to these limitations, it is possible that the analysis is performed observing the signal of interest in a time window where up to almost 50% of the window contains useless samples. Many DSOs are also complemented with math capabilities like Fast Fourier Transform (FFT) options that allow frequency domain analyses. In these applications the choice of the sample rate determines, in conjunction with the memory size, the frequency span and resolution settings; the limitations characterizing the sample rate selection lead to sub-optimal settings. Some DSOs allow the user applying an external clock signal to control the sampling rate. This option is not very common because of the following drawbacks: (i) the external path has a limited bandwidth, much inferior to that of the internal path, so that the operative range of the DSO is substantially reduced; (ii) some functionalities of the instrument, which cannot work with the external clock, are disabled; (iii) the precision specifications of the DSO, which are related to the operation with the internal clock, cannot be used to evaluate the accuracy of the measurement results.

In theory, fine control of the sampling rate in real-time DSOs can be obtained by resampling the ADC output stream by means of more effective methods alternative to hard decimation [7–9]. These methods can be inherited by digital signal processing theory, and rely either on the use of interpolation algorithms or polyphase filters [10,11]. The first method allows varying the sampling rate dynamically, and puts no restrictions on the selection of the output sampling rate. The second method is instead limited to decimation factors that are equal to $\frac{L}{M}$, where $L$ and $M$ are integers. Both methods counteract aliasing effects by means of low-pass filtering operations, which are implicit in the interpolation algorithm, and explicit in the processing scheme of polyphase filters [12–15]. In fact, the use of an interpolation function is equivalent to filtering the signal with a filter characterized by a frequency response where the number of taps is equal to the number of points used in interpolation. Unfortunately, the hardware implementation of both methods is difficult due to the strict requirements of seamless operation and fine resolution in sampling rate selection [16–18].

A method that shows a viable solution to finely control the sampling rate in DSOs has been presented in Reference [19], and a digital circuit that implements this method using field programmable gate array (FPGA) technology has been illustrated at the ApplePies 2019 Conference [20]. In detail, the digital circuit exploits a resampling method based on linear interpolation, which trades-off between accuracy and circuit complexity. It is designed to work between the ADC and the acquisition memory, and allows selecting sampling rates from the highest frequency, $f_{ck}$, down to its half value, $\frac{f_{ck}}{2}$. Choosing a sample rate lower than $\frac{f_{ck}}{2}$ is easily obtained by cascading the proposed circuit with a standard one that performs decimation by an integer value. The acquisition chain is made up of ADC, proposed digital circuit, and acquisition memory, all operating synchronously at the system clock rate $f_{ck}$. It provides samples that represent a version of the input signal characterized by a sample rate $f_s = C f_{ck}$, where $C$ is a fractional value in the interval $(\frac{1}{2}, 1)$. The defining resolution of $C$ is only limited by the number of bits adopted in its binary representation; the reciprocal of $C$ can be regarded as a non-integer decimation factor.

This work is an extended version of the article published in the Conference Proceedings [20]. It takes into consideration several different methods for DSOs sampling rate control, and, by evaluating their performance highlights how the proposed digital circuit represents a good compromise between achievable accuracy and circuit complexity. Starting from the primary version of the circuit, an improved version characterized by different pipeline levels is developed, and an application specific integrated circuit (ASIC) design of the proposed solution is also analyzed [21–23].

The paper discusses more about the resampling methods based on interpolation and polyphase filters in Section 2. The performance of different interpolators, which satisfy the requirements of effective hardware implementation and high resolution in sampling rate selection, is analyzed in Section 3 through simulations. Section 4 illustrates the design of the proposed digital circuit, and, finally, Section 5 gives concluding remarks.

## 2. Methods

In general, resampling a mono-dimensional signal, defined upon a sampling grid, aims at producing another representation of the same signal, referred to a different sampling grid. It basically requires gaining the samples referred to the output grid by processing the available ones. Resamplers manage a redundant representation of the signal, that includes both the input samples and the resampled ones; the second are the only ones returned by the circuit.

In the most common resampling applications both the input and output sampling grid are uniform and the circuit has to deal with samples that are streamed at regular time instants, such that a sampling rate is defined. Also, resampling has to be performed real-time seamlessly on the input stream, which is very challenging, especially in the presence of high-rate data streams [24–27].

Hereinafter, the attention is mainly paid to real-time seamless resampling of signals that are naturally defined in the time domain, for which the input sampling rate needs to be changed into a different sampling rate, lower than the input one. The methods that are illustrated can be adapted to other signals defined in different domains by exploiting the unique correspondence between the points of the sampling grids and the related time-stamps in their streamed form produced at the processing stage [28,29].

### 2.1. Resampling Based on the Use of Approximating Polynomials

The most straightforward resampling approach, capable of granting real-time seamless performance, exploits the zero-order interpolation process, which assumes the signal constant until the next sample is available. In other terms, the resampled value, $x(n + t)$, where $t$ is a fraction of the sampling period, $T_s$, (reciprocal of the sampling rate, $f_s$) of the input stream, is assumed equal to that of the most recent sample $x(n)$ [30,31].

Alternatively, the first-order or linear interpolator can be used to improve the accuracy of the resampling process. Linear interpolators wait for the subsequent sample of the input stream $x(n + 1)$ to compute the value of any sample at a time instant in the midst. Specifically, they compute it by adding to $x(n)$ a term equal to $t$ times the time derivative, which is estimated as first forward difference [32,33].

More generally, resampling can rely on interpolators that use a larger set of samples adjacent to the resampling instant to determine the resampled value. The samples of the set are processed to identify a polynomial of the $t$ variable, $P(t)$, that locally approximates the signal behavior. The value of the polynomial at the resampling instant provides the resampled value. The polynomial is identified imposing constraints that can involve the values of the signal and/or of its time derivatives. The most common solutions are:

- the approximating polynomial connects all the samples of the set (Lagrange polynomial) and is characterized by a degree equal to the number of samples of the set minus 1;
- the approximating polynomial is identified by fitting the samples in order to minimize the mean square error, and is characterized by a degree less than the number of samples of the set minus 1;
- the approximating polynomial connects a subset of the samples and has the same time derivative of the signal in those points (Hermite polynomial).

In all the aforementioned approaches the resampled value obtained using an approximating polynomial can be represented with a matrix formulation. For instance, for a 3-degree approximating polynomial, one as:

$$
\begin{aligned}
x(n + t) \quad &= \quad c_1(n) + c_2(n)t + c_3(n)t^2 + c_4(n)t^3 = \\[2mm]
&= \begin{bmatrix} 1 & t & t^2 & t^3 \end{bmatrix}
\begin{bmatrix}
a_{11} & a_{12} & a_{13} & a_{14} \\
a_{21} & a_{22} & a_{23} & a_{24} \\
a_{31} & a_{32} & a_{33} & a_{34} \\
a_{41} & a_{42} & a_{43} & a_{44}
\end{bmatrix}
\begin{bmatrix}
x_{n-1} \\
x_n \\
x_{n+1} \\
x_{n+2}
\end{bmatrix}
\end{aligned}
\tag{1}
$$

where $x(n + t)$ is the resampled value at time instant $n + t$, $t$ is within the interval $(0, 1)$, and each coefficient, $a_{ij}$, $i = 1, \ldots, 4$, is a linear combination of the values of the 4 consecutive samples $\{x_{n-1}, x_n, x_{n+1}, x_{n+2}\}$ with constant coefficients, namely:

$$
c_i(n) = \sum_{j=1}^{4} a_{ij} x_{n-2+j}.
\tag{2}
$$

The constant coefficients $a_{ij}$ can be determined imposing the constraints used to define the approximating polynomial. Hence, for a Lagrange polynomial, one can consider the system of equations obtained imposing that the polynomial connects the values $\{x_{n-1}, x_n, x_{n+1}, x_{n+2}\}$ characterized, respectively, by $t$ abscissas $\{-1, 0, 1, 2\}$:

$$\begin{bmatrix} 1 & -1 & 1 & -1 \\ 1 & 0 & 0 & 0 \\ 1 & 1 & 1 & 1 \\ 1 & 2 & 4 & 8 \end{bmatrix} \begin{bmatrix} c_1 \\ c_2 \\ c_3 \\ c_4 \end{bmatrix} = \begin{bmatrix} x_{n-1} \\ x_n \\ x_{n+1} \\ x_{n+2} \end{bmatrix} \tag{3}$$

from which the $a_{ij}$ values are determined by inverting the coefficient matrix in (3) as:

$$\{a_{ij}\} = inv \begin{bmatrix} 1 & -1 & 1 & -1 \\ 1 & 0 & 0 & 0 \\ 1 & 1 & 1 & 1 \\ 1 & 2 & 4 & 8 \end{bmatrix} = \frac{1}{6} \begin{bmatrix} 0 & 6 & 0 & 0 \\ -2 & -3 & 6 & -1 \\ 3 & -6 & 3 & 0 \\ -1 & 3 & -3 & 1 \end{bmatrix} \tag{4}$$

If an approximating polynomial of second degree is selected, then only three coefficients, $c_i(n)$, $i = 1, 2, 3$, that are still linear combination of the samples $\{x_{n-1}, x_n, x_{n+1}, x_{n+2}\}$ with constant coefficients are needed. These coefficients can be determined imposing the same constraints adopted to identify the Lagrange polynomial, i.e:

$$\begin{bmatrix} 1 & -1 & 1 \\ 1 & 0 & 0 \\ 1 & 1 & 1 \\ 1 & 2 & 4 \end{bmatrix} \begin{bmatrix} c_1 \\ c_2 \\ c_3 \end{bmatrix} = \begin{bmatrix} x_{n-1} \\ x_n \\ x_{n+1} \\ x_{n+2} \end{bmatrix} \tag{5}$$

but, since a second degree polynomial cannot in general grant the connection of more than 3 points, one has to accept an approximate solution that best fits the data according to a given cost function. The solution that grants the least mean square error, as well known, is obtained solving the system in (5) using the pseudo-inverse matrix method; in this case the $a_{ij}$ values, $i = 1, ..., 3$, $j = 1, ..., 4$, are:

$$\begin{aligned} \{a_{ij}\} &= inv \left\{ \begin{bmatrix} 1 & 1 & 1 & 1 \\ -1 & 0 & 1 & 2 \\ 1 & 0 & 1 & 4 \end{bmatrix} \begin{bmatrix} 1 & -1 & 1 \\ 1 & 0 & 0 \\ 1 & 1 & 1 \\ 1 & 2 & 4 \end{bmatrix} \right\} \begin{bmatrix} 1 & 1 & 1 & 1 \\ -1 & 0 & 1 & 2 \\ 1 & 0 & 1 & 4 \end{bmatrix} = \\ &= \frac{1}{20} \begin{bmatrix} 3 & 11 & 9 & -3 \\ -11 & 3 & 7 & 1 \\ 5 & -5 & -5 & 5 \end{bmatrix} \end{aligned} \tag{6}$$

The coefficients of a 3-degree Hermite polynomial are identified using also the time derivative of the approximating polynomial $P(t)$, namely:

$$\frac{dP}{dt} = c_2(n) + c_3(n)t + c_4(n)t^2 \tag{7}$$

to form a system of equations that imposes that the polynomial connects the central samples, referred to the $t$ abscissas equal to 0 and 1, and has the same derivative of the signal in those points. In matrix form, these constraints can be expressed by:

$$\begin{bmatrix} 1 & 0 & 0 & 0 \\ 1 & 1 & 1 & 1 \\ 0 & 1 & 0 & 0 \\ 0 & 1 & 2 & 3 \end{bmatrix} \begin{bmatrix} c_1 \\ c_2 \\ c_3 \\ c_4 \end{bmatrix} = \frac{1}{2} \begin{bmatrix} 0 & 2 & 0 & 0 \\ 0 & 0 & 2 & 0 \\ -1 & 0 & 1 & 0 \\ 0 & -1 & 0 & 1 \end{bmatrix} \begin{bmatrix} x_{n-1} \\ x_n \\ x_{n+1} \\ x_{n+2} \end{bmatrix} \tag{8}$$

where the time derivative of the signal is estimated in terms of finite central difference. The $a_{ij}$ values, $i, j = 1, \ldots, 4$, are obtained solving system (8) as:

$$\{a_{ij}\} = \frac{1}{2} inv \begin{bmatrix} 1 & 0 & 0 & 0 \\ 1 & 1 & 1 & 1 \\ 0 & 1 & 0 & 0 \\ 0 & 1 & 2 & 3 \end{bmatrix} \begin{bmatrix} 0 & 2 & 0 & 0 \\ 0 & 0 & 2 & 0 \\ -1 & 0 & 1 & 0 \\ 0 & -1 & 0 & 1 \end{bmatrix} = \frac{1}{2} \begin{bmatrix} 0 & 2 & 0 & 0 \\ -1 & 0 & 1 & 0 \\ 2 & -5 & 4 & -1 \\ -1 & 3 & -3 & 1 \end{bmatrix} \tag{9}$$

The equations that define the interpolation methods discussed above can be summarized as in Table 1.

**Table 1.** Lagrange, Hermite and best fitting polynomial (in the sense of least square error) adopted in resampling.

| Samples | Interpolation | Equation |
|---|---|---|
| 1 | zero order | $x(n+t) = x(n)$ |
| 2 | linear | $x(n+t) = x(n) + [x(n+1) - x(n)]t$ |
| 3 | linear (best fitting) | $x(n+t) = \frac{1}{3}[x(n-1) + x(n) + x(n+1)] + \frac{1}{2}[x(n+1) - x(n-1)]t$ |
| 3 | quadratic (Lagrange) | $x(n+t) = x(n) + \frac{1}{2}[x(n+1) - x(n-1)]t + \frac{1}{2}[x(n-1) - 2x(n) + x(n+1)]t^2$ |
| 4 | quadratic (best fitting) | $x(n+t) = \frac{1}{20}[3x(n-1) + 11x(n) + 9x(n+1) - 3x(n+2)] + \frac{1}{20}[-11x(n-1) + 3x(n) + 7x(n+1) + x(n+2)]t + \frac{1}{4}[x(n-1) - x(n) - x(n+1) + x(n+2)]t^2$ |
| 4 | cubic (Lagrange) | $x(n+t) = x(n) + \frac{1}{6}[-2x(n-1) - 3x(n) + 6x(n+1) - x(n+2)]t + \frac{1}{2}[x(n-1) - 2x(n) + x(n+1)]t^2 + \frac{1}{2}[-x(n-1) + 3x(n) - 3x(n+1) + x(n+2)]t^3$ |
| 4 | cubic (Hermite) | $x(n+t) = x(n) + \frac{1}{2}[x(n+1) - x(n-1)]t + \frac{1}{2}[2x(n-1) - 52x(n) + 4x(n+1) - x(n+2)]t^2 + \frac{1}{2}[-x(n-1) + 3x(n) - 3x(n+1) + x(n+2)]t^3$ |

## 2.2. Resampling Based on Polyphase Filters

Resampling with polyphase filters is commonly performed in a variety of systems, like multipurpose receivers, where several different sampling rates are supported to process signals characterized by different bandwidths, as well as in digital audio and video systems, and so forth [34,35]. In these systems the signal is initially sampled at a high sampling rate, then processed to modify the sampling rate by a factor $\frac{L}{M}$. Processing involves interpolation by $L$, low-pass filtering, and decimation by $M$. Low-pass filtering removes the image frequencies due to sampling rate changes; it is implemented using polyphase decomposition of both the input signal and filter coefficients [36,37].

For the sake of clarity, an example of a $\frac{3}{4}$-resampler that uses a short low pass filter with 9 coefficients, $h(n) = \{h(0), h(1), \ldots h(8)\}$, is shown in Figure 1. The input signal $y(n)$ is de-multiplexed in order to retrieve 4 consecutive samples and route them to 4 individual channels with a single operation. The output of the resampler, $z(m)$, is obtained by multiplexing the outputs produced by 3 filters, each filter defined in terms of 3 coefficients of $h(n)$ according to polyphase decomposition rules.

Polyphase filters are characterized by low requirements in terms of clock frequency and can be set to both up-sample and down-sample the input stream, but are not suitable for programmable resampling factors, because any polyphase structure is defined by the same ratio between the input and output sampling rate; consequently, any change of the resampling ratio implies modifying their structure [38].

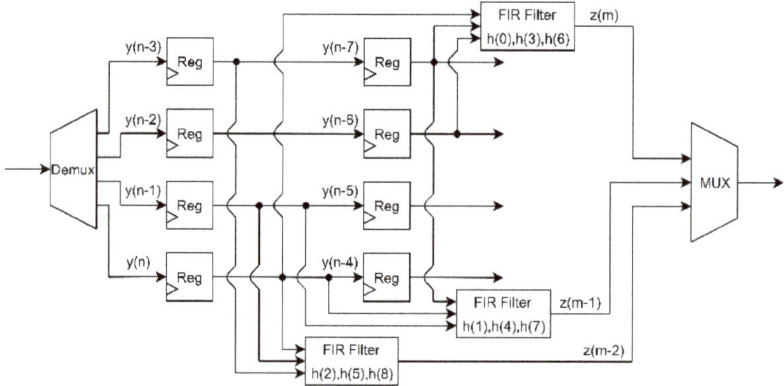


**Figure 1.** Schematic of a digital resampler implementation based on polyphase decomposition. Resampling factor equal to 3/4 low-pass filter with 9 taps.

### 2.3. Pro and Cons of Approximating Polynomials and Polyphase Filters

Resampling with polyphase filters straightforwardly changes the input sampling rate, $f_{ck}$ to the output one $f_s = \frac{L}{M} f_{ck}$. In fact, thanks to the use of a demultiplexer at the front-end, the polyphase filter processes any $\frac{T_{ck}}{M}$ seconds ($T_{ck} = f_{ck}^{-1}$) a set of $M$ input samples and returns a set of $L$ output samples, which are written in the acquisition memory with a single memory access, thus lowering the input sampling rate by a factor $\frac{L}{M}$. Unfortunately, any change of the sampling rate requires re-programming the digital circuit. Although, in theory, re-programming can be done, in case of sampling rates that involve very large $M$ and $L$ values, one should reserve sufficient hardware resources for huge polyphase structures, seldom required and largely unused; nonetheless, the responsiveness of the system would definitely slow-down.

Resamplers based on interpolators are instead less demanding in terms of hardware resources and allow controlling the sampling rate easily. They also require a suitable strategy for arranging the lower sampling rate output stream. Specifically, the digital resampler can take as input both a set of consecutive samples and the $t$ variable, as specified by the interpolation equations summarized in Table 1. It can run at a clock rate equal to the input sampling rate, quantifying the $t$ variable as the delay of the resampling instant with respect to the discrete time $n$. To this end, it can exploit an accumulator that increments by $\frac{T_s}{T_{ck}} - 1$ ($T_s = f_s^{-1}$) any $T_{ck}$ seconds. The accumulator represents the $t$ variable except when it overflows a unitary value. The overflow repeatedly occurs with a cadence related to the selected sampling rate. Overflow means that the resampling instant does not fall between the discrete time $n - 1$ and $n$, but is in the midst of $n$ and $n + 1$, such that it should be considered at the next processing step. At any occurrence of an overflow, the digital circuit skips the calculation of the resampled value, and performs a unitary decrement of the accumulator at the subsequent clock cycle, thus restoring $t$ between the expected discrete time instants.

The use of approximating polynomials is preferred in the development of a digital circuit aimed at granting fine control of DSOs sampling rate, because it has several interesting features. These include the capability of resampling even if the ratio of sample rates is not rational, as well as of seamlessly managing real-time streams even in the presence of time-varying sample rates.

## 3. Simulation Analyses

As well known, changing the sampling rate produces aliasing, which is usually counteracted by filtering the digital signal with a low pass filter before resampling. The performance of the resampling methods is affected by the presence of residual alias, thus the frequency response of the adopted anti-aliasing filter must be taken into account. The anti-aliasing filter is implicit in the resampling approach based on the use of an approximating polynomial, and its impulse response is given, in general, by a set of coefficients that depend on $t$; for instance, from Equation (1), one can obtain the coefficients of the 4-tap filter, $d_j, j = 1, ..., 4$ as:

$$d_j = \sum_{i=1}^{4} a_{ij} t^{j-1}. \tag{10}$$

An estimation of the residual alias can be approached taking into account that the spectrum of a digital signal is periodic with unitary period, and that lowering the sampling rate down to $f_s = C f_{ck}$ has the effect of replicating the spectrum at a pace equal to $C$. Moreover, since the $t$ variable changes during the resampling process, ranging in the interval $(0, 1)$, the features of the anti-aliasing filters change as well. Anyway, taking into account that $t$ is within $(0, 1)$ and, on average, $t = \frac{1}{2}$, one determines the average behavior of the filter. Using the frequency response, $H(v)$, of the filter that describes the average behavior, which is gained by taking the Fourier transform of the filter coefficients estimated with $t = \frac{1}{2}$, allows representing the spectrum of the resampled version as:

$$Z(v) = \sum_{p,q=-\infty}^{\infty} H(\frac{v}{C}) X(\frac{v-p}{C} - q), \tag{11}$$

where $v$ is normalized to the sampling rate $f_{ck}$. From Equation (11) the alias-free version of the resampled signal can be obtained using $p = q = 0$, whereas all the combinations satisfying $|p - Cq| < \frac{C}{2}$ identify the residual aliases that fall in the spectrum of the resampled signal. Figure 2 shows the frequency response of the anti-alias filters that are implicit in the 7 approaches detailed in Table 1. The different responses are characterized by specific markers and colors: 'plus' marker and blue color is for the linear interpolator, 'circle' marker and red color for the first-degree polynomial fitting 3 sample points, 'x' marker and green color for the second-degree Lagrange polynomial, 'star' marker and yellow color for the second-degree polynomial fitting 4 sample points, 'square' marker and magenta color for the third-degree Lagrange polynomial, and, finally, 'diamond' marker and cyan color for the third-degree Hermite polynomial (a suitable legend has been included to highlight these correspondences). For the sake of completeness also an additional graphic, related to the zero-order resampling approach, is shown using 'dot' marker and black color to highlight the all-pass nature of this approach, which is detrimental because it provides no mitigation of aliasing effects.

The frequency responses given in Figure 2 are obtained by Fourier transforming the impulse response estimated upon 50 points, and consist of 25 bins, equally spaced at a pace of 0.02; they show the behavior of the filters up to the normalized frequency 0.5, corresponding to $\frac{f_{ck}}{2}$ hertz. One can observe that the approximating polynomials with higher degree offer flatter gain and better selectivity. Also, the mean behavior of the anti-aliasing filters related to the use of the second-degree polynomial fitting 4 sample points, the third-degree Lagrange polynomial, and the third-degree Hermite polynomial are identical. The ideal frequency response behavior should exhibit unitary gain in the interval $(0, \frac{C}{2})$, to avoid undesired attenuation of the signal spectral content, and zero gain in $(\frac{C}{2}, \frac{1}{2})$ to cancel any alias contribution.

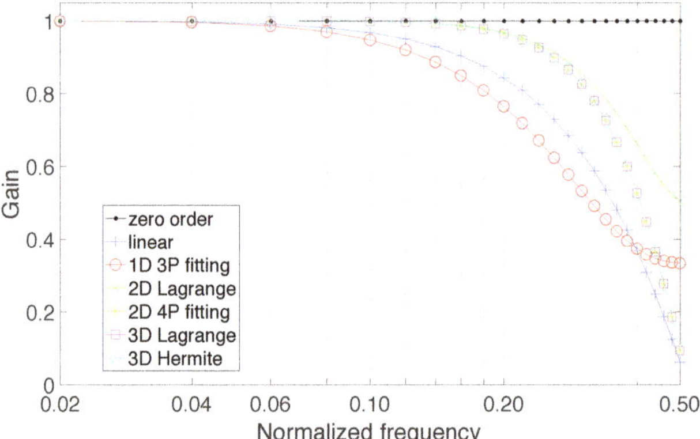

**Figure 2.** Frequency response of the anti-alias filters implicit in the resampling approaches based on the use of approximating polynomials.

Although the anti-aliasing filter plays an important role in the resampling process, the analysis of its mean behavior provides only partial insight, since the time-varying nature can play a role that cannot be analyzed using Equation (11). A deep insight in the performance of the resampling methods can instead be gained by using the standard test methods for ADC, detailed in Reference [39], such as the effective number of bits (ENOB) and the spurious-free dynamic range (SFDR). The first is a measure of the signal-to-noise and distortion ratio used to compare the actual ADC performance to an ideal one; the latter considers, in the presence of a pure sine-wave input, the ratio of the amplitude of the output spectral component at the input frequency, $f_0$, to the amplitude of the largest harmonic or spurious spectral component. Figure 3 shows the ENOB offered by the considered methods in the presence of test sine-waves.

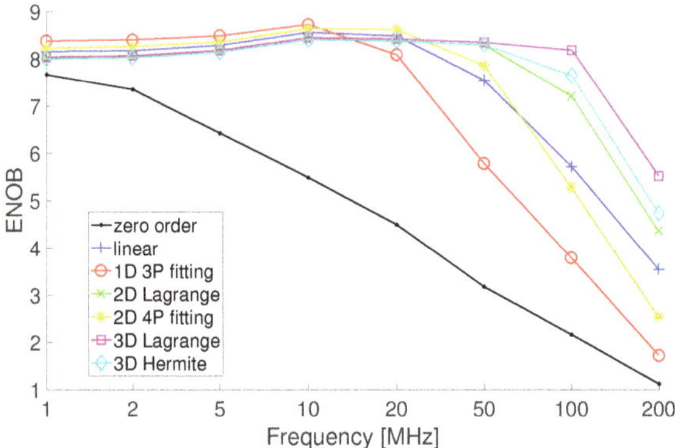

**Figure 3.** Effective number of bits (ENOB) offered by the resampling approaches based on the use of approximating polynomials.

The simulations have considered samples quantized by an 8 bit ADC. Quantization has been applied to a signal corrupted by white Gaussian noise, with rms value equal to 15% of the LSB of the ADC. The sampling rate of the input stream is $f_{ck} = 1$ GSa/s, that is resampled at $f_s = 743$ MHz, thus $C = 0.743$. The sine-waves adopted in the tests are characterized by the frequency values $\{1, 2, 5, 10, 20, 50, 100, 200\}$ MHz. The results show that ENOB obtained after resampling can even improve in the presence of the lower input frequencies of the considered set with respect to the nominal 8 bit. This is due to the anti-aliasing low-pass filter that reduces the acquired bandwidth and thus also the distortion due to quantization. As the input frequency approaches the upper limit of the Nyquist bandwidth, the performance of all methods rapidly decreases, and one can observe that the methods that use approximating polynomials with higher degree can grant ENOB close to the nominal number of bits on wider ranges. As expected, the effectiveness of interpolation algorithms diminishes as soon as the input sinusoidal signal is sampled collecting a few points per period, namely 7–8 points. This happens because the algorithms consider the local behavior of the signal, whereas the uniform sampling theorem claims for interpolation with *sinc* functions that consider the behavior on the whole time axis; unfortunately *sinc* interpolation is unfeasible and its straightforward approximations, like those based on the use of truncated *sinc* functions, are characterized by huge computational burden, which is not compatible with real-time execution. As a rule of thumb, suggesting some oversampling in the use of the acquisition mode with fine selection of the sample rate, avoids incurring in poor results.

The simulations have also estimated for the same test set-up the SFDR in order to highlight if the time-varying behavior of the anti-alias filters introduce relevant spurious; the obtained results are shown in Figure 4.

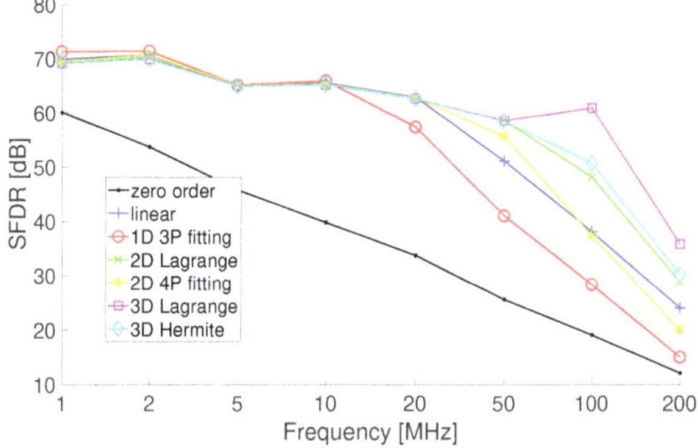


**Figure 4.** Spurious-free dynamic range (SFDR) offered by the resampling approaches based on the use of approximating polynomials.

The performance parameters highlight the convenience of using Lagrange or Hermite polynomials (the linear interpolation coincides with the adoption of a first-degree Lagrange polynomial) for interpolation rather then zero-order or fitting polynomials based methods.

Further simulations have been addressed to the analysis of any dependence of the performance on the output sampling rate. As an example, Figure 5 shows the ENOB obtained in the presence of a sine-wave input at 20 MHz when the 1 GHz input sampling rate is lowered down to the frequencies of the set $\{587, 641, 743, 797, 859, 907, 971\}$ MSa/s.

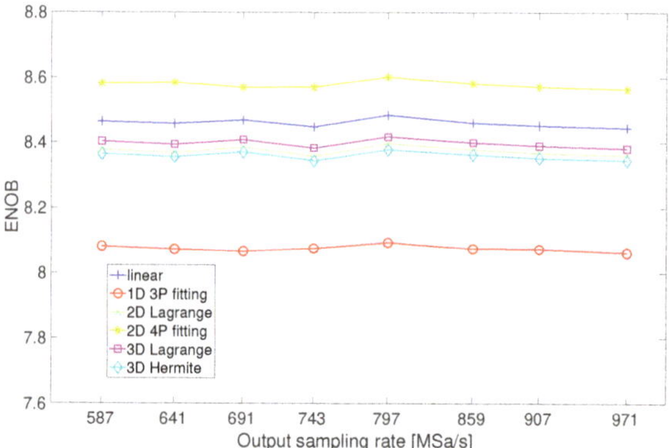

**Figure 5.** ENOB offered by the resampling approaches in the presence of an input sine-wave at 20 MHz when the 1 GHz input sampling rate is lowered down to frequencies of the set $\{587, 641, 743, 797, 859, 907, 971\}$ MSa/s.

The simulations highlight that the performance is unaffected by sampling rate changes; all the methods offer ENOB constant and above 8 bits, except for the zero-order method, the performance of which, although independent of the output sampling rate, is largely below the lower axis limit utilized in Figure 5.

## 4. Proposed Digital Circuit

### 4.1. Operation Details

The proposed digital circuit implements the linear interpolation method that represents a good compromise between accuracy and circuit complexity. It processes in real-time the signal $x(n)$ streaming out of the ADC, and returns the output, $y(n)$; both are characterized by the clock rate, $f_{ck}$, but $y(n)$ contains a resampled version of $x(n)$ characterized by a sampling rate $f_s = Cf_{ck}$.

More specifically, the value $y(n)$ is determined by combining the samples $x(n-1)$ and $x(n)$ returned by the ADC according to:

$$
\begin{aligned}
y(n) &= a(n)x(n-1) + (1 - a(n))x(n) = \\
&= a(n)x(n-1) + b(n)x(n)
\end{aligned}
\tag{12}
$$

where $a(n)$ is a time-varying coefficient, updated at every clock cycle by subtracting to its current value a quantity, chosen by the user, and related to the sampling factor as $\frac{1-C}{C}$. Notice that the aforementioned variable $t$ corresponds to the variable $b(n)$ of the Equation (12), and consequently $a(n) = 1 - t$. Subtraction is skipped if the current value of the coefficient $a(n)$ is negative, and in its place an addition by one is performed. Hence, the output of the digital circuit $y(n)$ contains, with some redundancy, the resampled version of $x(n)$.

The circuit also produces a signal $Ptr_X$, that indicates the memory location where $y(n)$ is stored. The generated sequence $y(n)$ is stored in memory at system frequency, $f_{ck}$ but, in order to cope with the lower sampling rate, $Ptr_X$ is not incremented when the $a(n)$ coefficient is incremented by one. In this way, two consecutive outputs share the same value of $Ptr_X$, which means that the second one overwrites the first.

An example will better clarify the meaning of $a(n)$. In Figure 6 a sinusoidal signal at 54 MHz is shown. It is sampled with the 1 GHz ($T_{ck}$ = 1.0 ns) system clock (sampling shown with circles). The result obtained resampling at 761 MHz ($T_s$ = 1.314 ns) is shown with red bullets. The resampling factor is $C = 0.761$, and the coefficient $a(n)$ is updated subtracting $\frac{1-C}{C} = 0.3141$ to the current value.

Variable $b(n)$ represents the point inside the sampling period where resampling must be performed. The bottom axis is the time while the top axis shows the increment of the memory pointer. When $a(n)$ is incremented (time: 6, 10, 14, 18 in Figure 6) the memory pointer is not updated.

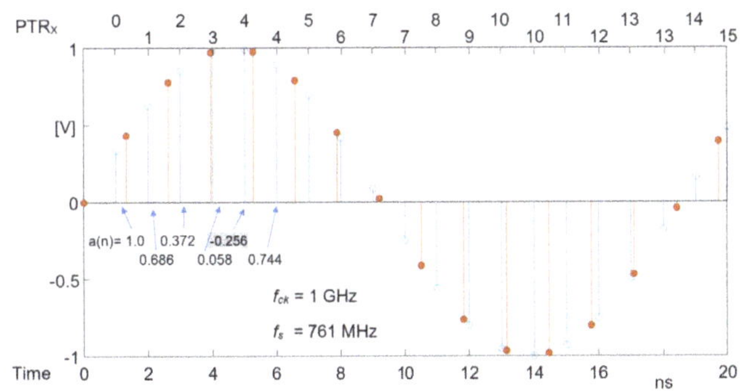


**Figure 6.** Example sequences for $a(n)$ and $Ptr_X$.

### 4.2. Design Details

A digital circuit for the implementation of the proposed resampling algorithm has been designed. The schematics before and after pipelining are in Figures 7 and 8.

Circuit input data are the signal to be resampled $x$, and factor $d = \frac{C-1}{C}$. The output data are the resampled stream $y$, and the memory pointer, $Ptr_X$. The number of bits for $x$, $d$, and $y$, is 8, while the memory pointer, $Ptr_X$ is represented with 32 bits.

The two complementary coefficients, $a$ and $b$, are multiplied by the previous value ($z$ signal in Figure 7) and the current value of the input signal ($x$ signal in Figure 7), respectively. Afterwards, the two products are summed, in order to produce the output signal, $y$, as indicated in Equation (12).

The updating of the coefficient $a$, relies on adding either the quantity $d$, or in the case of exception, a unitary value to the current value of $a$. In the case of exception, $a$ is negative, and the most significant bit (MSB) of the coefficient, is high, $a\,[9]=1$; otherwise, $a\,[9]=0$, and $d$ is added to the current value of $a$. This distinction is realized with the use of a multiplexer, controlled by the MSB of signal $a$. After the correct choice between "1" and "$d$", an accumulator is implemented for the updating of $a$.

A second accumulator is implemented, for the memory management. When $a$ is positive, $a\,[9]=0$, $g=1$, and $Ptr_X$ is incremented by a unitary value. In the case of exception, $a$ is negative, $a\,[9]=1$, $g=0$, and $Ptr_X$ remains unchanged. The above described memory management strategy allows to store only the resampled values. The fact that occasionally the memory pointer is not incremented reflects the fact that after resampling the number of samples is less than that of the input signal.

In Figure 8 the pipelined resampler can be observed. The four vertical dashed lines mark the four pipeline levels introduced to the circuit in order to isolate the combinational logic, thereby achieving a lower clock period and a higher throughput. On the other hand, latency and chip area are increased. The number of flip-flops (registers) used for the pipeline implementation is: $(8 + 8) + (10 + 10 + 8) + (12 + 1 + 12) + (8 + 32) = 109$.

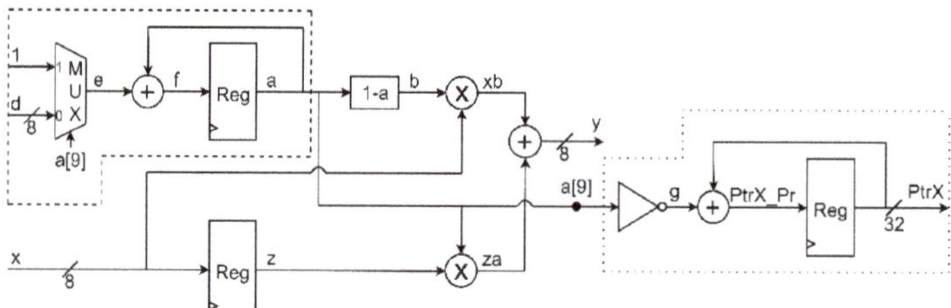

**Figure 7.** Circuital implementation of the proposed algorithm.

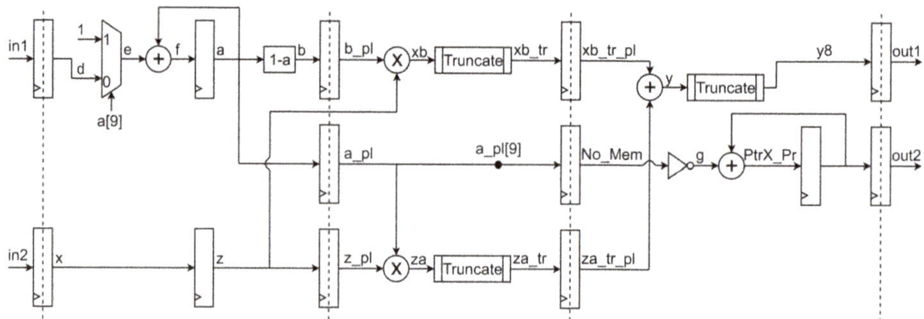

**Figure 8.** Circuital implementation of the proposed algorithm with pipeline registers (pipeline levels are highlighted with dashed lines).

### 4.3. Implementation and Performance

As mentioned earlier, 8 bits are used for the representation of $d$, where $d$ is within $(-1, 0)$. However, during the experimental procedure, a 16-bit signal was also tried out in order to test the performance of the circuit. For an $n$-bit signal, the resolution obtained for $d$ is constant and is equal to $2^{-n}$. The resolution obtained for $C$ can be derived from:

$$d(C) = \frac{d(C)}{d(d)}d(d) = \frac{d}{d(d)}\left(\frac{1}{1-d}\right)2^{-n}. \tag{13}$$

Given the fact that the relationship between $C$ and $d$ is not linear, the resulting resolution of $C$ (the actual resampling factor) differs for different $d$ values. In Table 2, some information related to the resolution of the resampling factor are presented. Assuming a 1 GHz clock frequency, using 8-bit for the $d$ signal allows a frequency resolution that ranges from 390 kHz to 97.5 kHz, while using 16 bit the frequency resolution can be as low as 4 kHz.

**Table 2.** Resampling factor resolution.

| Name | Value |
| --- | --- |
| Best C-Step (8 bits) | $9.76 \cdot 10^{-4} (C = 0.500976)$ |
| Worst C-Step (8 bits) | $39 \cdot 10^{-4} (C = 0.996094)$ |
| Best C-Step (16 bits) | $0.04 \cdot 10^{-4} (C = 0.500004)$ |
| Worst C-Step (16 bits) | $0.15 \cdot 10^{-4} (C = 0.999985)$ |

The circuit is described in hardware description language (HDL) and a first assessment of the performance has been conducted with a high-end FPGA as a target. This aims to demonstrate the available performances in a reconfigurable environment. In Table 3, some basic features and resources are presented for the implementation on a StratixIV GX FPGA device by Altera. The implemented design is the one depicted in Figure 8 with $d$ signal represented by 8 bits.

**Table 3.** Basic Features of the Resampler and FPGA resources (StratixIV-EP4SGX230KF40 implementation).

| Name | Value |
|---|---|
| Maximum Clock Frequency | 400 MHz |
| Combinational ALUTs | 532 (<1%) |
| Dedicated Logic Registers | 147 (<1%) |
| DSP Block 18-bit Elements | 3 (<1%) |

Tests similar to those considered in the simulation analyses have been repeated on sinusoidal signals, demonstrating that when the sampling frequency is at least ten times higher than the signal bandwidth, the results are satisfactory. For instance, in the presence of an input signal corrupted by white Gaussian noise (rms value equal to 15% the LSB of the ADC) and quantized by an 8 bit ADC, resampling at 743 MHz a 47.1 MHz signal converted with a 1GSs ADC has lowered the ENOB from 7.8 to 7.5 and left unaltered the SFDR, which is a quite limited degradation. The results do not exhibit recognizable changes if 50 kHz random deviations of the input frequency are considered.

An ASIC implementation has also been carried out. The circuit is synthesized by targeting a commercial standard-cell library in 14 nm fin field effect transistor (FinFET), from Global Foundry. Physical synthesis is performed by using Cadence Genus; no special cells are designed for the implementation and the circuit is automatically synthesized according to timing constraints. The considered technology corner is the typical one with 0.8 V of supply voltage and regular threshold voltage. The simulations, with delay and switching activity annotation, have been conducted with a suite of tools for the design and verification of ASICs and FPGAs, commonly referred to by the name NCSIM in reference to the core simulation engine. Power dissipation is computed by simulating the final netlist with 10,000 input vectors from an asynchronously sampled sinusoid to obtain the switching activity of each node.

While aiming for the highest frequency possible, several syntheses took place. Firstly, the circuit in Figure 8 was synthesized. Later the same circuit was synthesized, taking into account a retiming algorithm that moves the structural location of registers in order to improve the performance, while preserving the functional behavior at the outputs. Afterwards, two and three extra levels of pipeline were added to the design of Figure 8, and the synthesis was carried out with the retiming algorithm. The same syntheses were done for both an 8-bit and a 16-bit $d$ signal and the results are reported in Table 4, and Table 5, respectively. As expected, the maximum working frequency is largely increased with respect to the FPGA implementation. Moreover, there is a trade-off between maximum frequency and chip area as well as power consumption.

**Table 4.** ASIC implementation results for the resampler in 14 nm FinFET from Global Foundry technology using 8 bits for $d$ signal.

| | Clock Frequency [GHz] | Cell Count | Total Area [μm] | Flip Flops | Dynamic Power $\left[\frac{\mu W}{MHz}\right]$ | Leakage Power [μW] |
|---|---|---|---|---|---|---|
| basic | 3.03 | 1052 | 584 | 145 | 1.190 | 0.707 |
| retimed | 3.85 | 1035 | 620 | 239 | 1.184 | 0.568 |
| 2 extra PL | 5.26 | 1092 | 683 | 307 | 1.392 | 0.665 |
| 3 extra PL | 4.55 | 1066 | 712 | 341 | 1.524 | 0.693 |

**Table 5.** ASIC implementation results for the resampler in 14 nm FinFET GF technology using 16 bits for $d$ signal.

| | Clock Frequency [GHz] | Cell Count | Total Area [μm] | Flip Flops | Dynamic Power $\left[\frac{\mu W}{MHz}\right]$ | Leakage Power [μW] |
|---|---|---|---|---|---|---|
| basic | 2.70 | 1780 | 916 | 179 | 1.349 | 1.038 |
| retimed | 3.57 | 1636 | 932 | 314 | 1.451 | 0.846 |
| 2 extra PL | 5.00 | 1884 | 1077 | 398 | 1.791 | 1.082 |
| 3 extra PL | 4.55 | 1822 | 1120 | 457 | 1.791 | 1.059 |

A comparison between the FPGA implementation (Table 3) and the ASIC design (Tables 4 and 5) in this particular application allows the following considerations. The FPGA design is composed by quite large blocks and uses the digital signal processing (DSP) blocks to efficiently perform the binary multiplication. This is very useful for the FPGA that can reach a remarkable speed for a reconfigurable target but leaves very little space for the arithmetic optimization and for the introduction of pipeline levels (e.g., a pipeline is not possible inside the DSP blocks). On the other hand, the ASIC design exploits a standard cell library with very small granularity and can choose among various design techniques for the arithmetic blocks. Also, the pipeline level can be moved freely inside the arithmetic block if needed. As a consequence, retiming and pipelining allow a large leap in circuit clock frequency (from 3.03 GHz to 5.26 GHz in the d = 8 bit case and from 2.70 GHz to 5.00 GHz in the d = 16 bit case).

Implementation results show that the circuit is able to reach the 5 GHz target in both cases. The effect of the retiming and the presence of the additional pipeline levels is seen in an increase of both area (mainly due to the additional Flip Flops) and power dissipation.

## 5. Conclusions

The paper has reviewed the main digital signal processing based methods for controlling the sampling rate in DSOs by means of digital resampling approaches. A digital circuit that offers a promising solution to grant more control of the sampling rate, with respect to the existing approaches, has then been discussed. The circuit can be deployed in the acquisition channel of any DSO to interface the internal ADC and the acquisition memory. It has been implemented on FPGA and evaluated. Also, the performance of an ASIC design of the same circuit has been investigated. The proposed solution can be exploited to effectively improve the sampling rate selection capability of DSOs, especially when the instrument does not permit the use of an external clock to drive the internal ADC.

**Author Contributions:** Methodology, formal analysis, simulations and original draft preparation, M.D.; methodology, circuit design, supervision, E.N.; circuit design, validation, original draft preparation and review, E.Z. All authors have read and agreed to the published version of the manuscript.

**Funding:** This work is supported by the Project "Vision for Robotic Surgery (ViRoS)", funded by the Departmental of Electrical and Information Technology Engineering, University of Naples Federico II, Italy.

**Conflicts of Interest:** The authors declare no conflict of interest.

## Abbreviations

The following abbreviations are used in this manuscript:

| | |
|---|---|
| DSO | Digital Storage Oscilloscope |
| ADC | Analog to Digital Converter |
| FPGA | Field Programmable Gate Array |
| ASIC | Application Specific Integrated Circuit |
| ENOB | Effective Number Of Bits |
| SFDR | Spurious-Free Dynamic Range |
| MSB | Most Significant Bit |

finFET    fin Field Effect Transistor
DSP       Digital Signal Processing

## References


1. Oya, J.R.G.; Munoz, F.; Torralba, A.; Jurado, A.; Garrido, A.; Banos, J. Data acquisition system based on subsampling for testing wideband multistandard receivers. *IEEE Trans. Instrum. Meas.* **2011**, *60*, 3234–3237. [CrossRef]
2. Oya, J.R.G.; Munoz, F.; Torralba, A.; Jurado, A.; Marquez, F.J.; Lopez-Morillo, E. Data acquisition system based on subsampling using multiple clocking techniques. *IEEE Trans. Instrum. Meas.* **2012**, *61*, 2333–2335. [CrossRef]
3. D'Apuzzo, M.; D'Arco, M. A wideband DSO channel based on three time- interleaved channels. *JINST* **2016**, *11*, P08003. [CrossRef]
4. D'Apuzzo, M.; D'Arco, M. Sampling and time–interleaving strategies to extend high speed digitizers bandwidth. *Measurement* **2017**, *111*, 389–396. [CrossRef]
5. Monsurrò, P.; Trifiletti, A.; Angrisani, L.; D'Arco, M. Streamline calibration modelling for a comprehensive design of ATI-based digitizers. *Measurement* **2018**, *125*, 386–393. [CrossRef]
6. Monsurrò, P.; Trifiletti, A.; Angrisani, L.; D'Arco, M. Two novel architectures for 4-channel mixing/filtering/processing digitizers. *Measurement* **2019**, *142*, 138–147. [CrossRef]
7. Choi, H.; Gomes, A.; Chatterjee, A. Signal acquisition of high-speed periodic signals using incoherent sub-sampling and back-end signal reconstruction algorithms. *IEEE Trans. VLSI Syst.* **2011** *19*, 1125–1135. [CrossRef]
8. Angrisani, L.; D'Arco, M.; Ianniello, G.; Vadursi, M. An efficient pre-processing scheme to enhance resolution of band-pass signals acquisition. *IEEE Trans. Instrum. Meas.* **2012**, *61*, 2932–2940. [CrossRef]
9. Yuan, W.; Jiangmiao, Z.; Jingyuan, M. Correction of time base error for high speed sampling oscilloscope. In Proceedings of the 2013 IEEE 11th International Conference on Electronic Measurement & Instruments, Harbin, China, 16–19 August 2013; pp. 88–91.
10. Betta, G.; Capriglione, D.; Ferrigno, L.; Miele, G. Innovative methods for the selection of bandpass sampling rate in cost-effective RF measurement instruments. *Measurement* **2010**, *43*, 985–993. [CrossRef]
11. D'Arco, M.; Genovese, M.; Napoli, E.; Vadursi, M. Design and implementation of a preprocessing circuit for bandpass signals acquisition. *IEEE Trans. Instrum. Meas.* **2014**, *63*, 287–294. [CrossRef]
12. Porteous, M. Introduction to Digital Resampling. In *RF Engines White Paper*, 2011; lit.num.4216984. Available online: https://www.techonline.com (accessed on 13 March 2020).
13. Voronov, E.; Solodkov, A.; Belousov, E. Digital signal resampling device for self-organizing networks. In Proceedings of the 2015 Internet Technologies and Applications (ITA), Wrexham, UK, 8–11 September 2015; pp. 151–154.
14. Kirchner, M.; Bohme, R. Hiding Traces of Resampling in Digital Images. *IEEE Trans. Inform. Forensics Secur.* **2008**, *3*, 582–592. [CrossRef]
15. Popescu, C.; Farid, H. Exposing digital forgeries by detecting traces of resampling. *IEEE Trans. Signal Process.* **2005**, *53*, 758–767. [CrossRef]
16. Porwal, S.; Katiyar, S.K. Performance evaluation of various resampling techniques on IRS imagery. In Proceedings of the 2014 Seventh International Conference on Contemporary Computing (IC3), Noida, India, 7–9 August 2014; pp. 489–494.
17. Xu, T.; Fumagalli, A.; Hui, R. Real-time DSP-enabled digital subcarrier cross-connect based on resampling filters. *IEEE/OSA J. Opt. Commun. Netw.* **2018**, *10*, 937–946. [CrossRef]
18. Johansson, H.; Pillai, A.K.M. Lower bounds on the L2-norms of digital resampling filters with zero-valued input samples. In Proceedings of the 2016 IEEE International Conference on Acoustics, Speech and Signal Processing (ICASSP), Shanghai, China, 20–25 March 2016; pp. 4533–4537.
19. D'Arco, M.; Napoli, E.; Angrisani, L. A time base option for arbitrary selection of sample rate in digital storage oscilloscopes. *IEEE Trans. Instrum. Meas.* **2020**. [CrossRef]
20. D'Arco, M.; Napoli, E.; Zacharelos, E. Digital Circuit for the Arbitrary Selection of Sample Rate in Digital Storage Oscilloscopes. In Proceedings of the ApplePies Conference, Pisa, Italy, 12–13 September 2019.

21. Ruiz-Rosero, J.; Ramirez-Gonzalez, G.; Khanna, R. Field Programmable Gate Array Applications—A Scientometric Review. *Computation* **2019**, *7*, 63. [CrossRef]
22. Baugh, C.R.; Wooley, B.A. A Two's Complement Parallel Array Multiplication Algorithm. *IEEE Trans. Comput.* **1973**, *C-22*, 1045–1047. [CrossRef]
23. de Caro, D.; Petra, N.; Strollo, A.G.M.; Tessitore, F.; Napoli, E. Fixed-width multipliers and multipliers-accumulators with min-max approximation error. *IEEE Trans. Circuits Syst. I: Regul. Pap.* **2013**, *60*, 2375–2388. [CrossRef]
24. Wang, Q.; Zhou, W.; Feng, Y.T.; Ma, G.; Cheng, Y.; Chang, X. An adaptive orthogonal improved interpolating moving least-square method and a new boundary element-free method. *Appl. Math. Comput.* **2019**, *353*, 347–370. [CrossRef]
25. Stuart, R. A quarterly Phillips curve for Switzerland using interpolated data, 1963–2016. *Econ. Model.* **2018**, *70*, 78–86. [CrossRef]
26. Qaisar, S.M. A two stage interpolator and multi threshold discriminator for the Brain-PET scanner timestamp calculation. *Nucl. Instrum. Methods Phys. Res. Sect. A: Accelerat. Spectrom. Detect. Assoc. Equip.* **2019**, *922*, 364–372. [CrossRef]
27. Schmitter, D.; Fageot, J.; Badoual, A.; Garcia-Amorena, P.; Unser, M. Compactly-supported smooth interpolators for shape modeling with varying resolution. *Graph. Models* **2017**, *94*, 52–64. [CrossRef]
28. Mouri Zadeh Khaki, A.; Farshidi, E.; Hamid MD Ali, S.; Othman, M. An FPGA-Based 16-Bit Continuous-Time 1-1 MASH ΔΣ TDC Employing Multirating Technique. *Electronics* **2019**, *8*, 1285. [CrossRef]
29. Garofalo, V.; Petra, N.; Napoli, E. Analytical calculation of the maximum error for a family of truncated multipliers providing minimum mean square error. *IEEE Trans. Comput.* **2011**, *60*, 1366–1371. [CrossRef]
30. D'Apuzzo, M.; D'Arco, M.; Liccardo, L.; Vadursi, M. Modelling DAC output waveforms. *IEEE Trans. Instrum. Meas.* **2010**, *59*, 2854–2862. [CrossRef]
31. Angrisani, L.; D'Arco, M. Modelling timing jitter effects in digital to analog converters. *IEEE Trans. Instrum. Meas.* **2009**, *58*, 330–336. [CrossRef]
32. Martinek, R.; Rzidky, J.; Jaros, R.; Bilik, P.; Ladrova, M. Least Mean Squares and Recursive Least Squares Algorithms for Total Harmonic Distortion Reduction Using Shunt Active Power Filter Control. *Energies* **2019**, *12*, 1545. [CrossRef]
33. Szczepanska, A.; Gościewski, D.; Gerus-Gościewska, M. A GRID-Based Spatial Interpolation Method as a Tool Supporting Real Estate Market Analyses. *ISPRS Int. J. Geo-Inf.* **2020**, *9*, 39. [CrossRef]
34. Fiala, P.; Linhart, R. High performance polyphase FIR filter structures in VHDL language for Software Defined Radio based on FPGA. In Proceedings of the International Conference on Applied Electronics, Pilsen, Czech Republic, 9–10 September 2014; pp. 83–86.
35. D, S.; G, K. Polyphase Representation of Multirate Nonlinear Filters and Its Applications. *IEEE Trans. Signal Process.* **2007**, *55*, 2145–2157.
36. Haddad, F.; Rahajandraibe, W.; Zaid, L.; Frioui, O.; Bouchakour, R. Radio frequency tunable polyphase filter design. In Proceedings of the 16th IEEE Int. Conf. on Electronics, Circuits and Systems—(ICECS 2009), Yasmine Hammamet, Tunisia, 13–16 December 2009; pp. 21–24.
37. Johansson, H. Polyphase Decomposition of Digital Fractional-Delay Filters. *IEEE Signal Process. Lett.* **2015**, *55*, 1021–1025. [CrossRef]
38. Laddomada, M. On the Polyphase Decomposition for Design of Generalized Comb Decimation Filters. *IEEE Trans. Circuits Syst. I Regul. Pap.* **2008**, *55*, 2287–2299. [CrossRef]
39. IEEE Standard for Digitizing Waveform Recorders. *IEEE Stand.* **2007**, *1057*, 1–40.

*Article*

# Embedded Bio-Mimetic System for Functional ElectricalStimulation Controlled by Event-Driven sEMG [†]

**Fabio Rossi, Paolo Motto Ros, Ricardo Maximiliano Rosales and Danilo Demarchi ***

Dipartimento di Elettronica e Telecomunicazioni (DET), Politecnico di Torino, 10129 Torino, Italy; fabio.rossi@polito.it (F.R.); paolo.mottoros@polito.it (P.M.R.); s251237@studenti.polito.it (R.M.R.)

* Correspondence: danilo.demarchi@polito.it

† This paper is an extended version of our paper published in: Rossi, F., Rosales, M. R., Motto Ros, P., Danilo, D. Real-Time Embedded System for Event-Driven sEMG Acquisition and Functional Electrical Stimulation Control. In Proceedings of the 2019 International Conference on Applications in Electronics Pervading Industry, Environment and Society (ApplePies), Pisa, Italy, 11–13 September 2019.

Received: 7 February 2020; Accepted: 7 March 2020; Published: 10 March 2020

**Abstract:** The analysis of the surface ElectroMyoGraphic (sEMG) signal for controlling the Functional Electrical Stimulation (FES) therapy is being widely accepted as an active rehabilitation technique for the restoration of neuro-muscular disorders. Portability and real-time functionalities are major concerns, and, among others, two correlated challenges are the development of an embedded system and the implementation of lightweight signal processing approaches. In this respect, the event-driven nature of the Average Threshold Crossing (ATC) technique, considering its high correlation with the muscle force and the sparsity of its representation, could be an optimal solution. In this paper we present an embedded ATC-FES control system equipped with a multi-platform software featuring an easy-to-use Graphical User Interface (GUI). The system has been first characterized and validated by analyzing CPU and memory usage in different operating conditions, as well as measuring the system latency (fulfilling the real-time requirements with a 140 ms FES definition process). We also confirmed system effectiveness, testing it on 11 healthy subjects: The similarity between the voluntary movement and the stimulate one has been evaluated, computing the cross-correlation coefficient between the angular signals acquired during the limbs motion. We obtained high correlation values of 0.87 ± 0.07 and 0.93 ± 0.02 for the elbow flexion and knee extension exercises, respectively, proving good stimulation application in real therapy-scenarios.

**Keywords:** surface electromyography; event-driven; functional electrical stimulation; embedded system

---

## 1. Introduction

Nowadays, an increasing number of active rehabilitation techniques are moving to the bio-mimetic approach, which relies on the analysis of the surface ElectroMyoGraphy (sEMG) signal for, e.g., the application of Functional Electrical Stimulation (FES) [1], with the aim of physiologically controlling the muscle functional restoration as much as possible [2]. In particular, FES employs low energy current pulses to modulate the muscle contraction [3] where a complex stimulation pattern, useful to activate the group of muscles involved in a movement, is regulated by sEMG envelope evaluation or by muscle force indicators (e.g., Root Mean Square (RMS), Absolute Rectified Value (ARV)) [4].

In a practical application, the sEMG processing and FES control is a fundamental task to be carried out in real-time [5]. Since the run-time performance bottleneck could be easily related to the use of a general purpose computer for the FES control (often concurrently running, or loaded with, many other unrelated applications or functionalities, leading to unpredictable performances), here the idea is to replace it with a dedicated embedded system. In this regard, major concerns will be the effectiveness and safety of the stimulation and the resulting performance, i.e., a latency short enough to fulfill the real-time constraints and the quality of the stimulated movement.

We propose an embedded bio-mimetic FES system based on the Average Threshold Crossing (ATC) event-driven technique applied to the sEMG signal. The ATC essentially compares the sEMG signal with a threshold [6]: the Threshold Crossing (TC) events generate the quasi-digital TC signal, which is characterized by a digital waveform carrying analog (time-based) information. The ATC parameter is then computed by counting the number of TC events during a time window. In [7], we have demonstrated the correlation among ATC, ARV and the muscle force: in particular, having $0.95 \pm 0.02$ ATC-force w.r.t. $0.97 \pm 0.02$ ARV-force correlation, the ATC parameter can be used as indicator of muscle activity [8]. In this way, the event-driven approach enables the implementation of a low-complexity on-board feature extraction process, divided into two steps (TC generation and ATC computing), which can be directly performed in hardware [9,10], supporting, e.g., the recognition of different gestures [11–14]. While the theoretical background of ATC is quite similar to others common sEMG features, e.g., Zero-Crossing (ZC) or Wilson Amplitude (WAMP) [15], our event-based approach could overcome signal processing limitations for embedded feature extraction [16]. In particular, ZC and WAMP calculations are achieved by analyzing an already digitized signal, leading to high time, processing and power consumption, while by implementing in hardware our proposed ATC approach we are able to relax these issues.

Therefore, the minimal data size of the ATC information [10] and its sparsity (due to its event-driven nature) perfectly matches the low computational capabilities of an embedded system. Evolving from the architecture presented in previous works [10,17,18], with the aim of making the system portable and improving the run-time performance, we replaced the personal laptop, and the software based on the MATLAB® & SIMULINK® environment, with a Raspberry Pi 3 B+ as the processing and control core of the system, running a multi-platform software. Its main tasks are the management of the sEMG multi-channel wireless acquisition, the computation and update of the FES parameters from the ATC data, and the safe control of the stimulator. The software features a Graphical User Interface (GUI) as well, to monitor and control every aspect of the system, eventually guiding the user into setup different and personalized stimulation sessions.

From the application point of view, typical scenario consists in the reproduction of functional movements between two subjects in the therapist-patient rehabilitation context: The muscular activity monitored from an healthy subject (therapist), e.g., doctor, physiotherapist, during the execution of a movement, is processed in order to define the FES pattern to be applied to a second subject (patient) in order to induce the replication of the same movement.

This manuscript extends what already presented in [19] by discussing the design choices and details about the system architecture, focusing both on hardware and software aspects, and the related development approaches; finally, further system characterization and validation results are presented, as well as in-vivo experiments.

The paper is organized as follows: Section 2 presents the overall system architecture and details the design and development of both the hardware and software parts; Section 3 presents the results of the validation and characterization of the developed embedded system, while Section 4 the results of in-vivo experimental tests are reported; results, with particular emphasis on the feature of the embedded system, are then discussed and compared with related works in Section 5; in the end conclusion and future perspectives are outlined in Section 6.

## 2. System Architecture: Design and Development

### 2.1. Overview

A description of the proposed system can be conceptually schematized into inputs, control and output logical macro areas, as represented in Figure 1, according to the actions flow from signal acquisition to stimulation application. Data acquired by input devices (i.e., muscular activation and limbs motion) are processed by the control unit in order to drive the FES application through the output device.

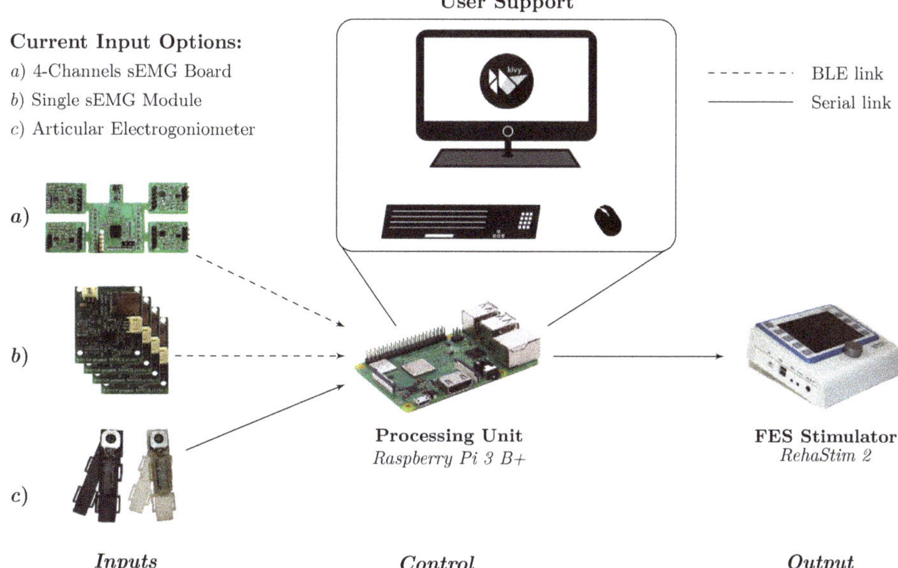


**Figure 1.** System hardware and user interface architecture: The Raspberry Pi acts as control logic linking the input devices (i.e., surface ElectroMyoGraphic (sEMG) acquisition board and electro-goniometer) with the output (Functional Electrical Stimulation (FES) stimulator).

We designed a flexible enough framework by developing a multi-platform software core, compatible with widespread Operating Systems (OSs) (such as Microsoft® Windows®, GNU/Linux and Android), able to run on commonly available devices, i.e., PC, laptop, tablet, smartphone as well as Raspberry Pi.

Among all the possibilities, we defined our optimized embedded version of the system as Reference Hardware Setup (RHS), which comprises individual acquisition channels for sEMG and electro-goniometers as inputs, Raspberry Pi as control logic and the RehaStim 2 FES stimulator as output. With respect to RHS, other configurations are characterized by changes in the inputs and control devices (i.e., a Microsoft® Windows® or GNU/Linux PC), which lead to slight variations in the wireless connectivity management and software structure.

*2.2. Hardware Platform*

The input devices are the sensors useful to record the signals of interest, i.e., the electrical signals produced by the muscles contraction (sEMG signal) during the execution of a movement and the angular signals representing the limbs motion of the human body.

In the first case, the employed device has to amplify and filter the muscular signal in order to allow its interpretation, since the raw signal amplitude varies between hundreds of µV and tens of mV [20]. Therefore, referring to the guidelines reported in [21], we developed an analog conditioning circuit for the bio-signal [9], which, using the three-electrodes differential approach (two as sources, one for reference), provides 1000 gain factor in the 30 Hz to 400 Hz bandwidth (obtained as a cascade of a differential first-order high-pass filter [22] and a second-order Sallen-Key low-pass filter [23]) in order to filter out electrode-skin movement artifact and high-frequency noise. Moreover, since the sEMG module has to be coupled with FES stimulator, we added overvoltage protection diodes on the channels input. As introduced, we carried out the first step of our event-driven signal processing by extracting the TC signal using an hysteresis voltage comparator (30 mV) so to avoid spurious glitches. The average counts of events (ATC) is then computed at the digital interface with the microcontroller (MCU): in [10], we demonstrated how to accomplish this task minimizing the MCU resources to a GPIO interrupt, which detects the TC digital events, and a timer, which defines the observation window. The length of the window is set to 130 ms as reported in the tests presented in [9], where this value has been proved to be an optimal trade-off between the time resolution of the muscle activation and the discrimination of different levels of generated muscular force.

We propose two solutions based on this acquisition and processing architecture, shown in Figure 2, depending on the user needs: The first option (a) is a complete four-channels board suitable for multiple-muscle monitoring on the same limb, e.g., extensor and flexor muscles of human forearm, while the second one (b) is a stand-alone single-channel module to be used independently when an individual detection is advantageous, e.g., biceps- and triceps- brachii muscles during the elbow flexion and extension. We equipped both solutions with wireless connectivity in order to improve a freedom movement executions, avoiding wiring hindrance, and to make the systems fully wearable: among the wide list of wireless (standard) option, we chose the Bluetooth Low Energy (BLE) protocol (stack 4.1 [24]) because of its low-energy features, which perfectly match with battery-device requirements. In particular, we equipped (a) with the Microchips RN4020 [25] module (with its own antenna), while in (b) the same MCU used for computing the ATC runs the BLE stack and directly feeds a PCB antenna, designed referencing to [26].

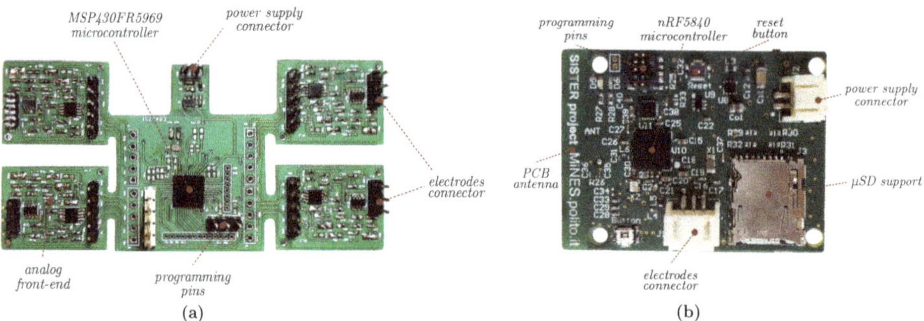


**Figure 2.** Custom sEMG acquisition device: (**a**) Four-channels board for multiple muscles monitoring, (**b**) independent single-channel module.

As second input typology, we developed custom electro-goniometers in order to record the limbs motion in form of electrical signals. Figure 3 shows their structure (very similar to standard goniometer's one), which basically consists of two parts fixed by a pivot at one extremity. Employing an absolute capacitive modular encoder, i.e., the AMT20 [27], and placing its center in correspondence of the pivot, we were able to detect the goniometer's angle decoding the encoder shaft position related to its inner capacitance changes. Angle values are represented on 12 bit, with a 0.2° accuracy and, since the AMT20 presents an SPI output line, we interfaced it with an Arduino micro MCU [28] in order to sample the signal at 80 Hz (appropriate w.r.t. human movement velocity [29]) and to transmit it to an external device (via USB cable) for graphical representation. The goniometer case has been manufactured by a 3D printing process, employing the Form 2 printer [30] with a bio-compatible photo-reactive resin, which allowed us to design an anatomical comfortable and lightweight structure. Four elastic strips secure the electro-goniometer in the proper location on the limb, ensuring its pivot to be in position with the rotation center of the articulation.

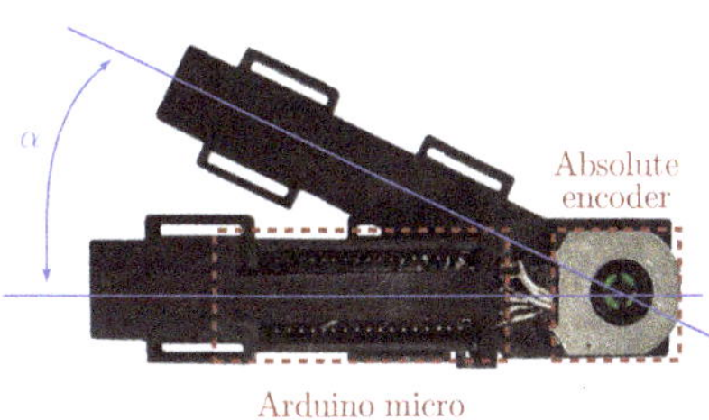


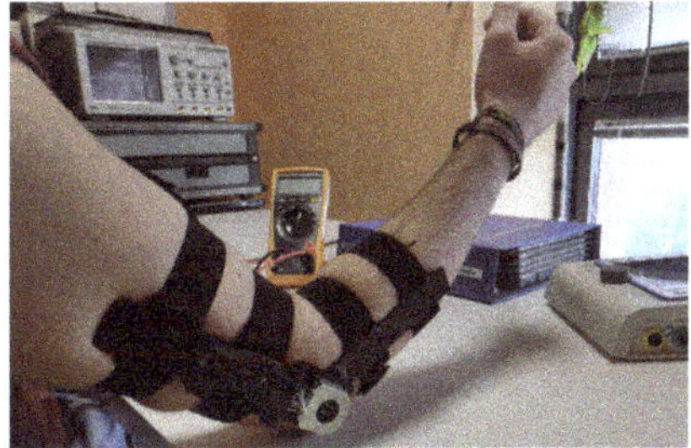

**Figure 3.** Wearable electro-goniometer developed to test the system performance and to provide an angular feedback on the ongoing stimulation.

The control of the induced FES pulses depends on how they are electrically generated and which pulse parameters can be modified during the stimulation. We decided for the medical-certified RehaStim 2 [31] because it allowed us to have an advanced control on the pulse definition per channel and the possibility to be easy interfaced with an external device by means of the ScienceMode2 bidirectional communication protocol [32].

The generated current pulses are characterized by a biphasic rectangular shape, shown in Figure 4, whose configurable parameters are the pulse amplitude, the stimulation frequency and the phase width, while the inter-phase interval is fixed to 150 µs guaranteeing a proper stimuli excitability [33].

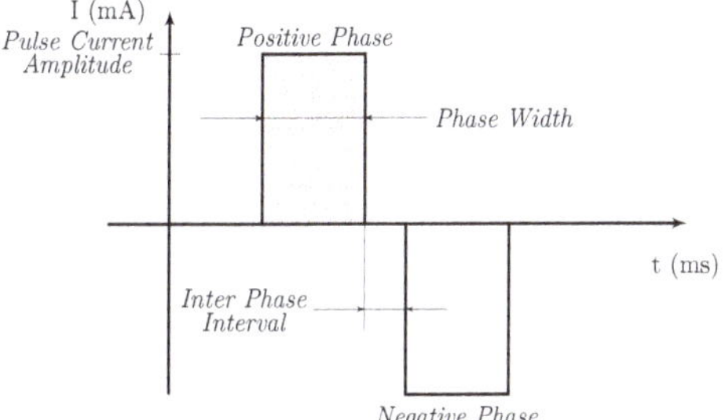


| FES params | Range | Control |
|---|---|---|
| amplitude | 0:1:130 mA | ATC |
| width | 20:10:500 µs | m. |
| interval | 150 µs (fixed) | - |
| frequency | 10:5:50 Hz | m. |

m.: physiotherapy manual

**Figure 4.** Rectangular biphasic current pulse generated by RehaStim 2 stimulator and its parameter.

Therefore, considering the ATC dependency on the muscle force (e.g., correlation between ATC and sEMG amplitude/energy indicators), our idea has been to modulate the FES pulses intensity on the basis of such parameter, while for the other settings we referred to the physiotherapy manual provided with the stimulator [34]. In this way, the modulation approach allows us to excite the muscle fibers with the proper amount of current during all the phases of a movement session (warm up, increasing force, relaxation as well as resting state) and for a wide list of exercises.

Last part of the system is represented by the Raspberry Pi, model 3 B+ [35], working as control logic which manages the entire system. Indeed, it runs the main software controlling the data acquisition, its processing, the stimulation definition and application. Moreover, since this Raspberry Pi is equipped with four USB ports and a full size HDMI, we improved the system usability developing a complete GUI and employing some peripherals, as keyboard, monitor and mouse.

As discussed in Section 2.1, different devices can act as control unit appropriately configuring the hardware: As an example, if a Microsoft® Windows® OS PC is used as control logic, the CC2540-Dongle [36] module is needed to communicate with the acquisition devices (limiting the maximum number of simultaneously connections to three) since Windows® machines do not allow an easy access to the Bluetooth interface.

## 2.3. Software Overview

As previously introduced in Section 2.1, although the project is finalized to the development of an embedded system, we want to provide a modular and flexible software core able to fulfill the compatibility requirements of different OSs. As previously introduced in Section 2.1, we provided a modular and flexible software core able to fulfill the compatibility requirements of different OSs. Consequently, from the development standpoint, we based the software on the Python language, because of its cross-platform nature, its widespread adoption, and the large availability of third-party multi-platform libraries (such as standard library for multi-threading features or Kivy library [37] for the GUI). Moreover, the embedded software has been based on an object-oriented (OO) framework in order to promote flexibility, modularity and robustness [38] (e.g., leveraging encapsulation, inheritance, and composition features), allowing a seamless integration and management of several devices (e.g., different input modules) along with the possibility of future integration of new processing algorithms. We also implemented a multi-threaded architecture in order to map the functional tasks onto different running threads [39], so to optimize the use of computational resources and to avoid complex (run-time) code interdependencies.

### 2.3.1. Classes Diagram Overview

As shown in the Unified Modeling Language (UML) diagram in Figure 5, the main System object is composed by four sub-objects: The FES class representing the stimulator, two Goniometer classes for the developed electro-goniometers and a Bluetooth class, which can have different implementations depending on the hardware configuration.

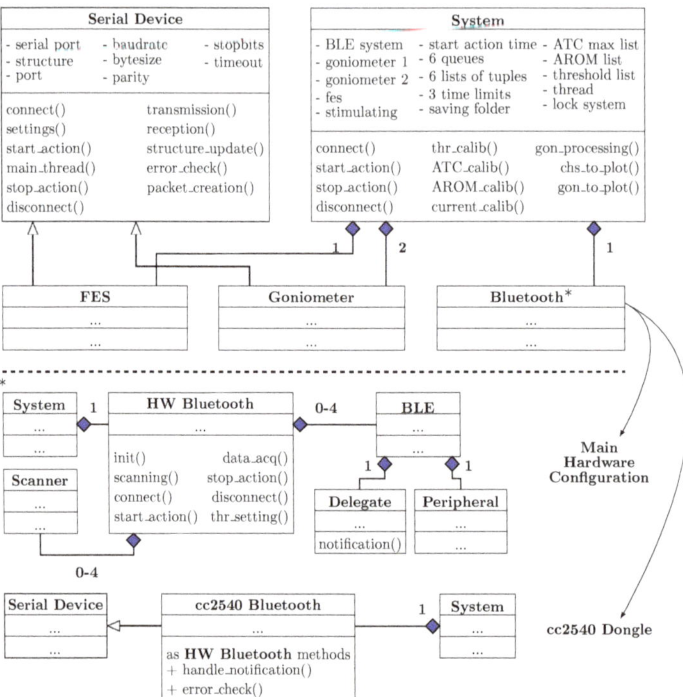


**Figure 5.** Classes diagram (UML) of our OO software organization. Bluetooth class implementation depends on system configurations.

Since both the goniometers and the stimulator are wired connected to the control unit, their classes inherit from an abstract custom Serial Device class, which provides a standard interface for every serial device (i.e., serial port, baudrate, stopbits, etc. attributes or connect(), settings(), transmission() methods and so on). Specific methods of different serial devices have been overwritten in order to provide the proper interfacing with the control unit.

Regarding the BLE software, it depends by system configurations: if the RHS is used, we combined the BlueZ [40] Linux Bluetooth stack with the bluepy [41] Python library (specific for low energy features). In particular, the HW Bluetooth class is composed by a variable number (zero to four) of BLE connections, which in turns consists of by a Delegate (notification data handling) and a Peripheral (bluepy instance for encapsulating BLE BlueZ connection) objects, and a Scanner, which seeks for advertising devices. On the other hand, if a common PC is employed, the CC2540-Dongle module is needed and, since it communicates through a serial port with the workstation, the Bluetooth class inherits from the Serial Device one.

### 2.3.2. Multi-Threading

Figure 6 shows the multi-threading structure of the system and the running state of the involved threads during a typical stimulation session.

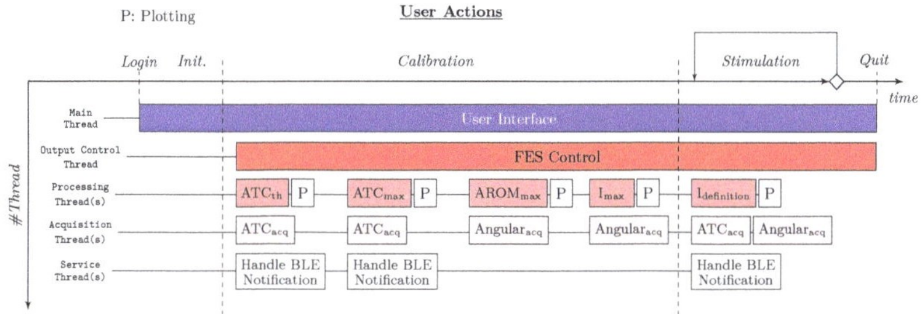


**Figure 6.** Multi-threading structure during a typical stimulation scenario.

The Main Thread starts after the user login and runs all along the session waiting for the user inputs, at which correspond the creation of child threads, handling the user interface. As primary sub-thread, the Output Control Thread manages the communication with the stimulator, e.g., watchdog timer, packet creation etc., during the calibration and stimulation phases. Moreover, the Main Thread runs all the calibration-step threads (i.e., $ATC_{th}$, $ATC_{max}$, $AROM_{max}$ and $I_{max}$, details in Section 2.4) during the settings and the $I_{definition}$ threads when the stimulation is applied, globally defined as Processing Threads. Each of them is also supported by a Plot thread, represented by white rectangle, which graphically represents the useful signals. Finally, we developed the Acquisition Threads, divided into $ATC_{acq}$ and $Angular_{acq}$ for the ATC and angular values acquisition, and Service Thread for BLE notifications managing. Data exchange among threads is organized with queue objects; therefore, each thread implements a specific method in order to continuously check the queue status.

### 2.3.3. Graphical User Interface

The GUI has been developed choosing the Kivy Python library [37], due to OS inter-compatibility, modern layout, open-access feature and optimized performance [42], in order to have an easy, intuitive, and practical high-level control of the application.

Figure 7 shows the main four screens of the GUI. In the Initialization one, the user inserts the personal information of therapist and patient, and chooses the system configuration (acquisition and stimulation channels) along with the movement that will be executed. The Calibration screen is properly designed to perform the calibration process, whereby the acquisition and stimulation parameters are optimized for the user-case. Subjects data are used to build up a database, useful to fast-configure application settings avoiding the calibration steps. In the `Main Stimulation` screen, the stimulation can be started and stopped, and the useful signals are graphically represented (i.e., pulse amplitude and angular signals) in order to provide a visual feedback for the therapist. Lastly, the `Parameters` screen allows the user to modify the parameters or save them if multi-session scenarios are expected. Transitions among the screens, represented by black arrows in Figure 7, have been arranged using the `Screen Manager` object, facilitating user navigation among sections.

From an OO prospective, all the screens directly inherit from the `Kivy Screen` class, with the exceptions of the ones containing graphs (i.e., `Main Stimulation` and `Calibration`) which are also defined by the `MyPlotScreen` class since it possesses Kivy plotting objects. Thus, the `System` is aggregated in every main screen where the system actions run through screens widgets.

Lastly, on the Raspberry Pi, we changed the RAM memory assigned to the Graphics Processing Unit (GPU) from 64 MB to 256 MB in order to execute the GUI without impacting on the graphical resources.

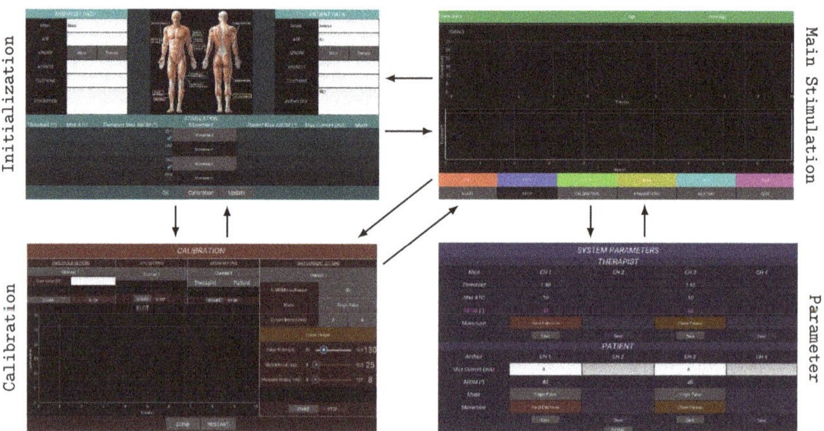

**Figure 7.** Kivy main screens: The `Initialization` one allows the user to store subjects information and set up the system; in the `Calibration` one the calibration process is achieved, and the optimized parameters can be visualized, saved and modified into the `Parameter` screen; finally, the `Main Stimulation` screen runs the stimulation and graphically represents the signals of interest. Arrows highlight transitions among screens.

### 2.4. ATC Dataflow: Processing and Calibration

The definition of the FES pulses amplitude dependent on the ATC values is the core of the FES control mechanism, linking the data acquisition with the stimulation one. Since embedded device has extremely low-computational power, we needed to implement this process trying to maintain the complexity lower as much as possible, also considering fast computing approach to respect real-time requirements. In this scenario, taking advantage of the sEMG-ATC (pre-)processing, our idea is to mimic the simplicity of a look-up table structure: Basically, we organized it as two matrices architecture, one for the ATC values and one for the FES current ones, with one-to-one cell correspondence between them.

Typical application scenario, considering $n$ active channels, is represented in Figure 8: Every time a new BLE packet arrives, containing the ATC data of n channels, the received data are appended to an $n \times 4$ matrix (ATC matrix), which also includes the three past ATC-window data. Then, the row-median operation is computed in order to obtain a robust ATC value without any noise corruptions. Since the ATC matrix is continuously updated (every ATC window), this operation basically represents a moving median. In this way, we obtain a $n \times 1$ array, whose values are interpreted as indexes pointing to the FES current values stored into the FES Current Matrix. Once the new stimulation data are defined through this algorithm, a FES data packet is built up and the command is transmitted to the stimulator.

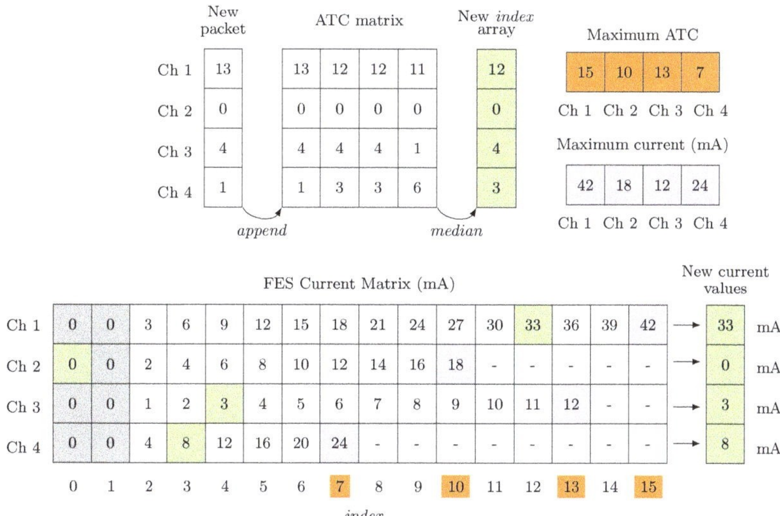

**Figure 8.** Average Threshold Crossing (ATC)-FES definition process: Green cells represents the links between inputs ATC data and outputs pulses current; orange labels identify the FES current Matrix indexes defined by the Maximal ATC calibration step; blue values correspond to the maximal FES current calibrated with the Current limitation process.

However, since different subjects could produce different ATC values or be stimulated by a diverse amount of current, a calibration process for the optimization of the acquisition (therapist) and stimulation (patient) parameters is fundamental, permitting us to develop a flexible system, able to suit different users, while maintaining the benefits of a proper and safe per-subject stimulation. Hence, we defined a four-steps calibration process as follows:

1. Threshold setting: The generation of the TC events strongly depends on the threshold value. Therefore, we tried to optimized the TCs setting the threshold just above the sEMG signal baseline in order to maximize the events with the minimal muscle effort. To accomplish this task, the therapist has to maintain a rest limb condition and, starting from an initial threshold value, we decrease it step-by-step until we find the baseline. Final threshold is set 30 mV above baseline reflecting voltage hysteresis comparator behavior.
2. Maximal ATC: The therapist has to repeat the movement to be calibrated at least four times. The maximal ATC value produced by the subject is calculated as the median value among the maximum of each repetition. This value limits the index dimension of the array, related to the calibrated channel, inside the FES Current Matrix, highlighted in orange in Figure 8.
3. AROM evaluation (optional): The maximal Absolute Range of Motion (AROM) of the involved articulation has been computed by processing the angular data of both therapist and patient. This measure standardizes the FES application and provides a comparison feedback between the voluntary movement and the stimulated one. We defined it as an optional step since the use of the electro-goniometers is not mandatory.
4. Current limitation: We define the maximal current, useful to properly reproduce the movement, as the 110% of the current able to produce a 30% AROM variation in the stimulated subject. If the goniometer is not used, this step can be visually performed. Maximal Current values, represented in blue in Figure 8, related to the indexes defined by the Maximal ATC, define the proper stimulation values inter-step.

Following this approach, we are able to set up our structure with a perfect matching between the muscular activation of the therapist and the pulses amplitude to adequately stimulate the patient limb. Looking at the example represented in Figure 8, the FES Current Matrix has a different column-dimension for each channel defined by the Maximum ATC values. In this way, setting the Maximum current values, we are able to define step and range of pulses amplitude. Concluding, simply controlling the lower values of the stimulation matrix (FES Current Matrix[(:, 1:2)], grey cells), combined with the moving median gate operation, we are able to implement a very low complex but efficient noise-gateway control.

## 3. System Validation and Characterization

The performances of the control unit have been studied by analyzing the real-time FES control processing, fundamental to achieve the proper online modulation of pulses amplitude, and examining how the developed software impacts the workstation resources, so evaluating memory, graphic and computational cost. Due to the multi-platform nature of our software, we carried out these tests comparing its behavior running on two different control units: in particular, we employed the Raspberry Pi 3 B+, equipped with a Cortex-A53 (ARMv8) 64 bit, running at 1 GHz, 1 GB RAM and Raspbian OS, to evaluate the performance of the embedded version; conversely, a Toshiba Satellite L830-14J PC, equipped with an Intel Core i3-3227U with 1.9 GHz clock frequency, 4 GB RAM and Microsoft® Windows® 10 OS, has been used to simulate personal laptop application scenario.

### 3.1. Latency Measurement

As mentioned in Section 2.1, the system can adopt different architectures depending on which sEMG acquisition device is used and by the employed processing unit. As a consequence, we defined five hardware configurations (**Cx**) to be tested, listed in Table 1. The latency has been evaluated for two crucial sections of the application: The FES current definition, which concerns the definition of the new pulses amplitude on the basis of the latest ATC values, and the Plotting, which regards the representation of both the angular signals and the FES currents over time. The duration of the test has been set to 3 min in order to obtain sufficient values ($180\,\text{s}/ATC_{window} = 180\,\text{s}/0.13\,\text{s} \simeq 1385$ measures) able to represent the system performance from the stimulation initialization to the stable working condition.

**Table 1.** Tested hardware configurations. Main differences concern the acquisition device (single channel or four-channel board) and the control unit (GNU/Linux Raspberry or Microsoft® Windows® PC).

| | Acquisition Device | | Control Unit | | BLE Module |
|---|---|---|---|---|---|
| Config. | Single Channel | 4-Channel Board | GNU/Linux Raspberry | Microsoft® Windows® PC | CC2540 * |
| C1 | ✓ | | ✓ | | |
| C2 | ✓ | | ✓ | | ✓ |
| C3 | | ✓ | ✓ | | ✓ |
| C4 | ✓ | | | ✓ | ✓ |
| C5 | | ✓ | | ✓ | ✓ |

* up to three concomitant connections.

### 3.1.1. FES Current Definition

This method represents the logical core that links the ATC values, describing the muscular activity, to the FES current values, which specify the amplitude of the incoming pulses. As described in Section 2.4, we implemented this functionality using a lookup table structure in order to minimize the complexity as well as the processing time. Indeed, the real-time FES definition is a fundamental task for a proper stimulation, avoiding any delay caused by data-queueing; in particular, our time constraint is directly related to the ATC window, which defines the time interval between two ATC values, and so the FES current definition processing time has to be lower than 130 ms. We tested this process studying how different methods split the workload among them and evaluating the total processing time. As details of the first test, the FES current definition is divided into five consequential methods: queue continuously checks the incoming of the new ATC values; when they are available, we append them to the ATC matrix and the median operation is performed. Then the FES_start method builds the FES packet to be transmitted to the stimulator, and in the meanwhile the plot thread is called for the graphical representation of the signals.

Table 2 reports the time profiling of the workload breakdown: As it can be observed, the majority of the time (around 90% in **C1**, **C2** and **C3**, and 97% for **C4** and **C5**) is spent inside the queue method waiting for the arrival of new ATC values, while the other sub-functions runs for a very short time. Therefore, from a methods breakdown point of view, this behavior confirms that the application works as expected, avoiding any queue formations caused by low computational processing.

**Table 2.** Time profiling results for the evaluation of the methods breakdown during the FES current definition process.

| Method | Configuration | | | | |
|---|---|---|---|---|---|
| | **C1** | **C2** | **C3** | **C4** | **C5** |
| queue | 90.71% | 92.78% | 88.03% | 97.41% | 97.25% |
| append | 1.43% | 1.03% | 2.35% | - | - |
| median | 1.65% | 1.40% | 2.24% | 0.53% | 0.52% |
| FES_start | 5.11% | 3.81% | 6.04% | 1.57% | 1.63% |
| plot | 0.68% | 0.62% | 0.81% | - | - |
| | 100% | 100% | 100% | 100% | 100% |

On the other hand, the real-time FES definition has been proved by looking at the delay data represented in the box plots in Figure 9. All the cases largely fulfill our time constraint, also considering the outliers, since none of the values is greater than 100 ms. In particular, in the Raspberry cases and PC ones the median values are below of 10 and 5 ms respectively, which avoid any possible delay between acquisition and stimulation caused by our FES definition processing. However, comparing the two OSs, laptop performances are considerably superior with respect the Raspberry ones since both hardware and software resources differ between the two architectures.

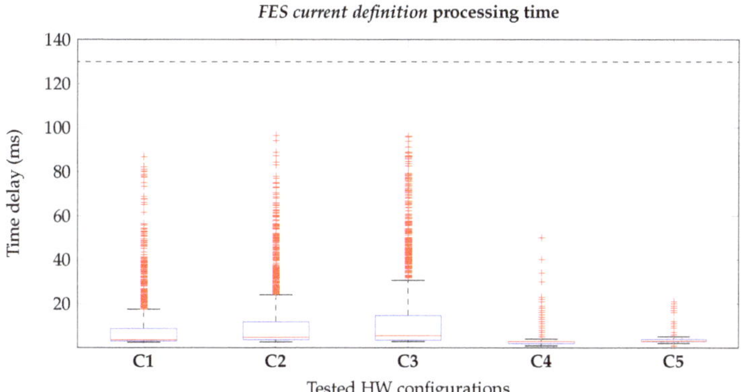



**Figure 9.** Box plots representing the time delays related to the FES current definition method for the different system configurations. Real-time constraints are respected, considering our time limitation of 130 ms (dashed black line), both in GNU/Linux Raspberry (**C1**, **C2** and **C3**) and Microsoft® Windows® PC (**C4** and **C5**) cases.

In conclusion, these results prove the low complexity implementation of our event-driven ATC-FES definition approach: considering the computational lightness of the ATC processing, based on simple mathematical relations (such as matrix and median operations) applied to a minimal data size, we were able to reach very fast pulses updates (lower than 10 ms) along with online modulation of FES parameters.

### 3.1.2. Plotting

The above measurements are repeated for the `Plotting` process in order to study if the graphical representation of the signals of interest can affect the run-time performances. The plotting is based on a `clock` object, whose methods are the `get_value` and the `sleep`: The former gets the new ATC and angular data, and represents them on the graphs; the latter puts the object into an idle state until new data are available. As in the previous test, this analysis allows us to detect the queues formation during the plotting process, but, looking at the Table 3, we can assume this critical condition has not been reached since the 99% of the time is spent in the `clock` sleep state. Moreover, we measured the time spent by the `Plotting` thread in order to verify the correctness of the data representation, discarding the possibility to misinterpret the stimulation visual feedback. This time, we report in Figure 10 our results only for the embedded configurations as worst case scenario: Again, since the 95% of the values (IQR) are lower than 2 ms for all the cases, we can assume the real-time behavior of our plotting method.

**Table 3.** Time profiling results for the `Plotting` process into its two sub-threads: `get_value` and `sleep` methods represent the active and inactive action, respectively, of the `clock` object which manages the plotting of the data.

| | Configuration | | | | |
|---|---|---|---|---|---|
| Method | C1 | C2 | C3 | C4 | C5 |
| `get_value` | 0.94% | 0.93% | 0.95% | 0.45% | 0.41% |
| `sleep` | 99.06% | 99.07% | 99.05% | 99.55% | 99.59% |
| | 100% | 100% | 100% | 100% | 100% |

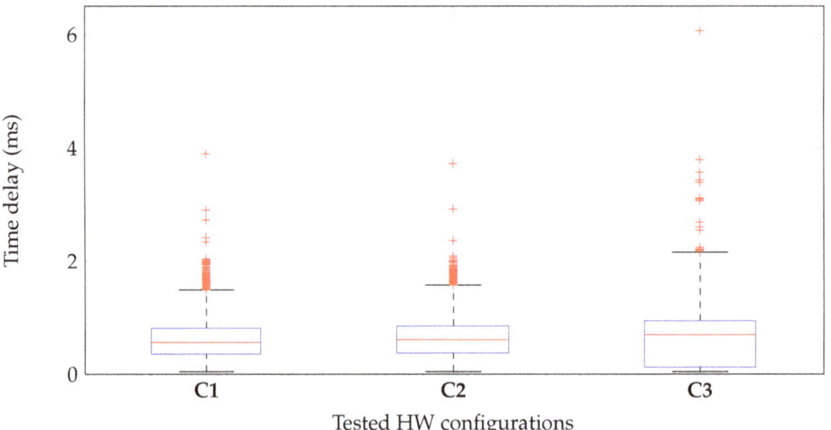


**Figure 10.** Plotting delay for the configurations employing the Raspberry Pi (**C1**, **C2** and **C3**). The very low plotting-time proves the real-time graphical representation by our application.

### 3.2. Computational Performance

The information about the usage of resources at the processing side, such as the Central Processing Unit (CPU) and the Random Access Memory (RAM), is of crucial importance for the evaluation of application fluency, performance and usability. From this perspective, we carried out this test using the RHS (**C2**), considering it as the worst case scenario in terms of processing power. Hence, we monitored the system processes through the `htop` GNU/Linux tool, running the system without the Raspbian GUI activated in order to take in consideration only the application and all its dependencies (i.e., BLE and Kivy).

Table 4 reports the results when four channels stimulation mode has been selected: As it can be observed, the most challenging CPU performance is reached in the main stimulation procedure, where the highest amount of threads are active. Therefore, focusing on the `Stimulation` stage, we repeated the measures studying two different but dependent cases: first, we varied the channels number from one to four looking at resources usage changes; second, with the same purpose, we analyzed our ATC-FES current definition process both in the `standard` situation (i.e., `append` and `median` methods), and when a direct equivalence between ATC and current values is performed (no data processing). From the results listed in Table 5, we can see that employing one single channel we are able to reduce the CPU usage of approximately the 20% w.r.t. the complete channels configuration, and it entirely depends on the fewer or higher number of co-running threads in the two situations. Instead, as it can be observed by the two main columns, the `FES current definition` implementation does not affect the CPU usage, which further proves the lightweight computational cost of our approach.

**Table 4.** CPU and RAM measurements of the application main stages, which have been evaluated testing the Reference Hardware Setup (RHS) with four working channels.

| Stages | CPU (%) | RAM (MB) |
|---|---|---|
| `Login` | 20 | 84 |
| `Initialization` | 21 | 85 |
| `Threshold calibration` * | 24.1 | 88.9 |
| `ATC maximum calibration` * | 26 | 87 |
| `AROM evaluation` * | 32 | 89 |
| `Maximum current calibration` * | 45.7 | 89 |
| `Stimulation` | 73.2 | 87.8 |
| `Parameters` | 15 | 88.3 |

* four steps of the `Calibration` process.

**Table 5.** RHS resources performance during the `Stimulation` stage depending on the number of working channels and the FES current definition implementation. The ATC processing has been enabled (standard flow) or disabled (direct ATC-$I_{FES}$ equivalence) in order to study if whether implementation affects the run-time system performance.

| | | 1 Ch. | 4 Ch. | 1 Ch. | 4 Ch. |
|---|---|---|---|---|---|
| **Process** | I/O operations | ✓ | ✓ | ✓ | ✓ |
| | ATC processing | ✓ | ✓ | x | x |
| **Resources** | CPU (%) | 53.8 | 73.2 | 53 | 74.4 |
| | RAM (MB) | 87.7 | 87.8 | 91 | 92 |

On the other hand, the dynamic memory suffers just low variations among the application stages and between one or four channel cases. This behavior is mainly due by the different amount and types of widgets the GUI owns, which are directly related to the number of active channels. Hence, since there is not any differences between single or four channels GUIs, the RAM consumption is almost constant (see Table 5).

## 4. In Vivo Experimental Tests and Results

As the system is intended for rehabilitative sessions, some tests have been carried out in order to prove the correctness and appropriateness of our approach in the control of the FES application. As introduced, typical scenario considers a first subject which performs an useful movement and whose muscle activity is monitored (therapist), and a second subject which replicates the movement as consequence of the stimulation application (patient). We studied the similarity between the two movements by analyzing the limb motion signals, acquired using the developed electro-goniometers which are worn from both the therapist and patient. We compared the angular signals calculating the maximum of the normalized cross-correlation coefficient ($\sigma$) as reported in the following formula:

$$\sigma = max(\sigma_{th\_pt,coeff}(m)) = \frac{1}{\sigma_{th\_th}(0) * \sigma_{pt\_pt}(0)} * \sigma_{th\_pt}(m) \tag{1}$$

where $m$ is the lag between the signals (th, pt), and the autocorrelation product normalization limits $\sigma$ values to 1 (perfect match between signals) and –1 (complete opposite signals).

We enrolled 11 healthy subjects, whose have been submitted their informed consent for our testing protocol (approved by the Bio-ethical Committee of the Università degli Studi di Torino, Italy), and we divided them into therapist-patient couples. In the next sections, we introduce the adopted methodologies for choosing electrode type and for preparing the subjects skin, and we report our complete results for upper limb exercise and the preliminary one for the lower limb exercise.

### 4.1. Electrodes and Skin Preparation

A proper treatment of the electrode-skin interface is essential in order to enhance the signal acquisition quality. Therefore, cleaning the skin surface with medical alcohol allows a removal of fat, dust and dead cell, and an increasing of the conductivity through the electrode [43]. We chose the Kendall™ Covidien H124G ⌀24 mm [44] for the sEMG signal acquisition due to the Ag/AgCl sensor, pre-gelled surface and long-term stability. Instead, for the stimulation, we employed the 5 cm × 9 cm RehaTrode [45] rectangular self-adhesive electrodes produced by the Hasomed®, which are perfectly designed to be coupled with the RehaStim 2 stimulator. The main difference between these two types regards the working area, having the acquisition electrodes a higher spatial resolution while the stimulation ones cover a bigger surface to properly induce the stimulation.

### 4.2. Upper Limb: Elbow Flexion

As upper limb benchmark exercise, we chose the Elbow Flexion (EF) movement, which consists in the forearm motion toward the upper arm rotating around the elbow join center. The active muscles of the arm are the brachialis, which attaches the humerus to the ulna, the brachioradialis, that connects humerus and radio, and the biceps brachii, which links the shoulder blade to the radius [46]. Since our idea was to perform this first tests with minimal complexity, we decided to monitor and to stimulate only the biceps brachii also due to its accessibility by surface electrodes. Therefore, we placed the couple of acquisition electrodes at 1/3 of the line between the fossa cubit and the medial acromion, with 20 mm inter-electrode distance, and the reference one on the back of the hand, as electric-neutral area [43]. In contrast, the FES electrodes position slightly differs from the previous ones, being located one on muscle belly and the other one closer to the crease of the elbow [47], in order to have a correct muscle fibers contraction. The experimental setup is shown in Figure 11a,b.

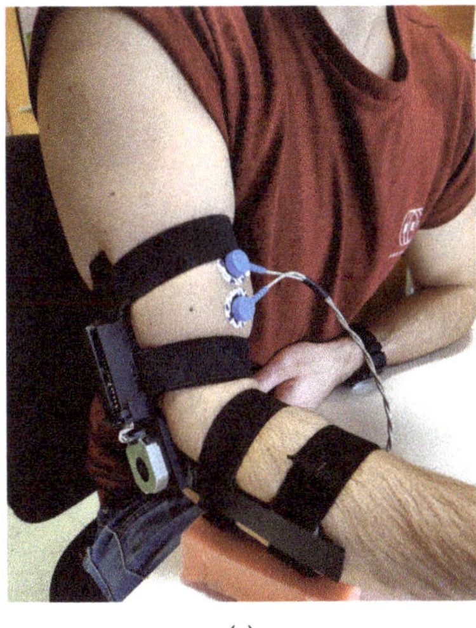 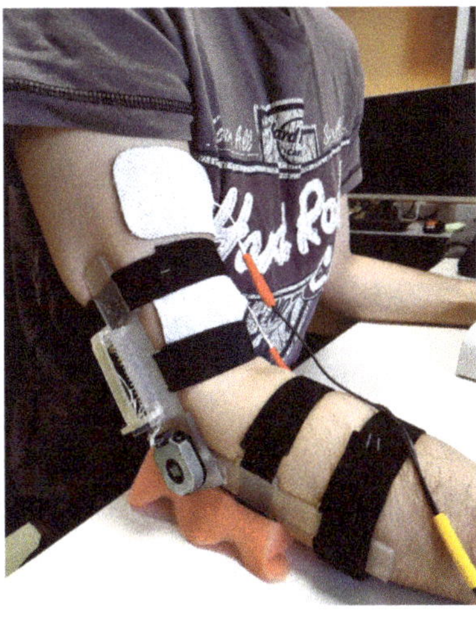

(a)        (b)

**Figure 11.** Acquisition (**a**) and stimulation (**b**) electrodes location, on the biceps femoris muscle, for the EF exercise.

Ten healthy subjects (five males and five females, 24–27 years old) took part to the testing phase: We divided them into five therapist-patient couples and, after the calibration of the acquisition and FES parameters, we asked them to repeat the EF exercise twelve times for each couple. A single repetition has to follow this flow: The starting position for both the therapist and patient is upright sitting, with their forearms and hands completely lean against the table, forming a 90° angle with the upper arm; then, the therapist performs the movement reaching her/him AROM, and finally returns to the starting position; once also the patient has finished the exercise, a short pause of at least 10 s prevents any muscle fatigue effects.

ATC, FES current, therapist and patient angular values have been collected during the entire session. They are successively processed in the MATLAB® environment in order to extract the useful information for the comparison between the voluntary and stimulated movement. The angular signal processing consists of the following steps:

1. Segmentation of the complete signals into single epochs representing one repetition.
2. Baseline removal, since it could be different depending on the limb starting position.
3. Signals normalization to the related AROM values.
4. Computing of the maximum of the cross-correlation coefficient for each epoch.

The box plot on the left of Figure 12 represents the entire dataset of $\sigma$ values (60 measures: Five couples per 12 repetitions each one) extracted during our test campaign. As it can be observed, the distribution is $Q_3$-skewered to the unity, which indicates a good reproduction of the movement, further confirmed by a median value above 0.8. Indeed, looking at the angular signals on the right graph of Figure 12, representing a single repetition, we can see how much the limb motion is similar between therapist and patient. Moreover, this graph also shows the on-line modulation of the stimulation current when the ATC values, directly proportional to the therapist limb angle, trigger the increasing, decreasing or plateau current phases. It is possible to notice that the total delay between the two movements is due to a first short processing phase, visible as distance between non-zero therapist angle and non-zero FES current, and a physiological longer one, distance between non-zero FES current and non-zero patient angle, which depends on the muscle mass and fibers contractions.

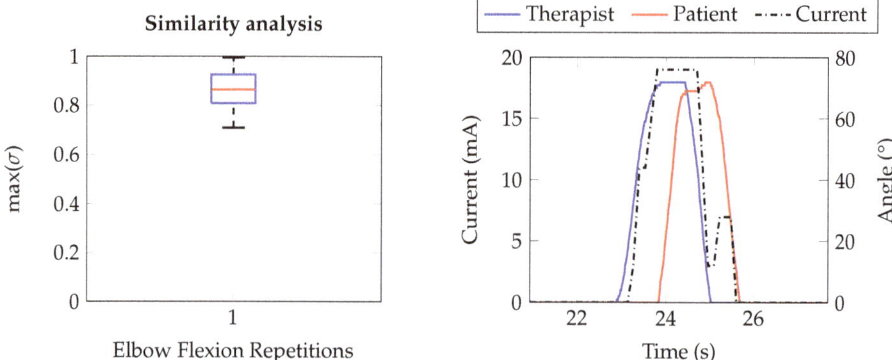

**Figure 12.** (**left**) Example of the stimulation application of a repetition of the Elbow Flexion (EF) movement: Blue and red are the angular signals of the therapist and patient, respectively; the dashed black line represents the FES current injected through the electrodes. (**right**) Similarity analysis using the maximum of the cross-correlation coefficient to compare the limb motion angular signals of the therapist and patient for the EF movement.

### 4.3. Lower Limb: Knee Extension

We also tried to replicate the Knee Extension (KE) movement, due to its largely employment as physiotherapy exercise. From a sitting initial position, the contraction of the quadriceps femoris muscle allows the extension of the leg with respect to the knee joint. This muscle is composed by four separate muscles: The rectus femoris, in the middle of the thigh; the vastus lateralis located on the lateral side of the femur; the vastus medialis on the medial side; and the vastus intermedius under the rectus femoris [48].

In our test, we decided to monitor the vastus lateralis and vastus medialis, setting up a two-channel stimulation layout: in the first case, the sEMG electrodes were placed at the 80% of the line from the anterior superior iliac spine and the medial side of the platella, while in the second case, they were put at 2/3 of the line connecting the anterior superior iliac spine with the superior lateral side of the patella [43], as shown in Figure 13a. Both reference electrodes were located on the patella. On the other hand, as reported in Figure 13b, the stimulation electrodes were placed along the muscle bellies in order to cover a surface including both the muscles and the rectus femoris.

Our preliminary results consist in 13 repetitions of the movement, performed by a single female subject (24 years old). As in the EF case, both the therapist and patient need to start from an initial position, which we defined as 135° between thigh and calf. Then, the therapist extends its leg until her/him AROM and, once the stimulation is completed, at least 10 s have to be waited before next repetition.

The cross-correlation results, calculated with the same method of the EF case, are reported on the box plot in Figure 14 (top left). However, these values are also represented by a time-graph (bottom left) in order to avoid any misinterpretations due to the low number of measures. Looking at the graphs, some considerations can be made: first of all, also for the KE movement, we obtained satisfactory results in term of similarity between the two signals, proved by the $\sigma$ values completely equal or greater to 0.9. Indeed, considering the repetition example on the left graph, we can observe the similar morphology among the therapist motion and the patient one. Anyway, the maximal angular value reached by the patient is lower than the therapist one. This behavior is possibly related to the muscle physiology activation and different fibers recruitment between voluntary and stimulated contraction. One cause could be associated to the stimulation of a healthy subject, with a normal muscles condition, that, by applying large values of stimulating current, could lead to sense of pain. Hence, we limited the current to the values represented by the dashed line in the graph avoiding this situation. A second possibility could be related to which muscles have been stimulated: A complete leg extension involves the total contraction of the quadriceps, while superficial electrodes could not result in the proper shortening of the deeper fibers, consequently producing a limited movement. Anyway, further studies will allow us to set up more complex stimulation scenarios, which will induce a better reproduction of the movement.

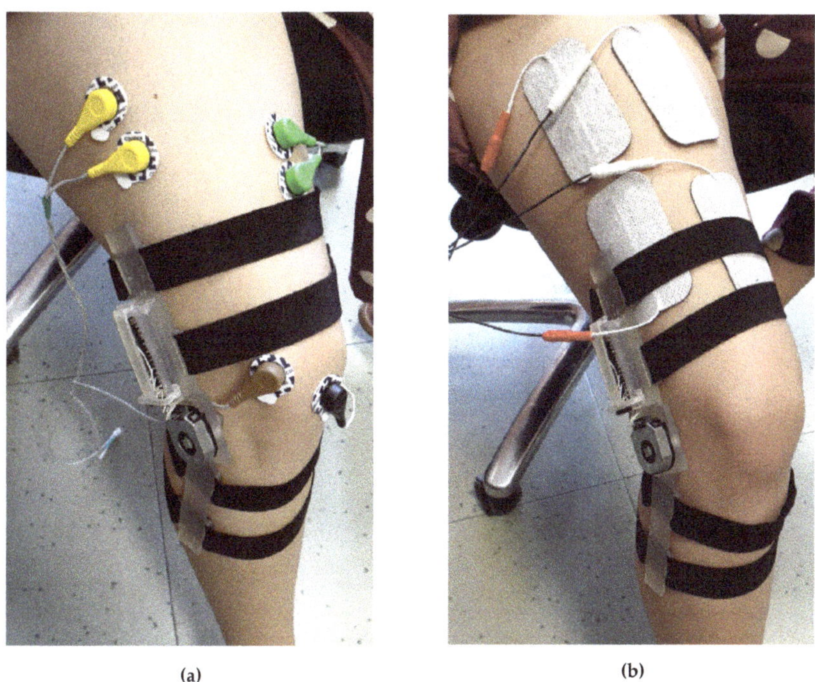

(a)        (b)

**Figure 13.** Knee extension exercise. (**a**) sEMG acquisition electrodes on the vastus lateralis and vastus medialis muscles. (**b**) the electrodes are directly placed on the muscle bellies of the vastus lateralis and vastus medialis. This locations and the electrodes dimension also contract the rectus femoris muscle improving the stimulation effectiveness.

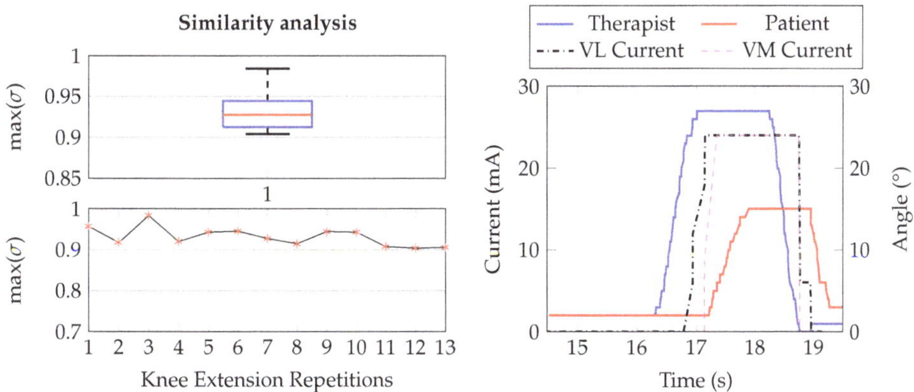



**Figure 14.** (**left**) Single repetition example showing the similarity between the two angular signals with respect to the vastus lateralis (VL) and vastus medialis (VM) stimulation currents. (**right**) Maximum of the cross-correlation coefficient for the 13 repetitions of the knee extension movement.

## 5. Discussion: sEMG-FES Systems Comparison

Table 6 reports a summary of literature works in the field of FES application triggered by sEMG signal analysis. The classification includes which control feature (e.g., RMS, envelope, ATC) has been used to online modulate one or more FES parameters, as well as the employed hardware, which summarizes the processing capability and the possibility to transfer the FES algorithm into an embedded device. Moreover, these systems could be also analyzed by considering additional features such as wireless connectivity, modularity and number of active channels which foster system wearability, future sensors and algorithms integration, and application typology. Lastly, since real-time behavior remains a major constraint, rightmost column shows the latency (FES pulses update period) measurements calculated as FES processing delay or (whenever available) the therapist-patient delay.

**Table 6.** sEMG-trigger-FES systems table comparison.

| Work | Control Feature | FES Parameter | Processing HW | Embedded | Wireless | Modular System | #Ch | Latency (ms) |
|---|---|---|---|---|---|---|---|---|
| [49] | RMS | intensity | MCU | ✓ | Bluetooth | x | 8 | 300 |
| [50] | envelope | intensity | n.a. | x | x | n.a. | 4 | n.a. |
| [51] | threshold crossing | frequency | MCU | ✓ | 335/433 MHz | x | 2 | 142 |
| [52] | force angle | intensity | PC | x | x | n.a. | 4 | n.a. |
| [53] | sEMG IMU | intensity width | MCU | ✓ | Bluetooth 2.1 | x | 4 | 21 |
| [54] | envelope | on/off stimuli | PC | x | x | x | 1 | 1600 |
| [55] | entropy | frequency width | MCU | ✓ | x | x | 1 | 300 [1] |
| [10,18] | ATC | intensity | PC | x | Bluetooth 4.2 | ✓ | 4 | 774.5 [1] 932 [1] |
| **This** | ATC | intensity | Raspberry | ✓ | Bluetooth 4.2 | ✓ | 4 | 140 |

[1] measured as therapist-patient delay.

A complete comparison between our system and those reported here is not straightforward due to the large variety of analyzed features; anyway, some considerations about different methods and performance can be carried out. In [51] authors used the threshold crossing feature extraction to modulate the stimulation frequency of the FES pulses, achieving a very promising FES definition latency of 142 ms, directly comparable to our outcomes. However, since the sEMG processing has been embedded in the MCU, a standard sampling approach is needed, which includes peripherals management and relative expensive processing power; on the other hand, with our event-based approach, MCU resources could be drastically reduced by implementing ATC in hardware. At the same time, linking the TC events with the stimulation frequency results particularly interesting in order to reproduce motor fibers firing rate; in our system we preferred, as a first step, to modulate the intensity, but a frequency-control approach could be easily implemented thanks to the flexible and modular architecture of our system. Another interesting frequency modulation has been presented in [54], which evaluates the sEMG entropy on an MCU architecture, but limiting the number of controlled channels to one.

With reference to the pulse amplitude control, it could be performed by extracting different features from the muscle signal (e.g., RMS [49], envelope [50] and force [52]) or using data-fusion techniques with different type of sensor (as Inertial Measurement Unit, IMU [53]): While the final latency among these works and our proposed system is quite similar, and respects the real-time constraints for such type of application, the processing methodology does not always allow to use an MCU [53] and, where it is possible, raw data acquisition seems to be the common adopted solution.

As another comparison point, to the best of our knowledge for this application, our architecture is the only one which presents a modular system structure, thanks to the combination of chosen programming language and strategies.

Lastly, looking at the ATC-FES system evolution from previous versions [10,18], we enhanced the real-time performance obtaining a total latency of about 140 ms, defined as the sum of the processing time (10 ms, RHS configuration) and the ATC widow (130 ms). Again, since the relevant upgrade in our architecture was to move towards an embedded device (i.e., Raspberry Pi), we confirm how the lightness and low-complexity of the ATC technique perfectly match with the low-processing capability of an embedded system, while maintaining adequate FES control, usability and power performance.

## 6. Conclusion and Future Perspectives

In this paper we presented our last prototype of the sEMG(ATC)-controlled-FES system, in which we have been replaced the previous MATLAB® & SIMULINK® software architecture [10] with the novel embedded version running on a Raspberry Pi, in order to overcome the performances limitation due to the use of a general purpose computer. Taking the advantages of the object-oriented and multi-threaded approach, along with the versatility of Python programming language, we developed a multi-platform software core able to work onto several devices and with different operating systems, also enhancing system usability by featuring a graphical user interface.

Since the main tasks of the system application concern the modulation of the FES pattern and its real-time application, we designed a processing structure able to match with the low computational power of an embedded device. We implemented a calibrated ATC-FES lookup table structure, combined with noise-gateway controls, which allows us to obtain a total processing time, defined as the delay between ATC data and the new FES parameters, below 30 ms (corner cases), obtained without substantially impact on the CPU and RAM usage, therefore demonstrating the lightness and responsiveness of the event-driven technique in the control of the stimulation.

We proved system efficiency by studying the similarity between voluntary and stimulated movements in therapist-patient real FES scenarios (healthy subjects): using the maximum of the normalized cross correlation coefficient, as comparison measurement between the signals of the involved limbs, we obtained a mean value of $0.87 \pm 0.07$ as result of 60 repetitions (5 therapist-patient couples per 12 repetition each one) during the reproduction of the elbow flexion movement. These promising outcomes allowed us also to preliminarily evaluate the FES performance for the knee extension movement: Adopting the same methodologies, we analyzed 13 exercise repetitions achieving a correlation value of $0.93 \pm 0.02$.

Future improvements, i.e., full FES parameters modulation, pre-FES movement recognition, will permit us to further optimize the ATC-based stimulation in order to extend our testing phase to a wide list of rehabilitation exercises, also performing some clinical trials with the support of medical staff.

**Author Contributions:** Investigation, R.M.R.; Project administration, D.D.; Supervision, P.M.R.; Writing—original draft, F.R. All authors have read and agreed to the published version of the manuscript.

**Funding:** This research received no external funding.

**Acknowledgments:** The authors would like to thanks the volunteers which kindly offered for the experimental tests useful for this work.

**Conflicts of Interest:** The authors declare no conflict of interest.

## References


1. Ferrante, S.; Chia Bejarano, N.; Ambrosini, E.; Nardone, A.; Turcato, A.M.; Monticone, M.; Ferrigno, G.; Pedrocchi, A. A Personalized Multi-Channel FES Controller Based on Muscle Synergies to Support Gait Rehabilitation after Stroke. *Front. Neurosci.* **2016**, *10*, 425. doi:10.3389/fnins.2016.00425. [CrossRef] [PubMed]
2. Perruchoud, D.; Pisotta, I.; Carda, S.; Murray, M.M.; Ionta, S. Biomimetic Rehabilitation Engineering: The Importance of Somatosensory Feedback for Brain–Machine Interfaces. *J. Neural Eng.* **2016**, *13*, 041001. doi:10.1088/1741-2560/13/4/041001. [CrossRef] [PubMed]
3. Reilly, J.P. Electrical Stimulation and Electropathology. In *Electrical Stimulation and Electropathology*, 1st ed.; Press, C.U., Ed.; Press Syndicate of the University of Cambridge: Cambridge, UK, 1992; Chapter 8: Skeletal Muscle Response to Electrical Stimulation by Sweeney, J. D.
4. Toledo, C.; Martinez, J.; Mercado, J.; Isabel Martín-Vignon-Whaley, A.; Vera-Hernández, A.; Leija-Salas, L. sEMG Signal Acquisition Strategy towards Hand FES Control. *J. Healthcare Eng.* **2018**, *2018*, 1–11. doi:10.1155/2018/2350834. [CrossRef] [PubMed]
5. Li, Z.; Guiraud, D.; Andreu, D.; Benoussaad, M.; Fattal, C.; Hayashibe, M. Real-time Estimation of FES-induced Joint Torque with Evoked EMG. *J. NeuroEng. Rehabil.* **2016**, *13*, 60. [CrossRef] [PubMed]
6. Shahshahani, A.; Shahshahani, M.;Motto Ros, P.; Bonanno, A.; Crepaldi, M.; Martina M; Demarchi, D.; Masera, G. An all-digital spike-based ultra-low-power IR-UWB dynamic average threshold crossing scheme for muscle force wireless transmission. In Proceedings of the 2015 Design, Automation & Test in Europe Conference & Exhibition (DATE), Grenoble, France, 9–13 March 2015; pp. 1479–1484.
7. Crepaldi, M.; Paleari, M.; Bonanno, A.; Sanginario, A.; Ariano, P.; Tran, D.H.; Demarchi, D. A quasi-digital radio system for muscle force transmission based on event-driven IR-UWB. In Proceedings of the 2012 IEEE Biomedical Circuits and Systems Conference (BioCAS), Hsinchu, Taiwan, 28–30 November 2012; pp. 116–119. doi:10.1109/BioCAS.2012.6418406. [CrossRef]
8. Sapienza, S.; Crepaldi, M.; Motto Ros, P.; Bonanno, A.; Demarchi, D. On Integration and Validation of a Very Low Complexity ATC UWB System for Muscle Force Transmission. In Proceedings of the IEEE Transactions on Biomedical Circuits and Systems, Paphos, Cyprus, 31 March–2 April 2016; Volume 10, pp. 497–506. doi:10.1109/TBCAS.2015.2416918. [CrossRef]
9. Guzman, D.; Sapienza, S.; Sereni, B.; Motto Ros, P. Very Low Power Event-Based Surface EMG Acquisition System with Off-the-Shelf Components. In Proceedings of the 2017 IEEE Biomedical Circuits and Systems Conference (BioCAS), Turin, Italy, 19–21 October 2017.
10. Rossi, F.; Motto Ros, P.; Sapienza, S.; Bonato, P.; Bizzi, E.; Demarchi, D. Wireless Low Energy System Architecture for Event-Driven Surface Electromyography. In *ApplePies 2018: Applications in Electronics Pervading Industry, Environment and Society*; Springer: New York, NY, USA, 2019; pp. 179–185. doi:10.1007/978-3-030-11973-7_21. [CrossRef]
11. Motto Ros, P.; Paleari, M.; Celadon, N.; Sanginario, A.; Bonanno, A.; Crepaldi, M.; Ariano, P.; Demarchi, D. A wireless address-event representation system for ATC-based multi-channel force wireless transmission. In Proceedings of the 5th IEEE International Workshop on Advances in Sensors and Interfaces IWASI, Bari, Italy, 13–14 June 2013; pp. 51–56. doi:10.1109/IWASI.2013.6576061 [CrossRef]
12. Sapienza, S.; Motto Ros, P.; Fernandez Guzman, D.A.; Rossi, F.; Terracciano, R.; Cordedda, E.; Demarchi, D. On-Line Event-Driven Hand Gesture Recognition Based on Surface Electromyographic Signals. In Proceedings of the 2018 IEEE International Symposium on Circuits and Systems (ISCAS), Florence, Italy, 27–30 May 2018; pp. 1–5.
13. Mongardi, A.; Rossi, F.; Motto Ros, P.; Sanginario, A.; Ruo Roch, M.; Martina, M.; Demarchi, D. Live Demonstration: Low Power Embedded System for Event-Driven Hand Gesture Recognition. In Proceedings of the 2019 IEEE Biomedical Circuits and Systems Conference (BioCAS), Nara, Japan 17–19 October 2019.
14. Mongardi, A.; Motto Ros, P.; Rossi, F.; Ruo Roch, M.; Martina, M.; Demarchi, D. A Low-Power Embedded System for Real-Time sEMG based Event-Driven Gesture Recognition. In Proceedings of the 2019 26th IEEE International Conference on Electronics, Circuits and Systems (ICECS), Genova, Italy, 27–29 November 2019; pp. 65–68.
15. Barioul, R.; Fakhfakh, S.; Derbel, H.; Kanoun, O. Evaluation of EMG Signal Time Domain Features for Hand Gesture Distinction. In Proceedings of the 2019 16th International Multi-Conference on Systems, Signals Devices (SSD), Istanbul, Turkey, 21–24 March 2019; pp. 489–493.

16. Toledo-Pérez, D.C.; Rodríguez-Reséndiz, J.; Gómez-Loenzo, R.A. A Study of Computing Zero Crossing Methods and an Improved Proposal for EMG Signals. *IEEE Access* **2020**, *8*, 8783–8790. [CrossRef]

17. Rossi, F.; Motto Ros, P.; Demarchi, D. Live Demonstration: Low Power System for Event-Driven Control of Functional Electrical Stimulation. In Proceedings of the 2018 IEEE International Symposium on Circuits and Systems (ISCAS), Florence, Italy, 27–30 May 2018; p. 1.

18. Rossi, F.; Motto Ros, P.; Cecchini, S.; Crema, A.; Micera, S.; Demarchi, D. An Event-Driven Closed-Loop System for Real-Time FES Control. In Proceedings of the 2019 26th IEEE International Conference on Electronics, Circuits and Systems (ICECS), Genova, Italy, 27–29 November 2019; pp. 867–870.

19. Rossi, F.; Rosales, M.R.; Motto Ros, P.; Demarchi, D. Real-Time Embedded System for Event-Driven sEMG Acquisition and Functional Electrical Stimulation Control. In Proceedings of the 2019 International Conference on Application in Electronics Pervading Industry, Environment and Society, Pisa, Italy, 11–13 September 2019.

20. De Luca, C.J. *Surface Electromyography: Detection and Recording*; DelSys Incorporated: Natick, MA, USA, 2002.

21. De Luca, G. *Fundamental Concepts in EMG Signal Acquisition*; DelSys Incorporated: Natick, MA, USA, 2001.

22. Spinelli, E.M.; Pallas-Areny, R.; Mayosky, M.A. AC-coupled front-end for biopotential measurements. *IEEE Trans. Biomed. Eng.* **2003**, *50*, 391–395. [CrossRef] [PubMed]

23. Franco, S. *Design with Operational Amplifiers and Analog Integrated Circuits*; Mac Graw Hill: New York, NY, USA, 2002.

24. (SIG), B.S.I.G. *BLUETOOTH SPECIFICATION Version 4.1 - Specification of the Bluetooth® System*. Available online: https://www.microchip.com/ (accessed on 3 December 2013).

25. Microchip. *RN4020 Bluetooth Low Energy Module DataSheet*. 2015. pp. 1–28. Available online: https://www.microchip.com/ (accessed on 3 December 2013).

26. Semiconductor, N. nRF52840 Dongle. Available online: https://www.nordicsemi.com/Software-and-tools/Development-Kits/nRF52840-Dongle (accessed on 9 March 2020).

27. Devices, C. *AMT20 Series Datasheet - Modular | Absolute | CUI Devices*. 2019. pp. 1–10. Available online: https://www.cuidevices.com/ (accessed on 21 December 2019).

28. Arduino Micro. Available online: https://store.arduino.cc/arduino-micro (accessed on 9 March 2020).

29. Jessop, D.; Pain, M. Maximum Velocities in Flexion and Extension Actions for Sport. *J. Human Kinet.* **2016**, *50*. doi:10.1515/hukin-2015-0139. [CrossRef] [PubMed]

30. Formlabs. Available online: https://formlabs.com/it/3d-printers/form-2/ (accessed on 9 March 2020).

31. GmbH, H. *Operation Manual RehaStim 2, RehaMove 2 - User Guide*. Available online: https://hasomed.de/ (accessed on 21 November 2012).

32. Kuberski, B. *ScienceMode2 - Description and Protocol*. Available online: https://hasomed.de/ (accessed on 12 December 2012).

33. Popovic, M.; Masani, K.; Micera, S. Functional Electrical Stimulation Therapy: Recovery of Function Following Spinal Cord Injury and Stroke. In *Neurorehabilitation Technology*; Springer: Cham, Switzerland, 2016.

34. GmbH, H. *RehaMove Functional Electrical Stimulation—FES Applications*. 2015. Available online: https://hasomed.de/ (accessed on 9 March 2020).

35. Foundation, R.P. Raspberry Pi 3 B+ Specifications. Available online: https://www.raspberrypi.org/products/raspberry-pi-3-model-b-plus/ (accessed on 9 March 2020).

36. Instruments, T. *CC2540 Bluetooth Low Energy USB Dongle*, 5 May 2015. Available online: http://www.ti.com/lit/ug/tidu977/tidu977.pdf (accessed on 9 March 2020).

37. Kivy. *Kivy Home*. Available online: https://kivy.org/#home (accessed on 9 March 2020)

38. Philips, D. *Python 3 Object-oriented Programming*, 2nd ed.; Packt Publishing: Birmingham, UK, 2015.

39. Lott, S. *Functional Python Programming*; Packt Publishing: Birmingham, UK, 2015.

40. BlueZ. BlueZ: Official Linux Bluetooth protocol stack. Available online: http://www.bluez.org/about/ (accessed on 9 March 2020).

41. Harvey, I. Bluepy: Python interface to Bluetooth LE on Linux. Available online: https://github.com/IanHarvey/bluepy (accessed on 9 March 2020).

42. Ulloa, R. *Kivy - Interactive Applications and Games in Python*, 2nd ed; Puckt Publishing: Birmingham, UK, 2015.

43. Stegeman, D.; Hermens, H. Standards for Suface Electromyography: The European Project Surface EMG for Non-Invasive Assessment of Muscles (SENIAM). 2007. Available online: https://www.researchgate.net/publication/228486725_Standards_for_suface_electromyography_The_European_project_Surface_EMG_for_non-invasive_assessment_of_muscles_SENIAM (accessed on 9 March 2020).

44. Covidien. *Kendall*™ *ECG Electrodes Product Data Sheet - Arbo*™ *H124SG*. Available online: https://www.medtronic.com/ (accessed on 9 March 2020).

45. FES STANDARD (5CM X 9CM) REHATRODE ELECTRODES. Available online: https://www.fescycling.com/shop/fes-rehatrode-electrodes-standard (accessed on 9 March 2020).

46. Kleiber, T.; Kunz, L.; Disselhorst-Klug, C. Muscular coordination of biceps brachii and brachioradialis in elbow flexion with respect to hand position. *Front. Physiol.* **2015**, *6*, 215. doi:10.3389/fphys.2015.00215. [CrossRef] [PubMed]

47. LTD, A.M.C. Education: Electrode Placement. Available online: https://www.axelgaard.com/Education/Elbow-Flexion (accessed on 9 March 2020).

48. Bordoni, B.; Varacallo, M. Anatomy, Bony Pelvis and Lower Limb, Thigh Quadriceps Muscle. Available online: https://www.axelgaard.com/Education/Elbow-Flexion (accessed on 9 March 2020).

49. Zhou, Y.; Fang, Y.; Zeng, J.; Li, K.; Liu, H. A Multi-channel EMG-Driven FES Solution for Stroke Rehabilitation. In Proceedings of the 11th International Conference Intelligent Robotics and Applications (ICIRA), Newcastle, NSW, Australia, 9–11 August 2018; pp. 235–243. doi:10.1007/978-3-319-97586-3_21. [CrossRef]

50. Fang, Y.; Chen, S.; Wang, X.; Leung, K.W.C.; Wang, X.; Tong, K. Real-time Electromyography-driven Functional Electrical Stimulation Cycling System for Chronic Stroke Rehabilitation. In Proceedings of the 2018 40th Annual International Conference of the IEEE Engineering in Medicine and Biology Society (EMBC), Honolulu, HI, USA, 18–21 July 2018; pp. 2515–2518. doi:10.1109/EMBC.2018.8512747. [CrossRef]

51. Wang, H.P.; Bi, Z.Y.; Zhou, Y.; Zhou, Y.; Wang, Z.G.; Lv, X.Y. Real-time and wearable functional electrical stimulation system for volitional hand motor function control using the electromyography bridge method. *Neural Regener. Res.* **2017**, *12*, 133. doi:10.4103/1673-5374.199216. [CrossRef]

52. Shima, K.; Shimatanu, K. A new approach to direct rehabilitation based on functional electrical stimulation and EMG classification. In Proceedings of the 2016 International Symposium on Micro-NanoMechatronics and Human Science (MHS), Nagoya, Japan, 28–30 November 2016; pp. 1–6. doi:10.1109/MHS.2016.7824200. [CrossRef]

53. Valtin, M.; Kociemba, K.; Behling, C.; Kuberski, B.; Becker, S.; Schauer, T. RehaMovePro: A versatile mobile stimulation system for transcutaneous FES applications. *Eur. J. Transl. Myol.* **2016**, *26*, 3. doi:10.4081/ejtm.2016.6076. [CrossRef] [PubMed]

54. Toledo Peral, C.L.; Nava, G.H.; Melía Licona, J.A.; Mercado Gutiérrez, J.A.; Aguirre Guemez, A.V.; Fresnedo, J.Q.; Hernández, A.V.; Salas, L.L.; Martínez, J.G. ON/OFF sEMG Switch for FES Activation. In Proceedings of the 2019 16th International Conference on Electrical Engineering, Computing Science and Automatic Control (CCE), Mexico City, Mexico, 11–13 September 2019; pp. 1–5. [CrossRef]

55. Bi, Z.; Bao, X.; Wang, H.; Lv, X.; Wang, Z. Prototype System Design and Experimental Validation of Gait-Oriented EMG Bridge for Volitional Motion Function Rebuilding of Paralyzed Leg. In Proceedings of the 2019 IEEE 7th International Conference on Bioinformatics and Computational Biology (ICBCB), Hangzhou, China, 21–23 March 2019; pp. 79–82. doi:10.1109/ICBCB.2019.8854632. [CrossRef]

*Article*

# Fast Approximations of Activation Functions in Deep Neural Networks when using Posit Arithmetic

**Marco Cococcioni [1], Federico Rossi [1], Emanuele Ruffaldi [2] and Sergio Saponara [1,***

[1] Department of Information Engineering, Università di Pisa, Via Girolamo Caruso, 16, 56122 Pisa PI, Italy; marco.cococcioni@unipi.it (M.C.); federico.rossi@ing.unipi.it (F.R.)
[2] Medical Microinstruments (MMI) S.p.A., Via Sterpulino, 3, 56121 Pisa PI, Italy; emanuele.ruffaldi@mmimicro.com
* Correspondence: sergio.saponara@unipi.it

Received: 25 January 2020; Accepted: 2 March 2020; Published: 10 March 2020

**Abstract:** With increasing real-time constraints being put on the use of Deep Neural Networks (DNNs) by real-time scenarios, there is the need to review information representation. A very challenging path is to employ an encoding that allows a fast processing and hardware-friendly representation of information. Among the proposed alternatives to the IEEE 754 standard regarding floating point representation of real numbers, the recently introduced Posit format has been theoretically proven to be really promising in satisfying the mentioned requirements. However, with the absence of proper hardware support for this novel type, this evaluation can be conducted only through a software emulation. While waiting for the widespread availability of the Posit Processing Units (the equivalent of the Floating Point Unit (FPU)), we can already exploit the Posit representation and the currently available Arithmetic-Logic Unit (ALU) to speed up DNNs by manipulating the low-level bit string representations of Posits. As a first step, in this paper, we present new arithmetic properties of the Posit number system with a focus on the configuration with 0 exponent bits. In particular, we propose a new class of Posit operators called L1 operators, which consists of fast and approximated versions of existing arithmetic operations or functions (e.g., hyperbolic tangent (TANH) and extended linear unit (ELU)) only using integer arithmetic. These operators introduce very interesting properties and results: (i) faster evaluation than the exact counterpart with a negligible accuracy degradation; (ii) an efficient ALU emulation of a number of Posits operations; and (iii) the possibility to vectorize operations in Posits, using existing ALU vectorized operations (such as the scalable vector extension of ARM CPUs or advanced vector extensions on Intel CPUs). As a second step, we test the proposed activation function on Posit-based DNNs, showing how 16-bit down to 10-bit Posits represent an exact replacement for 32-bit floats while 8-bit Posits could be an interesting alternative to 32-bit floats since their performances are a bit lower but their high speed and low storage properties are very appealing (leading to a lower bandwidth demand and more cache-friendly code). Finally, we point out how small Posits (i.e., up to 14 bits long) are very interesting while PPUs become widespread, since Posit operations can be tabulated in a very efficient way (see details in the text).

**Keywords:** alternative representations to float numbers; posit arithmetic; Deep Neural Networks (DNNs); neural network activation functions

---

## 1. Introduction

Due to the pervasivenss of real-time and critical systems like Internet of Things (IoT) platforms, automotives, and robotics, new types of requirements are being addressed in the use of Deep Neural Networks (DNNs).

The main challenges when dealing with DNNs are both the ubiquitous multiply-and-accumulate operations and the massive use of activation functions across the neural network layers. A big speed-up to these challenges is surely offered by parallelization of the workloads (e.g., Graphics Processing Units (GPUs) or Single-Instruction Multiple-Data (SIMD)) processors). However, these solutions are considerable demanding in terms of resources. Moreover, adding parallelization in critical systems may reduce the predictability of the said system (see References [1,2]). Furthermore, even the use of floating point SIMD engines is not always possible in embedded systems (e.g., ARM Cortex-M4 [3]). This means that we cannot always rely on high-performance processing units in critical and real time scenarios, thus needing to address new challenges.

Therefore, the challenging topic is to satisfy the real-time requirements while guaranteeing computational efficiency and lowering the power and the cost of such applications. One of the main paths to reduce the computational complexity when evaluating DNNs is stepping away from cumbersome arithmetic such as double-precision floats (represented on 64 bit). The basic idea is to use compressed formats that may save resources in terms of power consumption and computational efficiency. Great examples of compact formats are Brain Floats (BFLOAT) and Flexpoint [4,5] that consist in an optimized version of the 16-bit standard floating point number IEEE 754) used by Google for their TPU (tensor processing unit) engines. Other formats also come from the concept of transprecision computing [6,7] (NVIDIA Turing architectures allow computation with 4-, 8-, and 32-bit integers and with 16- and 32-bit floats). The up-and-coming Posit format has been theoretically [8–10] and practically [11] proven to be a perfect replacement for IEEE float numbers when applied to DNNs in terms of efficiency and accuracy.

Due to its novelty, this format lacks proper or standardized hardware support (e.g., a Posit Processing Unit (PPU)) to accelerate its computation, forcing the use of software implementations. However, in order to speed up the software emulation of Posits in DNNs, we present two different techniques. In this paper, we extend the work on deriving a fast and approximate version of the hyperbolic tangent (TANH) presented in Reference [12]. We introduce novel arithmetic properties of the Posit number system with a deep focus on the Posits with 0 exponent bits. This special case allows us to build common functions and arithmetic operators as simple bit manipulations on the bit-string representing a Posit number. This new class of functions (called L1 functions) has some interesting properties:

- The operation evaluation is faster than its counterpart with little to no degradation on accuracy.
- Since these operations only need integer arithmetic and logic operations, they are straightforwardly executed in the already existing ALU, also allowing a faster emulation of Posits
- Being able to write functions as a sequence of arithmetic-logic operations allows us to vectorize them exploiting already existing SIMD (Single Instruction–Multiple Data) engines.

In particular, in this extension, we also propose a new fast and approximated version of the Extended Linear Unit (ELU) activation function.

Moreover, if we consider really low-power devices that do not embed a floating point unit but only an arithmetic logic unit, the approach proposed can become very interesting to enable DNN processing even in this class of devices (although for inference only, not for training).

Furthermore, we investigate operator tabulation as a different approach to speed up Posit emulation without constraints on the exponent configuration. This allows us to accelerate basic arithmetic operators like sum and multiplication that are not suitable for being implemented as L1 functions. Although very powerful, this approach has clear limitations to its scalability, having a considerable spatial complexity.

*Paper Structure*

The paper is organized as follow: Section 2 introduces the Posit format, proposing novel approaches to approximation and speed-up of Posit arithmetic, exploiting the 0-bit exponent Posit configuration. Section 3 describes the cppPosit library implemented in Pisa for the computation of the new numerical format. Section 4 introduces the hyperbolic tangent and ELU activation functions along with their approximations. Section 5 shows the results of our approach with DNN and common benchmarking datasets. Finally, Section 6 provides the conclusions.

## 2. Posit Arithmetic

The Posit format has been introduced by John L. Gustafson in Reference [8] and was further investigated in Reference [9,10,12]. The format is a fixed-length one with up to 4 fields as also reported in Figure 1:

- Sign field: 1-bit
- Regime field: variable length, composed of a string of bits equal to 1 or 0 ended respectively by a 0 or 1 bit.
- Exponent field: at most *es* bits
- Fraction field: variable length mantissa

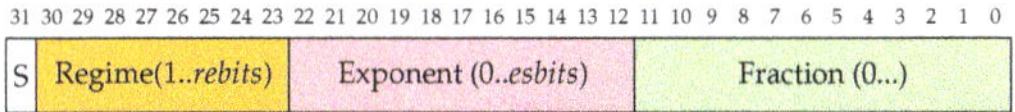

**Figure 1.** Illustration of the of 32-bit Posit data type.

Given a Posit on *nbits*; *esbits*, represented by the integer $X$, and $e$ and $f$ respectively as the exponent and fraction values, the real number $r$ represented by that encoding is as follows:

$$r = \begin{cases} 0, \text{ if } X = 0 \\ \text{NaN, if } X = -2^{(nbits-1)} \\ sign(X) \times useed^k \cdot 2^e \cdot (1 + f), \text{ otherwise} \end{cases}$$

An example of Posit decoding operation is shown in Figure 2.

15 14 13 12 11 10 9 8 7 6 5 4 3 2 1 0

1 0 0 1 1 1 0 1 1 1 0 1 1 0 0 1

15 14 13 12 11 10 9 8 7 6 5 4 3 2 1 0

| S | R | E | F |
|---|---|---|---|
| 1 | 1 1 0 | 0 0 1 | 0 0 0 1 0 0 1 1 1 |

**Figure 2.** An example of a 16-bit Posit with 3 bits for the exponent (*esbits* = 3): Given the sequence on top of the figure, after detecting that it starts with one 1, we have to compute the 2's complement of all the remaining bits (passing from 001-110-111011001 to 110-001-000100111). Then, we can proceed to decode the Posit. The associated real value is therefore $-256^1 \cdot 2 \cdot (1 + 39/512)$. The final value is therefore $-512 \cdot (1 + 39/512) = -551$ (exact value, i.e., no rounding, for this case).

The design of a hardware Posit Processing Unit (PPU) as a replacement for the FPU has already started on several universities worldwide, but it will take time for their availability on real platforms. Fortunately, we can still do many things related to DNNs even in the absence of a hardware PPU. Furthermore, when DNN weights can be represented with less than 14-bit Posits, we can tabulate some core operations like sum and multiplication (see Section 3.1) and can use the ALU for other operations that will be shown hereafter in order to reduce the number of tables.

As reported above, the process of decoding a Posit involves the following steps: obtaining regime value by reconstructing the bit-string, building exponent, and extracting fraction. We can make use of C low-level building blocks to speed up the decoding:

- Count leading zeros: using the embedded `__builtin_clz` C function that several CPU families provide in hardware [13].

- Next power of two: used to extract the fraction. An efficient way to obtain the next power of two, given a representation $X$ on 32 bit, is the following:

```
next_p2(X) -> Y
      Y = X - 1
      Y = Y | X >> 1
      Y = Y | X >> 2
      Y = Y | X >> 4
      Y = y | X >> 8
      Y = Y | X >> 16
      Y = Y + 1
```

This approach copies the highest set bit to all the lower bits. Adding one to such a string will result in a sequence of carries that will set all the bits from the highest set to the least significant one to 0 and the next (in order of significancy) bit of the highest set to 1, thus producing the next power of two. Let us use an example. Suppose $X = (5)_{10} = (0101)_2$. At the first step, $Y = (0100)_2$. At the second step, $Y = (0100)_2|(0010)_2 = (0110)_2$. At the next step, $Y = (0110)_2|(0001)_2 = (0111)_2$. From now on, $Y$ will remain set to $Y = (0111)_2$. At the last step, $Y = (0111)_2 + (0001)_2 = (1000)_2 = (8)_{10}$, that is the next power of two starting from 5.

### 2.1. The Case of No Exponent Bits (esbits = 0)

When using a Posit configuration with zero exponent bits (*esbits* = 0), some interesting properties arise. In this case, we can express the real number represented by the Posit as follows:

$$x = 2^k \cdot (1 + \phi \cdot 2^{-F}) \tag{1}$$

where $\phi$ is the fraction field and $F$ the fraction length. The value of $k$ depends on the regime length R. In particular, $k = -R$ for $x < 1$ (from now on $x^-$) and $k = R - 1$ for $x >= 1$ (from now on $x^+$). If we denote the bit immediately following the regime bit string (stop-bit) as $\sigma$, we can express the value of $R$ as $R = N - 2 - \sigma_p$, where $\sigma_p$ is the position of the stop-bit in the Posit bit-string. For $x^-$, we can note that, substituting the expression for $F = N - 2 - R$ in (1), we get the following expression:

$$x^- = 2^{-R} + \phi \cdot 2^{-(N-2)} = 2^{N-2} \cdot (2^{-F} + \phi) \tag{2}$$

Moreover, we can link $x^-$ with its representation $X$ using Equation (3), obtaining:

$$x^- = X \cdot 2^{-(N-2)} \tag{3}$$

A particular property emerges with 0-bit exponent Posits when considering the $[0, 1]$ range. In fact, if we plot the resolution (that is the decimal difference between the real numbers represented by two consecutive bit-strings) of a Posit$\langle X, 0 \rangle$ in $[0, 1]$ we obtain the resolution of a fixed-point format. This property is visualized in Figure 3. This is a very important property that will be exploited below.

As we will see below, the novel equations introduced above for the first time play an important role for deriving fast approximation of activation functions in DNNs. Equation (3) says also that a Posit with zero exponent bits can be interpreted as a fixed point number with a shift of $(N - 2)$ bits. This has implications on the accuracy and further operations.

An example of how to exploit the expressions discovered in the previous section is building a fast approximated inversion operator. Given $x$, we want to find a fast and efficient way to compute $y$ such that $x \cdot y \approx 1$. In the following, we will consider only positive values of x. The simplest case is when $f - 0$. Let us consider $x > 1$; we simply need to apply a reduction of the regime length by 1 as in Equation (4).

$$x \cdot y = 2^{R_x - 1 - R_y} = 1 \rightarrow R_y = R_x - 1 \tag{4}$$

A trickier case is when $f > 0$. Here, we can easily see that $k_x + 1 = k_y$, that implies $R_x = R_y$. Therefore, we get Equation (5).

$$x \cdot y = 2^{-1} \cdot \left(1 + f_x \cdot f_y \cdot 2^{-2F_x} + (f_x + f_y) \cdot 2^{-F_x}\right) = 1 \tag{5}$$

Then, discarding the term $f_x \cdot f_y \cdot 2^{-2F_x}$, we obtain Equation (6):

$$1 + (f_x + f_y) \cdot 2^{-F_x} = 2 \rightarrow f_y = 2^{F_x} - f_x \tag{6}$$

The latter can be obtained by simply bitwising-not $f_x$ and by adding 1, thus obtaining Equation (7):

$$Y = X \oplus (\neg signmask) \tag{7}$$

where $\oplus$ is the exclusive or (XOR) operator, $\neg$ is the bitwise negation operator, and *signmask* is the a mask obtained as shown in the following pseudo-code. For example, given a 5-bit Posit, the signmask is simply $(10000)_2$. The pseudocode also takes into account the holder size; in fact, a 5-bit Posit may be held by an 8-bit integer. This means that, for this holder type, the signmask produced by the pseudocode is $(11110000)_2$.

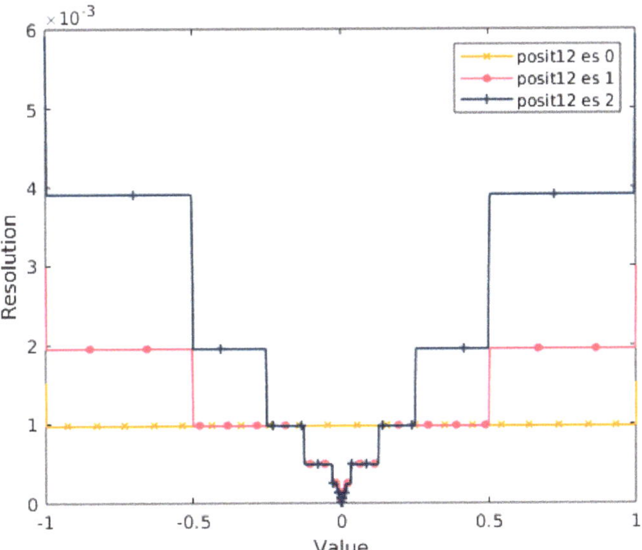


**Figure 3.** Resolution of a 12-bit Posit when varying the exponent size. With a 0-bit exponent, the Posit resolution in the $[0, 1]$ range is the one of a 12-bit fixed point format.

A pseudo-code implementation for $f > 0$ (otherwise, we simply invert the sign) is as follows:

```
inv(x) -> y
    X = x.v      // 'v' field: bit-string representing the Posit
    msb = 1 << (N-1)
    signmask = ~((msb | msb -1) >> 1)
    Y = X ^ (~signmask)     // negation operator followed by XOR operator (C-style)
    y(Y)
```

Another useful function to implement as bit manipulation is the one's complement operator (8)

$$y = 1 - x \tag{8}$$

This is of interest when $x \in [0, 1]$. In this case, $y \in [0, 1]$, of course. From Equations (1) and (2), we can rewrite the operator as in Equation (9).

$$y = 1 - 2^k \cdot (1 + \phi \cdot 2^{-F}) = 1 - 2^{N-2} \cdot (2^{-F} + \phi) \tag{9}$$

Since we can link $x$ to its representation $X$, we obtain (10).

$$y = 1 - X \cdot 2^{-(N-2)} \tag{10}$$

Then, we can also link $y$ to $Y$, obtaining (11).

$$Y = 2^{N-2} \cdot \left[ 1 - X \cdot 2^{-(N-2)} \right] = 2^{N-2} - X \tag{11}$$

The latter can be obtained easily with an integer subtraction only using the ALU. A pseudo-code implementation is the following:

```
comp_one(x) -> y
    X = x.v     // 'v' field: bit-string representing the Posit
    invert_bit = 1 << (N-2)
    Y = invert_bit - X
    y(Y)
```

when $esbits = 0$, we know that $x = 2^k \cdot (1 + \phi \cdot 2^{-F})$. when doubling/halving $x$, we simply increment/decrement the exponent $k$ by 1. For 0-bit exponent Posits, this operation corresponds to one left shift for doubling and one right shift for halving the number. For instance, let us take a Posit$\langle 5, 0 \rangle$ with the value $3/4$. The correspondent bit-string will be $(00110)_2$. If we shift it by one position right, we will get $(00011)_2$, that is the bit-string corresponding to a Posit with value $3/8$.

### 2.2. FastSigmoid

As pointed out in Reference [8], if we plot the 2's complement value of the signed integer representing the Posit against the real number obtained from Equation (2), we obtain an S-shaped function very similar to the sigmoid curve. What we need to do is to rescale it to have the co-domain $\in [0, 1]$ and to shift it in order to center it in 0. To bring the Posit in $[0, 1]$, we must notice that the quadrant is characterized by having the two most significant bits set at 00 (see Figure 4).

Moreover, we can notice that adding the *invert bit* seen in previous sections to the Posit representation means moving it a quarter of the quadrant. In fact with $esbits = 0$, when adding the invert bit, we are adding $2^{N-2}$, that is equal to $L = \frac{1}{minpos}$, which is the number of Posits that fit in a single quarter of a ring. This means moving $L$ times along the Posit ring, thus skipping a quarter of it. A pseudo-code implementation of this transformation is the following:

```
fastSigmoid(x) -> y
    X = x.v     // 'v' field: bit-string representing the Posit
    Y = (invert_bit + (X >> 1)) >> 1
    y(Y)
```

In order to understand how this code works, we need to separate the analysis for $x^-$ and $x^+$, considering only positive values, since the reasoning is symmetric for negative ones. Figure 5 shows the behaviour of the two sigmoid versions.

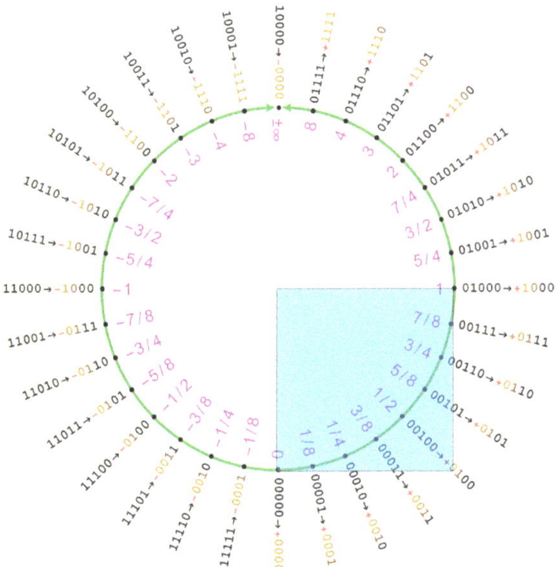

**Figure 4.** The $[0, 1]$ quadrant in the Posit ring.

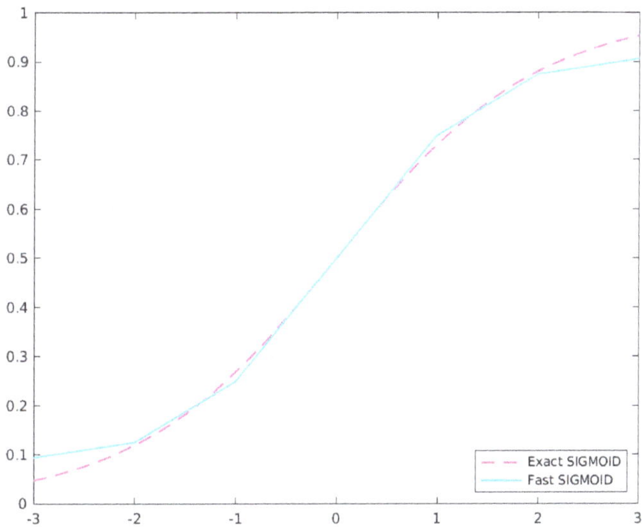


**Figure 5.** Accuracy comparison between the exact and approximated versions of the Sigmoid function.

We know that, for values of $x \in [0, 1]$, the behaviour of $x$ is like the one of a fixed point representation, so the first right shift is simply a division by two. When we add the *invert bit*, we move the Posit in the northeast ring quarter ($[1, +NaR)$). After this addition, the last shift can be considered as a division by two as well, thus obtaining the following:

$$y = \frac{x}{4} + \frac{1}{2} \tag{12}$$

Equation (12) is also the first-order Taylor expansion of the Sigmoid function in $x_0 = 0$.

With $x$ represented as the bit-string $X = (0, 1[R_x], 0, \phi_x)$, the right shift will produce $X' = (0, 0, 1[R_x - 1], 0, \phi_x')$. Now, with some computation, we can express $x'$ as a function of $x$ and $R_x$, obtaining (13).

$$\frac{x}{2^{2 \cdot R_x}} + \frac{2^{2 \cdot R_x} - 3 \cdot 2^{R_x - 1}}{2^{2 \cdot R_x}} \tag{13}$$

when adding the *invert bit*, we obtain $X'' = (0, 1[R_x + 1], 0, \phi_x)$. Finally, with the last right shift, we obtain (14).

$$\frac{x}{2^{2 \cdot R_x + 1}} + 3 \cdot \frac{2^{2 \cdot R_x} - 2^{R_x}}{4 \cdot 2^{2 \cdot R_x}} \tag{14}$$

We know that we can approximate $R_x \sim \log_2(x) \rightarrow x \sim 2^{R_x}$. If we substitute it back in Equation (14), we obtain Equation (15), close to $sigmoid(R_x)$:

$$\frac{3 \cdot 2^{R_x} - 1}{4 \cdot 2^{R_x}} \tag{15}$$

## 3. CppPosit Library

For this paper, we employ our software implementation of Posit numbers developed at the University of Pisa, called cppPosit. As already described in References [9,12], the library classifies Posit operations into four different classes (from L1 to L4), with increasing computational complexity.

Among the others, L1 operations are the ones we want to focus on, since they can be fully emulated with an ALU. For this reason, they provide means to produce very efficient operators, as reported in Table 1.

This level supports Posit reciprocation and sign-negation as well as one's complement. Furthermore, when dealing with 0 exponent-bit configuration, they provide the fast and approximated sigmoid function (FastSigmoid) as described in Reference [8] and the fast approximation of the hyperbolic tangent (FastTanh) investigated in Reference [12]. Other interesting operators that require 0 exponent bits are the double and half functions. It is clear that, given these requirements, it is not always easy to derive a simple expression for a particular function that can be implemented in an L1 way. However, the effort put in this step is completely rewarded since it brings both faster execution both in a emulated and hardware Posit Processing Unit (PPU) and reduction of transistor occupation when dealing with hardware implementation of the unit.

**Table 1.** Most interesting L1 operators implemented in cppPosit and their requirements to be applied on the argument $x$.

| Operation | Approximated | Requirements |
|:---:|:---:|:---:|
| $2 \cdot x$ | no | $esbits = 0$ |
| $x/2$ | no | $esbits = 0$ |
| $1/x$ | yes | $esbits = 0$ |
| $1 - x$ | no | $esbits = 0, x \in [0, 1]$ |
| FastSigmoid$(x)$ | yes | $esbits = 0$ |
| FastTanh$(x)$ | yes | $esbits = 0$ |
| FastELU$(x)$ | yes | $esbits = 0$ |

### 3.1. Tabulated Posits

In the absence of proper hardware support of a Posit Processing Unit (PPU), there still is the need for speeding up the computation. An interesting mean to cope with this problem is the pre-computation of some useful Posit operators in look-up tables. These lookup tables (LUTs) become useful when the number of bits is low (e.g., $nbits < 12$). The core idea is to generate tables for the most important arithmetic operations (addition/subtraction and multiplication/division) for all combinations of a given Posit configuration $nbits, esbits$. Moreover, some interesting functions can be tabulated in order to speed up their computation, like *logarithm* or *exponentiation*. Given an *nbits* bit Posit with a naive approach, a table will be $T \in P^{R \times C}$ where $R = C = 2^{nbits} - 1$.

Depending on the underlying storage type T, each table entry will occupy b=sizeof(T) bits. Typically, there will be between $N = 8$ and $N = 10$ tables for a Posit configuration. This means that the overall space occupation will be $S = N \cdot (R \cdot C) \cdot b$.

Table 2 shows different per-table occupations of different Posit configurations. As reported, only Posits with 8 and 10 bits have reasonable occupation, considering current generation of CPUs. In fact, we can obtain a considerable speed-up when one or more tables can be entirely contained inside the cache.

**Table 2.** Table occupation for various configurations.

| Total Bits (X) | Storage Type Bits (b) | Per-Table Occupation |
|:---:|:---:|:---:|
| 8 | 8 | 64 KB |
| 10 | 16 | 2 MB |
| 12 | 16 | 32 MB |
| 14 | 16 | 512 MB |
| 16 | 16 | 8 GB |

In order to reduce both LUT size and their number, we can exploit some arithmetic properties:

- Addition and subtraction are respectively symmetric and antisymmetric. The two tables can be merged into one, and only one half of it is required (above or below the main diagonal).
- Multiplication and division can be simplified through logarithm properties. Given $p = x \cdot y$, we can apply *log* operator on both sides (see Reference [14]), thus obtaining $\log(p) = \log(x \cdot y)$. From logarithm properties, this results in $\log(p) = \log(x) + \log(y)$ Finally, going back with exponentiation, we get $p = e^{\log(x) + \log(y)}$. Since tabulation of single operators scales linearly with the Posit size, it is feasible only to store $\exp, \log$ instead of multiplication and division, thus exploiting addition/subtraction LUT for the computation.
- We can compact multiplication tables even more by exploiting the fast inversion (L1) shown in Section 2. Suppose to have two Posit numbers $x, y$ and their reciprocates, if we want to provide every multiplication or division combination, we would build a LUT like in Table 3. This table would result in 16 entries for only 4 numbers, hence not manageably growing with Posit size. If we apply the L1 inversion and symmetry of negative values, we only need to store the operations for $x \cdot y$ and $x/y$, thus resulting in a LUT size of only 2 elements for the same amount of numbers, as shown in Table 4.

**Table 3.** All the possible combinations for multiplying and dividing two Posit numbers.

| | 1/x | x | −1/x | −x |
|:---:|:---:|:---:|:---:|:---:|
| **1/y** | $1/xy$ | $x/y$ | $-1/xy$ | $-x/y$ |
| **y** | $y/x$ | $xy$ | $-y/x$ | $-xy$ |
| **−1/y** | $-1/xy$ | $-x/y$ | $1/xy$ | $-x/y$ |
| **−y** | $-y/x$ | $-xy$ | $-y/x$ | $xy$ |

**Table 4.** All the possible combinations for multiplying and dividing two Posit numbers: all the cells in italics correspond to the same LUT entry, and all the remaining ones correspond to another LUT entry.

| | 1/x | x | −1/x | −x |
|:---:|:---:|:---:|:---:|:---:|
| **1/y** | *1/xy* | $x/y$ | *−1/xy* | $-x/y$ |
| **y** | $y/x$ | *xy* | $-y/x$ | *−xy* |
| **−1/y** | *−1/xy* | $-x/y$ | *1/xy* | $-x/y$ |
| **−y** | $-y/x$ | *−xy* | $-y/x$ | *xy* |

### 3.2. Type Proxying

When dealing with Posit configuration with *esbits* $\neq 0$ it is not possible to exploit fast approximation of operators that relies on this property. A possible solution is to switch to a different Posit configuration with 0 exponent bits and higher total number of bits to exploit a fast approximation and to then switch back to the original one.

Increasing the number of bits is also useful when the starting Posit configuration has already 0 exponent bits. In fact, increasing *nbits* for the operator computation increases the accuracy of the computation, avoiding type overflows.

Given a Posit configuration $P1 \langle X, Y \rangle$, the basic idea is to proxy through a configuration $P2 \langle Z, 0 \rangle$ with $Z \gg X$. The core step in the approach is the Posit conversion between different configurations. The base case is converting $P1 \langle X, 0, T1 \rangle \rightarrow P2 \langle Z, 0, T2 \rangle$, with $Z \gg X$ and `sizeof(T2)`$\gg$`sizeof(T1)`. In this case, the conversion operation is the following:

```
convert0(p1) -> p2
    v1 = p1.v     // 'v' field: bit-string representing the Posit
    v2 = cast<T2>(v1) << (Z - X)
    p2.v2 = v2
```

### 3.3. Brain Posits

The idea behind Brain Floats is to define a Float16 with the same number of bits for the exponents of an IEEE 754 Float32. BFloat16 is thus different from IEEE 754 Float16, and the rationale of its introduction is that, when we have a DNN already trained with IEEE Float32, we can perform the inference with a BFloat16 and we can expect a reduced impact on the accuracy due to the fact that the dynamic range of a BFloat16 is the same as that of IEEE Float32. Following the very same approach, we can define *Brain Posits* to be associated to the Posit16 and Posit32 that will be standardized soon. In particular, BPosit16 can be designed in such a way that it has the same dynamic range of a standard Posit32, which will be the one with 2 bits of exponent. Since we are using the Posit format, we can define the BPosit16 as the 16-bit Posit having a number of bits for the exponent such that its dynamic range is similar to the one of Posit<32,2>. Using the same approach, we will define BPosit8, where the number of bits for the exponent, in this case, must be the one that allows the BPosit8 to cover most of the dynamic range of the standard 16-bit Posit, which is the Posit<16,1>. In the following, we will perform some computations to derive the two number of exponents. Indeed, another interesting aspect of type proxying is that we can also reduce the total number of bits while increasing the exponent ones and still being able to accommodate the entire dynamic range. In doing so, we need to know the minimum number of exponent bits of the destination type. Suppose we are converting from Posit $P1 \langle X_1, Y_1 \rangle$ to Posit $P2 \langle X_2, Y_2 \rangle$, with $X_1 > X_2$. We know that the maximum value for $P_1$ (similarly, it holds for $P_2$ as well) is $max_1 = \left[ 2^{2^{Y_1}} \right]^{X_1 - 2}$ If we set the inequality $max_2 \geq max_1$ and we apply logarithms to both sides, we get $(X_2 - 2) \cdot 2^{Y_2} \geq (X_1 - 2) \cdot 2^{Y_1}$ From this, we obtain the rule for determining the exponent bits of the destination type:

$$Y_2 \geq \log_2\left(\frac{X_1 - 2}{X_2 - 2}\right) + Y_1 \tag{16}$$

From Equation (16), we can derive some interesting cases. A Posit $P1 \langle 16, 1 \rangle$ can be transformed into a Posit $P2 \langle 8, 2 \rangle$ without a significant loss in the dynamic range. Furthermore, the same holds for a Posit $P1 \langle 32, 2 \rangle$, which can be approximated using Posit $P1 \langle 16, 3 \rangle$.

For all this reasons, the Brain Posits proposed in Table 5 might deserve a hardware implementation too.

**Table 5.** Brain Posits.

| Standard Posits | Brain Posits |
| --- | --- |
| Posit<16,1> | Posit<8,2> |
| Posit<32,2> | Posit<16,3> |
| Posit<64,3> | Posit<32,4> |

## 4. Hyperbolic Tangent, Extended Linear Unit, and their Approximations

The hyperbolic tangent (*tanh* from now on) is a commonly used activation function. Its use over the sigmoid function is interesting since it extends the sigmoid codomain to the interval $[-1, 1]$. This allows both the dynamic range of the sigmoid in the output to be exploited twice and the negative values in classification layers during training to be given meaning. The first advantage is particularly important when applied to Posit, especially to small-sized ones. In fact, when considering the sigmoid function, if we apply it to a $\mathrm{Posit}\langle X, Z \rangle$, we practically obtain in the output the dynamic range of a $\mathrm{Posit}\langle X/2, Z \rangle$, that is, for instance, quite limiting for Posits with 8 to 14 number of bits. Figure 6 stresses this point, highlighting how the tanh function insists on the two most dense quarters of the Posit circle (the interval $[-1, 1]$ occupies half of the Posit circle).

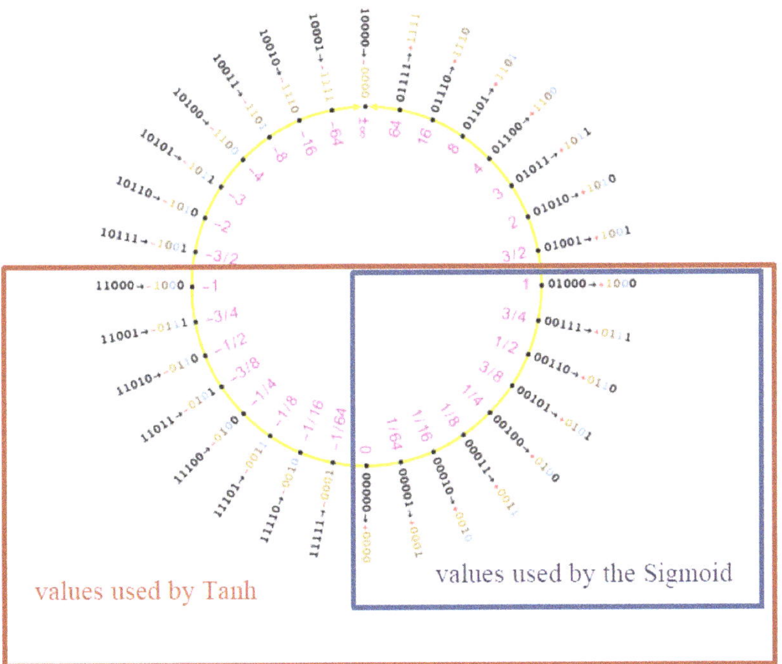


**Figure 6.** Five-bit Posit mapping to the Posit circle: As reported, the tanh function manages to cover the lower half of the circle while the sigmoid one covers only the quarter $[0, 1]$.

However, the sigmoid function has an important property, as shown in Table 1 and in Reference [8]: it can be implemented as L1 function, thus having a fast and efficient approximation only using integer arithmetics. The idea is to use the sigmoid function as a building block for other activation functions, only using a combination of L1 operators. We know that the sigmoid function is:

$$sigmoid(x) = \frac{1}{e^{-x} + 1} \tag{17}$$

Now, we can scale and translate (17) to cover the desired range $[-1, 1]$ on the output obtaining the scaled sigmoid:

$$sSigmoid_k(x) = k \cdot sigmoid(k \cdot x) - k/2 \tag{18}$$

Equation (18) is useful when setting $k = 2$, thus obtaining the tanh expression in (19):

$$sSigmoid_2(x) = (e^{2x} - 1)/(e^{2x} + 1) = tanh(x) = 2 \cdot sigmoid(2 \cdot x) - 1 \tag{19}$$

From this formulation, we want to build an equivalent one that only uses L1 operators to build the approximated hyperbolic tangent, switching from sigmoid to the fast approximated version called FastSigmoid. Since we are dealing with 0 exponent bit Posits, the operations of doubling the Posit argument, computing the FastSigmoid, and doubling again is just a matter of bit manipulations, thus efficiently computed. However, the last step of subtracting 1 to the previous result is not an L1 operator out-of-the-box; thus, we reformulate the initial expression obtaining (20):

$$tanh(x) = -(1 - 2 \cdot sigmoid(2 \cdot x)) \tag{20}$$

If we consider only negative arguments $x$, we know that the result of the expression $2 \cdot sigmoid(2 \cdot x))$ is always in the unitary region. This, combining with the 0 exponent bit hypothesis allows us to implement the inner expression with the 1's complement L1 operator seen in Table 1. The last negation is obviously an L1 operator; thus, we have the L1 fast approximation of the hyperbolic tangent in (21):

$$FastTanh(x) = -(1 - 2 \cdot FastSigmoid(2 \cdot x)) \tag{21}$$

Finally thanks to the antisymmetry of the tanh function, we can extend what we have done before to positive values. The following is a pseudo-code implementation:

```
FastTanh(x) -> y
    x_n = x > 0 ? -x:x
    s = x > 0
    y_n = neg(compl1(twice(FastSigmoid(twice(x_n)))))
    y = s > 0 ? -y_n:y_n
```

As already described, tanh and sigmoid functions can be implemented in their fast approximated version. However, the use of such kinds of shapes presents the well-known behaviour of vanishing gradients [15]; for this reason, ReLU -like functions (e.g., ELU, Leaky-ReLU, and others) are preferable when dealing with a large number of layers in neural networks. As in Reference [15], the ReLU activation function is defined as in (22):

$$ReLU(x) = \begin{cases} 0, \text{ if } x \leq 0 \\ x \text{ otherwhise} \end{cases} \tag{22}$$

Its use is important in solving the vanishing gradient problem, having a non-flat shape towards positive infinity. However, when used with Posit numbers, this function can only cover $[0, \inf)$, ignoring the very dense region $[-1, 0]$.

In order to provide a more covering function with similar properties, we switch to the Extended Linear Unit (ELU) (23):

$$ELU(x) = \begin{cases} \alpha \cdot (e^x - 1), \text{ if } x \leq 0 \\ x \text{ otherwhise} \end{cases} \tag{23}$$

This function is particularly interesting when $\alpha = 1$ (24), covering the missing dense region from the ReLU one:

$$\text{ELU}(x) = \begin{cases} e^x - 1, \text{ if } x \leq 0 \\ x \text{ otherwhise} \end{cases} \tag{24}$$

Figure 7 shows the difference in Posit ring region usage of ELU and ReLU functions. It is remarkable how the ELU function manages to cover all the high density regions of the Posit ring. Moreover, the ELU function brings interesting normalization properties across the neural network layers as proven in Reference [16]. This helps in keeping stable the range of variation of the weights of the DNN.

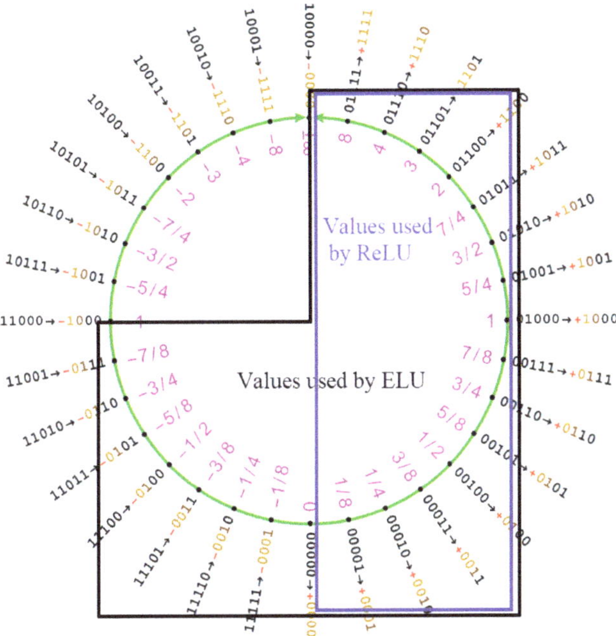


**Figure 7.** The Posit circle when the total number of bits is 5: The extended linear unit uses all the numbers in $[-1, \inf)$, while the ReLU function uses only the ones in $[0, \inf)$.

From Equation (24), we can build a L1 approximation exploiting operators in Table 1. The $ELU(x)$ behaviour for $x > 0$ is the identity function, that is L1 for sure. The first step for negative $x$ values is seeing that the ELU expression is similar to the reciprocate of Sigmoid function (17). We can manipulate (17) as follows:

$$\text{Sigmoid}(-x) = \frac{1}{1 + e^x} \tag{25}$$

$$1/\text{Sigmoid}(-x) = 1 + e^x \tag{26}$$

$$1/(2 \cdot \text{Sigmoid}(-x)) = \frac{1 + e^x}{2} \tag{27}$$

$$1/(2 \cdot \text{Sigmoid}(-x)) - 1 = \frac{1 + e^x}{2} - 1 = \frac{e^x - 1}{2} \tag{28}$$

$$2 \cdot [1/(2 \cdot \text{Sigmoid}(-x)) - 1] = e^x - 1 \tag{29}$$

We need to prove that the steps involved are L1 operations. The step in Equation (25) is always L1 for $esbits = 0$ thanks to fast Sigmoid approximation. The result of this step is always on $[1, 2]$. The step in Equation (26) is always L1, and the output is on $[1/2, 1] \in [0, 1]$. The step in Equation (27) is always L1 for *esbits* $= 0$, and the output is on $[0, 1/2]$. The step in Equation (28) is L1 since the previous step output is in the unitary range $[0, 1]$. The output of this step is in $[0, 1]$ as well. Finally, the last step is L1 for *esbits* $= 0$. Expression (29) is exactly the ELU expression for negative values of the argument.

A pseudo-code implementation of the FastELU using only L1 operations is shown below:

```
FastELU(x) -> y
    y_n = neg(twice(compl1(half(reciprocate(FastSigmoid(neg(x)))))))
    y = x > 0 ? x:y_n
```

Figure 8 shows the behaviour of the two functions when approximated with our approach.

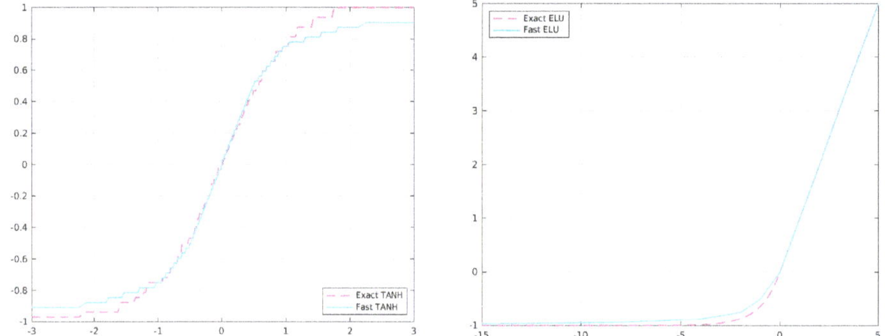


**Figure 8.** Comparison between the exact and approximated versions of hyperbolic tangent (TANH) and extended linear unit (ELU).

## 5. Implementation Results

In this section, the different proposed activation function performances are analyzed in both the exact and approximated fashions when used as activation function in the LeNet-5 neural network model [17]. As shown in Figure 9, the neural network is trained with the MNIST digit recognition benchmark (GTRSB) [17] and the German Traffic Road Sign Benchmark [18] datasets using the Float32 type. The performance metrics involved are the testing accuracy on said datasets and the mean sample inference time. Testing phase is executed converting the model to *Posit* $\langle X, Y \rangle$ type and to SoftFloat32 (a software implementation of floats). We used SoftFloats in order to ensure a fair comparison between the two software implementations due to the absence of proper hardware support for Posit type.

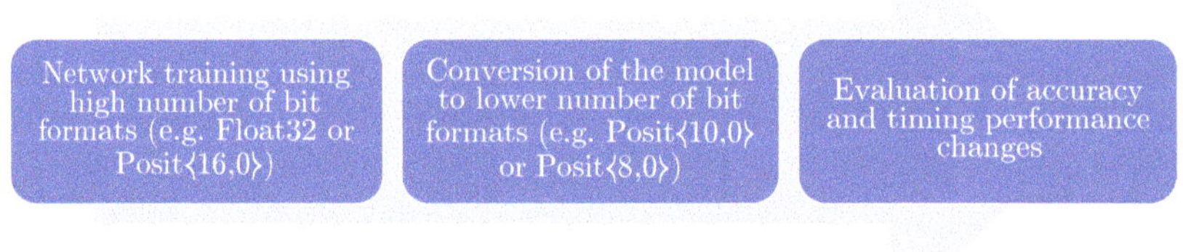


**Figure 9.** Flowchart for the proposed method: models are trained using formats with high bit count like Float32 or, in the future, Posit⟨16,0⟩. The models obtained this way are then converted to formats with lower bit count (e.g., Posit⟨8,0⟩) to increase space efficiency and bandwidth.

Benchmarks are executed on a 7th generation `Intel i7-7560U` processor, running Ubuntu Linux `18.04`, equipped with `GCC 8.3`. Benchmark data is publicly available in References [17]. The C++ source code can be downloaded from Reference [19].

As reported in Tables 6 and 7, the approximated hyperbolic tangent can replace the exact one, with a small degradation in accuracy but improving the inference time of about 2 ms in each Posit configuration. Moreover, the performance of FastTanh also overcome FastSigmoid in terms of accuracy. Furthermore, as reported in Tables 8 and 9, the approximated ELU function can replace the exact one, with little-to-no accuracy degradation, improving the inference time of about 1 ms in each Posit configuration. Moreover, performance of FastELU also overcomes the ReLU in terms of accuracy, showing the benefits of covering the additional region in $[-1, 0]$. At the same time, the FastELU is not much slower than ReLU, thus being an interesting replacement to increase accuracy of Posits with few bits (e.g., Posit$\langle 8, 0 \rangle$) without losing too much in time complexity.

**Table 6.** Comparison using Posits for the MNIST dataset for three different activation functions: fast approximated version of Tanh (FastTanh), exact Tanh, and FastSigmoid. Accuracy of the neural network and mean sample inference time are reported.

| Activation | FastTanh (This Paper) | | Tanh | | FastSigmoid | |
|---|---|---|---|---|---|---|
| | Acc. (%) | Time (ms) | % | ms | % | ms |
| SoftFloat32 | - | - | 99.4 | 8.3 | 97.1 | - |
| Posit$\langle 16, 0 \rangle$ | 99.1 | 3.2 | 99.4 | 5.28 | 97.1 | 3.31 |
| Posit$\langle 14, 0 \rangle$ | 99.1 | 2.9 | 99.4 | 4.64 | 97.1 | 3.09 |
| Posit$\langle 12, 0 \rangle$ | 99.1 | 2.9 | 99.4 | 4.66 | 97.1 | 3.04 |
| Posit$\langle 10, 0 \rangle$ | 99.1 | 2.9 | 99.3 | 4.62 | 96.9 | 3.08 |
| bottomrule Posit$\langle 8, 0 \rangle$ | 98.6 | 2.9 | 98.5 | 4.84 | 94.2 | 3.01 |

**Table 7.** Comparison using Posits for the GTRSB dataset (see Table 6).

| Activation | FastTanh (This Paper) | | Tanh | | FastSigmoid | |
|---|---|---|---|---|---|---|
| | Acc. (%) | Time (ms) | % | ms | % | ms |
| SoftFloat32 | - | - | 94.2 | 15.2 | 82.0 | - |
| Posit$\langle 16, 0 \rangle$ | 93.5 | 5.3 | 93.5 | 6.2 | 81.9 | 5.0 |
| Posit$\langle 14, 0 \rangle$ | 93.4 | 4.2 | 93.5 | 5.1 | 81.9 | 4.3 |
| Posit$\langle 12, 0 \rangle$ | 93.4 | 4.2 | 93.4 | 5.1 | 81.9 | 4.3 |
| Posit$\langle 10, 0 \rangle$ | 93.4 | 4.2 | 93.3 | 5.1 | 81.0 | 4.2 |
| Posit$\langle 8, 0 \rangle$ | 93.0 | 4.0 | 92.3 | 5.0 | 72.1 | 4.0 |

**Table 8.** Comparison using Posits for the MNIST dataset for three different activation functions: fast approximated version of ELU (FastELU), exact ELU, and ReLU. Accuracy of the neural network and mean sample inference time are reported.

| Activation | FastELU (This Paper) | | ELU | | ReLU | |
|---|---|---|---|---|---|---|
| | Acc. (%) | Time (ms) | % | ms | % | ms |
| SoftFloat32 | - | - | 98.6 | 8.8 | 89.1 | 6.3 |
| Posit$\langle 16, 0 \rangle$ | 98.5 | 3.2 | 98.6 | 3.9 | 89.1 | 2.0 |
| Posit$\langle 14, 0 \rangle$ | 98.5 | 2.4 | 98.6 | 3.1 | 89.05 | 2.0 |
| Posit$\langle 12, 0 \rangle$ | 98.5 | 2.3 | 98.6 | 3.1 | 89.0 | 2.0 |
| Posit$\langle 10, 0 \rangle$ | 98.3 | 2.3 | 98.5 | 3.0 | 89.0 | 1.9 |
| Posit$\langle 8, 0 \rangle$ | 91.1 | 2.2 | 90.1 | 3.0 | 88.4 | 1.9 |

**Table 9.** Comparison using Posits for the GTRSB dataset (see Table 8).

| Activation | FastELU (This Paper) | | ELU | | ReLU | |
| --- | --- | --- | --- | --- | --- | --- |
| | Acc. (%) | Time (ms) | % | ms | % | ms |
| SoftFloat32 | - | - | 94.2 | 15.86 | 92.0 | 8.2 |
| Posit$\langle 16,0 \rangle$ | 94.0 | 5.8 | 94.2 | 6.37 | 92.0 | 5.0 |
| Posit$\langle 14,0 \rangle$ | 94.0 | 4.6 | 94.2 | 5.21 | 92.0 | 4.3 |
| Posit$\langle 12,0 \rangle$ | 94.0 | 4.6 | 94.2 | 5.08 | 92.0 | 4.3 |
| Posit$\langle 10,0 \rangle$ | 94.0 | 4.6 | 94.2 | 5.0 | 92.0 | 4.2 |
| Posit$\langle 8,0 \rangle$ | 92.0 | 4.6 | 91.8 | 5.0 | 86.8 | 4.0 |

If we compare FastELU and FastTanh, their performance are quite similar in the benchmarks provided. However as already said in Section 4, increasing the number of layers in the neural network model can lead to the so called "vanishing gradient" problem; s-shaped functions like sigmoid and hyperbolic tangent are prone to this phenomenon. This has been proven not to hold for ReLU-like functions.

The results highlight how Posits from Posit $\langle 16,0 \rangle$ to Posit $\langle 10,0 \rangle$ are a perfect replacement for float numbers; Posit $\langle 10,0 \rangle$ is a particularly interesting format since it offers the best data compression without any drop in accuracy. This reasonably makes Posit $\langle 10,0 \rangle$ the configuration of choice for low-precision inference when using Posits.

## 6. Conclusions and Future Work

In this work, we have introduced some interesting properties of Posit format for the specific configuration having zero exponent bits (*esbit* $= 0$), that allows building fast arithmetic operators that only requires ALU support. In particular, we have derived two novel fast approximated versions of two important activation functions in neural networks: the hyperbolic tangent and the extended linear unit. These approximations are fast since they involve only bit manipulations (at the so-called "L1 level"). This means that such functions do not need to be implemented in hardware within the so-called Posit processing unit. Instead, they can be efficiently computed using the ALUs of most of the current CPUs. We have used this approximation to speed up the inference phase of deep neural networks. The proposed approximations have been tested on common deep neural network benchmarks. The use of this approximations resulted in a slightly less accurate neural network with respect to the use of the (slower) exact version but with better performance in terms of mean sample inference time of the network. In our experiment, the FastTanh and FastELU functions also outperform both the ReLu and the FastSigmoid (a well-known approximation of the sigmoid function), a de facto standard activation function in neural networks. Future developments of the work will include porting the Posit format inside the Apollo Autonomous Driving Framework to test it on the assisted/autonomous driving scenario; this will allow us to test our approach in object detection and semantic segmentation tasks. We plan to implement a Field Programmable Gate Array (FPGA) based Posit Processing Unit (PPU) in order to evaluate real-world hardware performance of our library. Furthermore, we are actively working to port the cppPosit library for the new RISC-V processor architecture; we plan to develop both a software-accelerated version using the vector extension of the RISC-V Instruction Set Architecture (ISA) and an intellectual property (IP) core for the RISC-V hardware architecture.

**Author Contributions:** The four authors have equally contributed to all the phases of this work. All authors have read and agreed to the published version of the manuscript.

**Funding:** This work is partially funded by H2020 European Processor Initiative (grant agreement No. 826647) and partially by the Italian Ministry of Education and Research (MIUR) in the framework of the CrossLab project (Departments of Excellence).

**Conflicts of Interest:** The authors declare no conflict of interest. The founding sponsors had no role in the design of the study; in the collection, analyses, or interpretation of data; in the writing of the manuscript; and in the decision to publish the results.

## References


1. Alcaide, S.; Kosmidis, L.; Tabani, H.; Hernandez, C.; Abella, J.; Cazorla, F.J. Safety-Related Challenges and Opportunities for GPUs in the Automotive Domain. *IEEE Micro* **2018**, *38*, 46–55. doi:10.1109/MM.2018.2873870. [CrossRef]
2. Benedicte, P.; Abella, J.; Hernandez, C.; Mezzetti, E.; Cazorla, F.J. Towards Limiting the Impact of Timing Anomalies in Complex Real-Time Processors. In Proceedings of the 24th Asia and South Pacific Design Automation Conference (ASPDAC'19), Tokyo, Japan, 21–24 January 2019; pp. 27–32. doi:10.1145/3287624.3287655. [CrossRef]
3. 10 Useful Tips for Using the Floating Point Unit on the Cortex-M4. Available online: https://community.arm.com/developer/ip-products/processors/b/processors-ip-blog/posts/10-useful-tips-to-using-the-floating-point-unit-on-the-arm-cortex--m4-processor (accessed on 4 March 2020).
4. Köster, U.; Webb, T.; Wang, X.; Nassar, M.; Bansal, A.K.; Constable, W.; Elibol, O.; Gray, S.; Hall, S.; Hornof, L.; et al. Flexpoint: An adaptive numerical format for efficient training of deep neural networks. In Proceedings of the 31st Conference on Neural Information Processing Systems (NIPS'17), Long Beach, CA, USA, 4–9 December 2017; pp. 1742–1752.
5. Popescu, V.; Nassar, M.; Wang, X.; Tumer, E.; Webb, T. Flexpoint: Predictive Numerics for Deep Learning. In Proceedings of the 25th IEEE Symposium on Computer Arithmetic (ARITH'18), Amherst, MA, USA, 25–27 June 2018; pp. 1–4. doi:10.1109/ARITH.2018.8464801. [CrossRef]
6. NVIDIA Turing GPU Architecture, Graphics Reinvented. Available online: https://www.nvidia.com/content/dam/en-zz/Solutions/design-visualization/technologies/turing-architecture/NVIDIA-Turing-Architecture-Whitepaper.pdf (accessed on 4 March 2020).
7. Malossi, A.C.I.; Schaffner, M.; Molnos, A.; Gammaitoni, L.; Tagliavini, G.; Emerson, A.; Tomás, A.; Nikolopoulos, D.S.; Flamand, E.; Wehn, N. The transprecision computing paradigm: Concept, design, and applications. In Proceedings of the Design, Automation Test in Europe Conference Exhibition (DATE'18), Dresden, Germany, 19–23 March 2018; pp. 1105–1110. doi:10.23919/DATE.2018.8342176. [CrossRef]
8. Gustafson, J.L.; Yonemoto, I.T. Beating Floating Point at its Own Game: Posit Arithmetic. *Supercomput. Front. Innov.* **2017**, *4*, 71–86.
9. Cococcioni, M.; Rossi, F.; Ruffaldi, E.; Saponara, S. Novel Arithmetics to Accelerate Machine Learning Classifiers in Autonomous Driving Applications. In Proceedings of the 26th IEEE International Conference on Electronics, Circuits and Systems (ICECS'19), Genoa, Italy, 27–29 Noember 2019; pp. 779–782. doi:10.1109/ICECS46596.2019.8965031. [CrossRef]
10. Cococcioni, M.; Ruffaldi, E.; Saponara, S. Exploiting Posit arithmetic for Deep Neural Networks in Autonomous Driving Applications. In Proceedings of the 2018 IEEE International Conference of Electrical and Electronic Technologies for Automotive (Automotive '18), Milan, Italy, 9–11 July 2018; pp. 1–6. doi:10.23919/EETA.2018.8493233. [CrossRef]
11. Carmichael, Z.; Langroudi, H.F.; Khazanov, C.; Lillie, J.; Gustafson, J.L.; Kudithipudi, D. Deep Positron: A Deep Neural Network Using the Posit Number System. In Proceedings of the 2019 Design, Automation Test in Europe Conference Exhibition (DATE), Florence, Italy, 29 March 2019; pp. 1421–1426. doi:10.23919/DATE.2019.8715262. [CrossRef]
12. Cococcioni, M.; Rossi, F.; Ruffaldi, E.; Saponara, S. A Fast Approximation of the Hyperbolic Tangent when Using Posit Numbers and its Application to Deep Neural Networks. In Proceedings of the International Workshop on Applications in Electronics Pervading Industry, Environment and Society (ApplePies'19), Pisa, Italy, 18 September 2019. doi:10.1007/978-3-030-37277-4_25. [CrossRef]
13. Other Built-in Functions Provided by GCC. Available online: https://gcc.gnu.org/onlinedocs/gcc/Other-Builtins.html (accessed on 4 March 2020).
14. Arnold, M.G.; Garcia, J.; Schulte, M.J. The interval logarithmic number system. In Proceeding of the 16th IEEE Symposium on Computer Arithmetic (ARITH'03), Santiago de Compostela, Spain, 15–18 June 2003; pp. 253–261. doi:10.1109/ARITH.2003.1207686. [CrossRef]
15. Nair, V.; Hinton, G.E. Rectified Linear Units Improve Restricted Boltzmann Machines. In Proceedinsgs of the 27th International Conference on International Conference on Machine Learning (ICML'10), Haifa, Israel, 21–24 June 2010; pp. 807–814.

16. Klambauer, G.; Unterthiner, T.; Mayr, A.; Hochreiter, S. Self-Normalizing Neural Networks. In *Advances in Neural Information Processing Systems 30*; Guyon, I., Luxburg, U.V., Bengio, S., Wallach, H., Fergus, R., Vishwanathan, S., Garnett, R., Eds.; Curran Associates, Inc.: Long Beach, CA, USA, 4–9 December 2017; pp. 971–980.

17. LeCun, Y.; Bottou, L.; Bengio, Y.; Haffner, P. Gradient-based learning applied to document recognition. *Proc. IEEE* **1998**, *86*, 2278–2324. doi:10.1109/5.726791. [CrossRef]

18. Stallkamp, J.; Schlipsing, M.; Salmen, J.; Igel, C. The German Traffic Sign Recognition Benchmark: A multi-class classification competition. In Proceedings of the IEEE International Joint Conference on Neural Networks (IJCNN'11), San Jose, CA, USA, 31 July–5 August 2011; pp. 1453–1460.

19. European Processor Initiative: Posit-Based TinyDNN. Available online: https://gitlab.version.fz-juelich.de/epi-wp1-public/tinyDNN (accessed on 4 March 2020).

*Article*

# Steerable-Discrete-Cosine-Transform (SDCT): Hardware Implementation and Performance Analysis †

**Riccardo Peloso *, Maurizio Capra, Luigi Sole, Massimo Ruo Roch, Guido Masera and Maurizio Martina**

Department of Electronics and Telecommunication (DET), Politecnico di Torino, C.so Duca degli Abruzzi 24, 10129 Turin, Italy; maurizio.capra@polito.it (M.C.); luigi.sole@studenti.polito.it (L.S.); massimo.ruoroch@polito.it (M.R.R.); guido.masera@polito.it (G.M.); maurizio.martina@polito.it (M.M.)
* Correspondence: riccardo.peloso@polito.it
† This paper is an extended version of our paper published in Applications in Electronics Pervading Industry, Environment and Society, APPLEPIES 2019.

Received: 30 January 2020; Accepted: 28 February 2020; Published: 4 March 2020

**Abstract:** In the last years, the need for new efficient video compression methods grown rapidly as frame resolution has increased dramatically. The Joint Collaborative Team on Video Coding (JCT-VC) effort produced in 2013 the H.265/High Efficiency Video Coding (HEVC) standard, which represents the state of the art in video coding standards. Nevertheless, in the last years, new algorithms and techniques to improve coding efficiency have been proposed. One promising approach relies on embedding direction capabilities into the transform stage. Recently, the Steerable Discrete Cosine Transform (SDCT) has been proposed to exploit directional DCT using a basis having different orientation angles. The SDCT leads to a sparser representation, which translates to improved coding efficiency. Preliminary results show that the SDCT can be embedded into the HEVC standard, providing better compression ratios. This paper presents a hardware architecture for the SDCT, which is able to work at a frequency of 188 MHz, reaching a throughput of 3.00 GSample/s. In particular, this architecture supports 8k UltraHigh Definition (UHD) (7680 × 4320) with a frame rate of 60 Hz, which is one of the best resolutions supported by HEVC.

**Keywords:** video coding; discrete cosine transform; directional transform; VLSI

---

## 1. Introduction

In recent years, high-resolution multimedia content has fostered research in the field of video compression. Indeed, in 2013 the Joint Collaborative Team on Video Coding (JCT-VC) released the High-Efficiency Video Coding (HEVC) standard, also referred to as H.265 [1].

Interestingly, the HEVC standard improved the coding efficiency gain by reaching 50% of bit-rate reduction (for the same quality level) with respect to the previous Advanced Video Coding (AVC)/H.264 standard. Noticeably, HEVC not only improved the compression capability, but it effectivelyh andles high-quality video resolutions, enhanced frame rates, and increased dynamic range. In particular, the HEVC standard relies on coding tree units (CTUs) to improve transform coding and prediction. Each CTU contains two coding tree blocks (CTBs), one for the luma component and one for the chroma components. CTBs are partitioned into smaller blocks called coding units (CUs) along with a tree-based coding structure that includes prediction units (PUs). PUs exploit the temporal and spatial redundancies present in video streams leading to inter-frame and intra-frame prediction. The sizes of PUs vary from 8 × 4 and 4 × 8, to 64 × 64 pixels for inter-frame, while for intra-predicted PUs size goes from 4 × 4 to 32 × 32 pixels. As PUs are coded without including neighboring blocks, blocking artifacts due to discontinuous block boundaries can occur. To reduce these artifacts and to improve the quality of the decoded frames, the HEVC standard includes two in-loop filters: the deblocking filter (DBF) and the sample adaptive offset (SAO), as depicted in Figure 1.

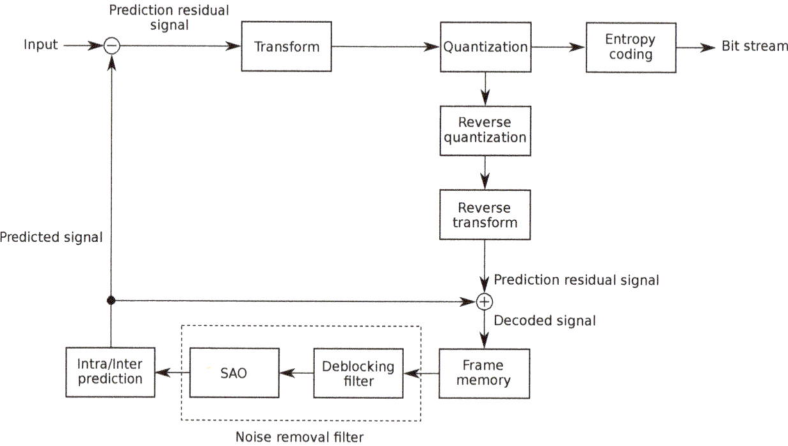


**Figure 1.** HEVC basic structure.

During the prediction, for each PU, the difference between the predicted block and the current block (*residual*) is lossly coded by the means of transform and quantization. The transform stage can be either the Discrete Sine Transform (DST) or Discrete Cosine Transform (DCT). While the DST is used only for the smallest block size, namely $4 \times 4$ pixels, the DCT is used for all the other sizes, up to $32 \times 32$. For this reason, some works pointed out that the complexity of the transform stage in the HEVC context is particularly relevant [2,3]. This motivated several researchers to propose dedicated architectures for variable size DCTs, such as [4–6]. Recently, G. Fracastoro et al. [7] proposed the Steerable DCT (SDCT) and showed that it can give some coding advantages when embedded in the HEVC standard [8]. Such a directional transform is not tailored to any specific one, but it can be potentially applied to any two-dimensional separable transform. Moreover, it can be oriented in any selected direction, providing a more scattered representation depending on the chosen orientation. Unfortunately, such enhancements in HEVC lead to further complexity increases. These features interfere with battery-powered platforms and real-time applications, since the higher the complexity, the higher the power consumption. This current paper details the hardware accelerator for the SDCT described in [9,10]. Such an accelerator is able to support the ultimate video coding resolutions like the 8k UltraHigh Definition ($7680 \times 4320$ pixels). After a brief introduction on the SDCT in Section 2, Section 3 analyses the proposed architecture and finally Section 4 presents the implementation results discussing possible trade-offs. Lastly, Section 5 offers an overview of the entire work by providing some results about the effectiveness of the SDCT in comparison to other canonical solutions.

## 2. Background

HEVC is a block-based video compression algorithm and, like similar compression schemes, it employs spatial transforms. In particular, the 2-D DCT is the main one, which acts along the horizontal and vertical directions. The 2-D DCT is defined as

$$
\begin{aligned}
X_{k_1,k_2} &= \sum_{n_1=0}^{N_1-1} \left( \sum_{n_2=0}^{N_2-1} x_{n_1,n_2} \cos\left[ \frac{\pi}{N_2}\left(n_2 + \frac{1}{2}\right)k_2 \right] \right) \cos\left[ \frac{\pi}{N_1}\left(n_1 + \frac{1}{2}\right)k_1 \right] \\
&= \sum_{n_1=0}^{N_1-1} \sum_{n_2=0}^{N_2-1} x_{n_1,n_2} \cos\left[ \frac{\pi}{N_1}\left(n_1 + \frac{1}{2}\right)k_1 \right] \cos\left[ \frac{\pi}{N_2}\left(n_2 + \frac{1}{2}\right)k_2 \right]
\end{aligned}
\tag{1}
$$

which is, by definition, a separable transform. The DCT deals better than the DFT (Discrete Fourier Transform) with the borders of the coding blocks. This allows higher energy compaction with reduced sensitivity to quantization. It is also a real transform, thus, computations on complex numbers are not required. The operation can be stated as a convolution, leading to a compact and efficient implementation.

It is possible to demonstrate that for blocks that include diagonal edges, a directional transform will be better suited, leading to a higher compression ratio. The work of B. Zeng et J. Fu [11] presents a mathematical framework about directional DCT (DDCT). This transform is difficult to handle as it requires non-canonical DCT lengths and complex reshaping of the blocks. Recently, G. Fracastoro et al. [7] proposed the Steerable DCT (SDCT). It employs the graph Fourier transform from [12] to obtain an easier-to-handle directional DCT. The SDCT kernels still retain a square shape so that computation remains easy to perform, even though this 2-D transform is not separable in two 1-D operations as for the classic 2-D DCT. Lately, the work in [8] demonstrated that it is possible to split the steerable cosine transform into a traditional DCT followed by a geometrical rotation. The resulting kernels are the same as the SDCT but the computation workload is reduced by exploiting the 2-D DCT separability. Section 3 will better deal with this issue. Figure 2 shows different kernels obtained by the SDCT, the DCT being a special case of the SDCT with a rotation by zero degrees.

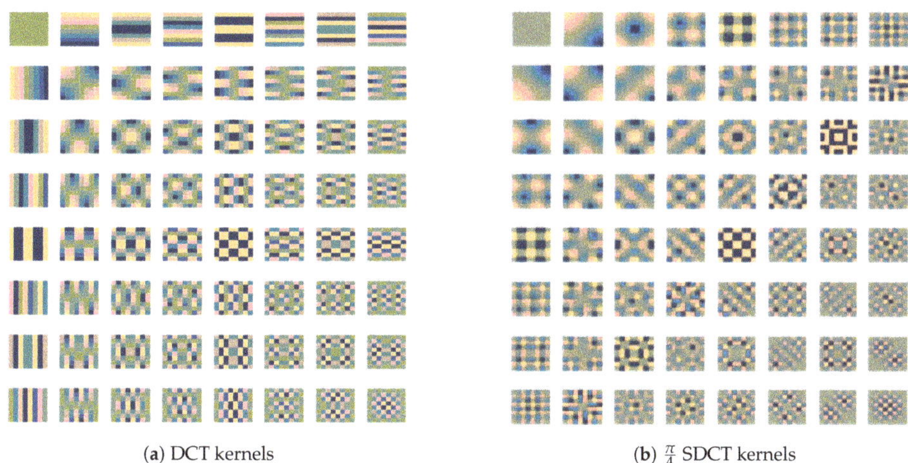

(**a**) DCT kernels            (**b**) $\frac{\pi}{4}$ SDCT kernels

**Figure 2.** Example of Discrete Cosine Transform (DCT) and Steerable Discrete Cosine Transform (SDCT) kernels.

## 3. Architectural Implementation

### 3.1. Datapath

While the 2D-DCT employed in HEVC is an inherently separable operation, the SDCT must be computed all at once. The complexity of a transform that is not separable is far greater than a separable one, so this may be a big drawback for the implementation. However, the complexity can be decreased drastically by splitting the SDCT into two parts, namely, a separable 2D DCT followed by some rotations, and then by computing the separable transform before applying rotations, as reported in [8]:

$$\tilde{x} = T(\boldsymbol{\theta})x = R(\boldsymbol{\theta})Tx = R(\boldsymbol{\theta})\hat{x} \tag{2}$$

where $x$ are the input samples, $\hat{x}$ are the results obtained by applying the $T$ transform matrix, $R(\boldsymbol{\theta})$ is the rotation matrix, while $\tilde{x}$ is the result of the SDCT. Thus, the SDCT can be decomposed in a DCT followed by a steering transformation. The DCT part can be implemented as suggested in the literature using a folded architecture [13]. When all the samples returned by the 2D-DCT are available, the rotations must be applied to obtain a steering transform. Since the DCT works exploiting a sliding window approach on the data, the process takes several steps to complete. However, the results will be provided all at once. This means that the steering part of the architecture has to work faster than the DCT. This issue has been tackled in this work by defining two clock regimes, one for the 2D-DCT and one, four times faster, for the steering part, to comply with the throughput offered by the 2D-DCT transform block. A FIFO memory between the two parts acts as a buffer memory. The whole structure is depicted in Figure 3.

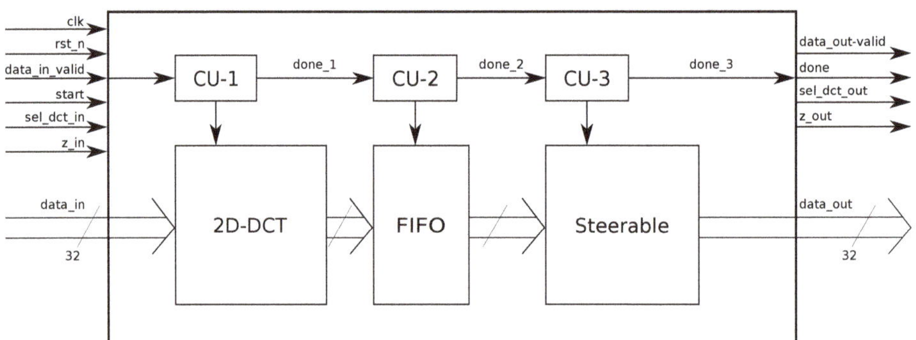


**Figure 3.** Whole SDCT structure.

The 2D-DCT block is based on the architecture proposed in [13] by Meher et al., which is very flexible and efficient, especially when dealing with folded transforms of size 4, 8, 16 and 32. The steerable part is shown in Figure 4. It is composed of an input memory (IM), an output memory (OM) and the lifting blocks that perform the rotation [14]. Some multiplexers are used to bypass the lifting blocks for the case of no rotation, returning directly the result given by the DCT. Despite the possiblity to bypass the IM and OM blocks when no rotation has to be applied, such an alternative leads to different latency of the architecture as a function of the rotation angle. Thus, in order to simplify the interface of the architecture, we decided to only bypass the lifting blocks. The IM is required also to reorder the samples as the steering process is computed on the custom zig-zag order given in Figure 5; this is different from the classic zig-zag ordering, as the vectors are rotated in pairs with respect to the diagonal elements. Rotation by lifting scheme:

$$\begin{pmatrix} \cos\theta & \sin\theta \\ -\sin\theta & \cos\theta \end{pmatrix} = \begin{pmatrix} 1 & \frac{1-\cos\theta}{\sin\theta} \\ 0 & 1 \end{pmatrix} \begin{pmatrix} 1 & 0 \\ -\sin\theta & 1 \end{pmatrix} \begin{pmatrix} 1 & \frac{1-\cos\theta}{\sin\theta} \\ 0 & 1 \end{pmatrix} \tag{3}$$

The rotation matrix is decomposed in the multiplication of three other rotation matrices, in such a way that the resulting structure shown in Figure 6, presents a lower complexity. Indeed, this implementation requires only three multipliers, one less concerning the original implementation, leading to a reduction of the 25% of the computational area, shorter latency and less power consumption. To further simplify the architecture, the multiplication for P and U coefficients from Equation (3)

$$P = \frac{1 - \cos\theta}{\sin\theta} \tag{4}$$

$$U = -\sin\theta \tag{5}$$

in Figure 6 are implemented as shift and add, as the number of possible rotation angles has been fixed to 8 (from 0, no rotation, to 7), as reported as optimum in [8] by M. Masera et al. The steerable block thus introduces $2 \times N$ clock cycles of latency for the reordering stage plus 4 clock cycles due to the internal pipeline. Therefore, in the event that all the SDCTs have a length N = 32, the latency is equal to 68 clock cycles, which corresponds to the worst case.

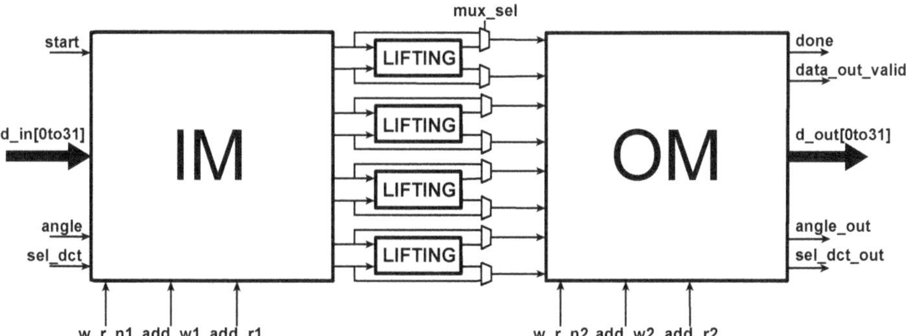


**Figure 4.** Steerable block structure.

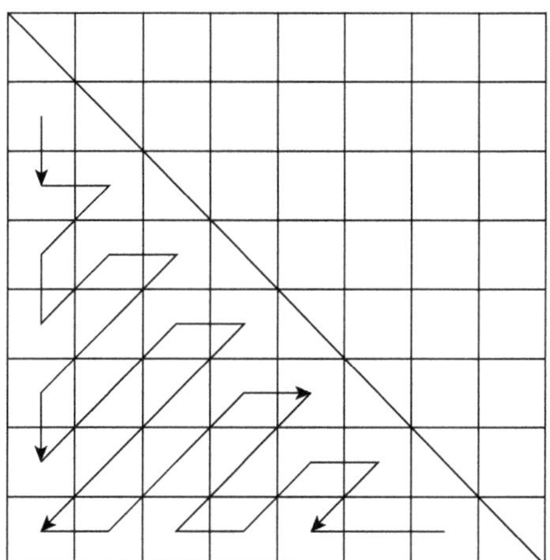

**Figure 5.** Zig-zag scanning order.

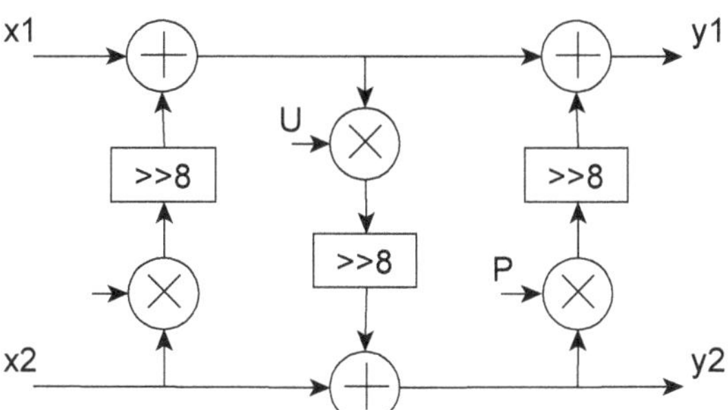


**Figure 6.** Lifting-based rotation.

## 3.2. Control Unit

The design requires two Control Units (CUs), one for the DCT part and one for the steering part. The 2D-DCT block is managed by its control unit, which generates all the control signals and the required memory addresses. It is composed of a Finite State Machine (FSM) and a counter. The FSM is composed of two states (FWR1 and FWC1), plus an IDLE state. When the external starting input signal is received, the FSM switches from IDLE to FWR1. The counter starts to increase its value and the write_enable signal is raised so that the partial 1D-DCT results are stored in the transposition buffer at the position indicated by the counter address value. The input signal itself encapsulates the length of the current DCT and consequently the value to be reached by the counter. Once the maximum counter value (cnt_max) is reached, the FSM switches from FWR1 to FWC1. In this state, the FSM is responsible for the read memory address generation and the assertion of the data_out_valid signal. The maximum counter value in this state remains the same as the previous one. Once cnt_max is reached, the two-dimensional transformation is completed, and the FSM evolves to a new FWR1 state if the start signal is asserted again, otherwise, it returns to the IDLE state.

For what concerns the steerable block, its control unit generates all the signals needed to manage the datapath and to address the two buffers. This unit is made up of an FSM and four counters. The FSM is composed of 14 states and an IDLE state, divided into 5 functional groups. Table 1 reports all groups functionalities. All the states belonging to the same group are similar, they are distinguished only by the different behavior of the output signals and the counter threshold.

State A coincides with the start of the steering process. Here, the 2D-DCT results are written into the input buffer. After that, the FSM switches to the B state, where the data is read from the input buffer and is written to the output one after being rotated. Then the results must be removed from the output buffer. However, as the video coding application requires to process a continuous stream of data, every time the previous results are completely written in the output buffer, new values need to be fetched and stored in the input one. State C handles such a situation, allowing the architecture to provide uninterrupted input/output data flow.

**Table 1.** Steerable Control Unit FSM states.

| | | |
|---|---|---|
| write input buffer | $A$ | START |
| read input & write output buffer | $B$ | WAIT |
| write input & read output buffer | $C, F, I, L$ | WB |
| read input & write output & read output buffer | $D, G, H, M$ | RWB |
| read output buffer | $E, E_1, E_2, E_3$ | RB |

In principle, these three states plus E are enough to execute the steerable operation but the execution of multiple steerable with different lengths must be considered. The FSM complexity grows with the number of different supported SDCT lengths. As stated before, this unit supports lengths of N = 4, 8, 16 and 32. Consequently, many different states are required. For instance, Table 2 shows one simple FSM execution, in which a steerable operation with length N = 16 is followed by a new operation with a length of N = 8. In this case, after the eight columns of new data are written in the input buffer, it is necessary to read and rotate them. The first N = 16 columns of the output buffer are filled with previous data, but not all of them have been read. Thus, the FSM introduces an offset in the writing address to avoid the overwrite of previous results. At this point, new data can be stored in the output buffer, while the old ones are read at the same time. In the opposite situation there are no problems: the new execution is longer than the previous one, so temporary storage is not needed.

**Table 2.** Example of FSM state evolution.

| Memory Operation | Number of Cycles | | | | | | | | | | | |
|---|---|---|---|---|---|---|---|---|---|---|---|---|
| Write | 16 | X | 16 | X | 8 | X | 4 | X | 4 | 12 | X | X |
| Read & Write | X | 16 | X | 16 | X | 8 | X | 4 | X | X | 16 | X |
| Read | X | X | 16 | X | 8 | 8 | 4 | 4 | 4 | X | X | 16 |
| State | A | B | C | B | C | D | C | D | C | A | B | E |
| | | | 16 | | 16 | | 8 | 4 | | | | 16 |

The four counters are responsible for the generation of the double buffer addresses and to control the FSM evolution from state to state. Two counters are necessary to decide the next state: while the first one takes into account the previous SDCT length, the second one deals with the current SDCT length. A third counter generates the addresses for the input buffer and the coefficients Read-Only Memory (ROM). Finally, the last counter is used to point to the SDCT results in the output memory. Figure 7 visually represents the simplified evolution of states in the control unit. The states are grouped as in Table 1:

- START: write input buffer
- WAIT: read input & write output buffer
- WB (Write Buffer): write input & read output buffer
- RWB (Read and Write Buffer): read input & write output & read output buffer
- RB (Read Buffer): read output buffer

The decision about which will be the next state depends on the the current SDCT computation phase and the size of the next SDCT to be computed, this is why the states RB, RWB and RB are so thightly interconnected.

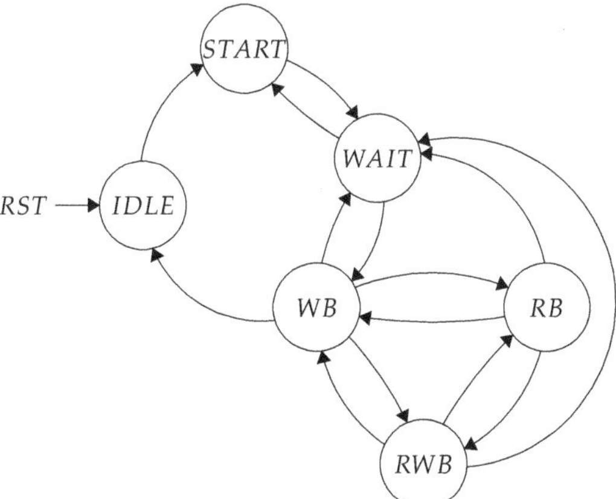


**Figure 7.** Simplified FSM diagram.

### *3.3. Reduced SDCT Architectures*

The unit presented so far can compute SDCT of lengths 4, 8, 16 and 32. This type of structure has been designed to be implemented inside the HEVC standard while providing maximum flexibility. This algorithm could be also used for video compression standards with lower constraints and image compression algorithms, such as JPEG. As these cases do not require the whole range of SDCT lengths, two reduced SDCT units have also been developed. The first can compute SDCT of length 4, 8 and 16 (called SDCT-16), while the second is capable of computing SDCT of length 4 and 8 (called

SDCT-8). These two units have a reduced throughput of 50% and 75%, respectively, with a parallelism of 16 or 8 data samples instead of 32. This leads to a consistent reduction of the memory sizes. In particular, the length of both rows and columns of all memories is halved in the SDCT-16 unit, while it is four times lower in the SDCT-8 unit with respect to SDCT-32. As a result, the area occupation of these units is much lower than the SDCT-32 one, providing suitable solutions tailored to the final application. Moreover, since the throughput is reduced, just one clock domain has been used for both DCT and steerable block. In this way it is possible to remove the FIFO memory interface and lower the design complexity.

## 4. Results

In order to satisfy the HEVC stream requirements for a video resolution of $7680 \times 4320$, frame rate of 60 fps, with the YUV 4:2:0 image coding, the proposed structure needs a throughput of almost 3 GSample/s. As discussed in Section 3, the folded version presented in [13] has been chosen since this approach guarantees the required throughput. This structure has a processing rate of 16 pixels per cycle, therefore the architecture needs a frequency of at least 187 MHz ($2.99 \times 10^9/16$ MHz). Furthermore, clock gating has been included during the synthesis process, leading to a power consumption reduction of about 58% as shown in Table 3. The technology employed for the synthesis is the UMC 65 nm. The following architectures have been considered and synthesized:

- two-dimensional DCT
- SDCT
- reduced SDCT-16
- reduced SDCT-8

Concerning the steering part, several clocks have been tested, namely $1\times$, $2\times$, $4\times$ and $8\times$ (with respect to the DCT clock frequency). By increasing the Steerable unit frequency it is possible to decrease the parallelism and consequently the number of input/output ports of the buffers.

**Table 3.** Estimated power consumption at 188 MHz.

| Power | Internal | Switching | Total Dynamic | Leakage |
|---|---|---|---|---|
| basic DCT | 36.55 mW | 17.72 mW | 54.47 mW | 33 μW |
| clock gated DCT | 21 mW | 12.52 mW | 33.52 mW | 30 μW |
| basic SDCT | 290.47 mW | 60.33 mW | 350.88 mW | 106 μW |
| clock gated SDCT | 88.71 mW | 59.85 mW | 148.67 mW | 94 μW |
| clock gated SDCT-16 | 27.86 mW | 28.97 mW | 56.85 mW | 27 μW |
| clock gated SDCT-8 | 6.56 mW | 7.20 mW | 14.17 mW | 7 μW |

It can be noticed in Table 4 that by reducing the data parallelism of the Steerable unit, the size of the input memory (IM) and output memory (OM) decreases considerably, while the size of all the other sub-blocks slightly increases, due to the synthesizer constraints with different clock regimes.

**Table 4.** SDCT area occupation for different clock regimes.

| Cell | $1\times$ Total Area | $2\times$ Total Area | $4\times$ Total Area | $8\times$ Total Area |
|---|---|---|---|---|
| SDCT | $4,337,744$ μm$^2$ | $3,042,226$ μm$^2$ | $1,608,759$ μm$^2$ | $1,301,522$ μm$^2$ |
| 2D-DCT | $438,866$ μm$^2$ | $601,970$ μm$^2$ | $455,150$ μm$^2$ | $474,167$ μm$^2$ |
| IM | $1,401,523$ μm$^2$ | $820,032$ μm$^2$ | $495,856$ μm$^2$ | $335,932$ μm$^2$ |
| OM | $2,377,837$ μm$^2$ | $1,418,162$ μm$^2$ | $482,048$ μm$^2$ | $319,037$ μm$^2$ |
| FIFO | $86,542$ μm$^2$ | $110,594$ μm$^2$ | $113,008$ μm$^2$ | $110,604$ μm$^2$ |
| ROM | $5895$ μm$^2$ | $22,228$ μm$^2$ | $13,227$ μm$^2$ | $33,223$ μm$^2$ |

Table 5 presents an overview of the obtained results, comparing the DCT baseline with the SDCT proposed.

**Table 5.** Overview of the obtained architectures.

| Architecture | DCT | SDCT | SDCT-16 | SDCT-8 |
|---|---|---|---|---|
| Technology (nm) | 65 | 65 | 65 | 65 |
| Frequency (MHz) | 188 | 188 | 188 | 188 |
| Power (mW) | 33.52 | 148.67 | 56.85 | 14.17 |
| Throughput | 2.992 G | 2.992 G | 1.496 G | 0.748 G |
| Area (mm$^2$) | 0.321 | 1.427 | 0.444 | 0.110 |

As it can be noticed, the area and power results of the SDCT-16 are around 60% smaller than the complete SDCT. On the other hand, the SDCT-8 area is around 75% smaller than the SDCT-16 and 90% smaller than the complete SDCT while the throughputs are reduced respectively by 50% and 75%. Finally, comparing the DCT and the SDCT architecture we can observe that the hardware overhead to support up to N = 32 is very large. However, removing the hardware support for the steering part with N = 32 (SDCT-16), the area becomes comparable with the one of the DCT. As a consequence, this solution can be of interest to increase the rate-distortion performance [8].

*4.1. Reduced SDCT Compression Savings*

The performance of the proposed encoder with a DCT directional transform is analyzed using the metric gauge Bjøntegaard Delta Bit-Rate (BDBR) [15], using the original HEVC encoder HEVC test Mode (HM-16.6) as the reference method. The full SDCT requires on average 22% more time to be executed with respect to plain DCT on an modified HM version, while SDCT-16 took on average 18% more time and SDCT-8 only 15% more time. By further optimization this overhead could be reduced to make the execution times closer to the DCT case. On one hand, negative values of BDBR stand for bit-rate savings, thus improved coding efficiency, while, on the other hand, positive values denote loss of rate-distortion.

Kimono, ParkScene, Cactus, BQTerrace and BasketballDrive are standard sequences employed to assess the encoder performances. The BDBR has been measured and the compression results are presented in Table 6 and Figure 8. As expected, the full SDCT presents a BDBR reduction but with a high computational cost. Reduced SDCTs are still able to maintain an average reduction, superior with respect to plain DCT compression. All the sequences have been compressed as *all intra* with default settings with Constant QP (*Quantization Parameter*) of values 22, 27, 32 and 37 for BDRD computation. Even when using only small SDCT transforms, the quality of the output is still better than the plain DCT. This is to be expected as the DCT can be seen as a special case of SDCT with steering angles of integer multiples of $\frac{\pi}{4}$.

**Table 6.** BDBR [%] for implemented reduced SDCT sizes versus DCT-only.

| Sequence | SDCT [8] | SDCT-16 | SDCT-8 |
|---|---|---|---|
| Kimono | −0.795 | −0.144 | −0.020 |
| ParkScene | −0.617 | −0.500 | −0.128 |
| Cactus | −0.485 | −0.392 | −0.209 |
| BQTerrace | −0.265 | −0.267 | −0.193 |
| BasketballDrive | −0.199 | −0.174 | −0.112 |
| **Average** | −0.472 | −0.295 | −0.132 |

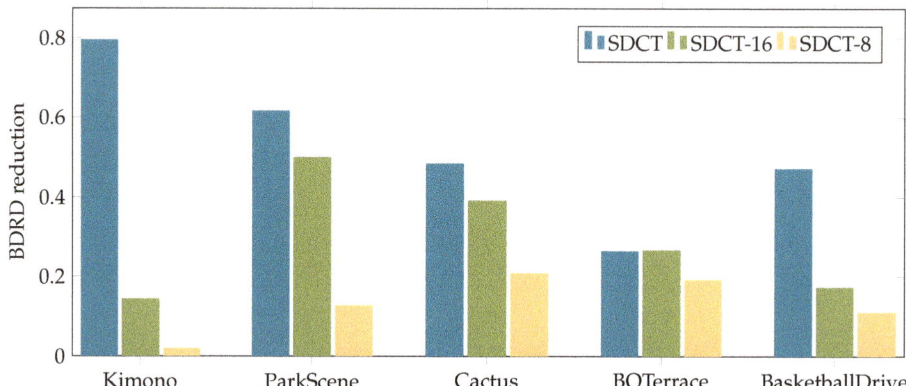

**Figure 8.** Histogram of obtained BDBR saving with respect to DCT.

*4.2. Comparison with Previous Works*

Since the Steerable-DCT is a new approach, it is not easy to make a fair comparison with other architectures found in the literature. However, for the sake of completeness, Table 7 proposes a comparison between the proposed SDCT architecture and some state of the art DCT ones. Zhao et al. [16] proposed an architecture able to support transform sizes from 4 × 4 to 32 × 32 with an implementation policy that reuses structure parts in order to contain the final dimension. Moreover, multiplications are substituted by shift and sum operations. Even though it uses a smaller technology compared to SDCT (45 nm vs. 65 nm) that grants a faster clock frequency (1.7×), the SDCT presents 4.7× higher throughput. Ahmed et al. [17] designed a folded structure that decomposes the DCT matrices into sparse submatrices to reduce the multiplications. Moreover, these last are eliminated thanks to a lifting scheme. Albeit such scheme supports 1080P HD video codec, its throughput is more than 12 times lower than the SDCT as well as the worst of those presented in Table 7 in terms of samples per second. Meher et al. [13] describe two versions of a pruned design: folded and full-parallel. Both present a working frequency equal to the SDCT, however, while the folded has also the same throughput, the full-parallel outperforms the rest since the hardware is replicated many times. Despite the SDCT follows a low-power paradigm with its folded-based structure, the hardware overhead needed to decompose the 2D-DCT transform results in superior power consumption. As a consequence, the pruned approach used in [13] grants a higher energy efficiency. Finally, Masera et al. [18] outline a folded approximated architecture with a just 7% higher throughput than SDCT, but with an energy per sample (EPS) comparable to the SDCT Folded-8 version.

**Table 7.** Comparison of 2D-DCT and SDCT Architectures.

| Design | | Technology [nm] | Frequency [MHz] | Throughput [Gsps] | Power [mW] | EPS [pJ] |
|---|---|---|---|---|---|---|
| Zhao et al. [16] | | 45 | 333 | 0.634 | - | - |
| Ahmed et al. [17] | | 90 | 150 | 0.246 | - | - |
| Meher et al. [13] | Folded | 90 | 187 | 2.992 | 40.04 | 13.38 |
| | Full-parallel | 90 | 187 | 5.984 | 67.57 | 11.29 |
| Masera et al. [18] | Architecture 1 | 90 | 250 | 3.212 | 51.72 | 16.10 |
| SDCT | Folded | 65 | 188 | 2.992 | 148.67 | 49.69 |
| | Folded-16 | 65 | 188 | 1.496 | 56.85 | 38 |
| | Folded-8 | 65 | 188 | 0.748 | 14.17 | 18.94 |

## 5. Conclusions

The most recent state-of-the-art compression technique is the HEVC, which almost doubles the performance in terms of rate-distortion compared to the H.264/AVC. Nevertheless, the continuous development of new High Definition (HD) or Ultra-HD (UHD) techniques introduces high requirements concerning the storing and the transmission of such sequences of frames. Thus, researchers and companies are trying to push further the HEVC boundaries.

This paper provides an efficient and compact hardware architecture accelerator for the SDCT algorithm to be used in the HEVC algorithm. Many of the design choices explained above present an optimized approach, such as the lifting-based approach, in which the hardware resources are reduced to a minimum. Moreover, the flexibility showed by this architecture makes it appealing for a wide range of applications, being able to work with different coding formats. The proposed SDCT framework is able to cope with 8k UltraHigh Definition (UHD) (7680 × 4320 pixels) with a frame rate of 60 Hz for the 4:2:0 YUV format, which is one of the highest resolution supported by HEVC. The steerable DCT is a viable solution to improve compression efficiency, as reported in [8]. Further work will cover the integration of the proposed accelerator in a complete HEVC framework to validate the performances in a real case scenario.

**Author Contributions:** Conceptualization, L.S.; methodology, L.S.; software, L.S. and M.M.; validation, R.P., M.C. and M.M.; formal analysis, M.M.; investigation, L.S.; resources, G.M., M.R.R. and M.M.; data curation, L.S.; writing—original draft preparation, R.P., M.C. and M.M.; writing—review and editing, M.M., G.M. and M.R.R.; visualization, R.P., M.C., M.M., G.M. and M.R.R.; supervision, M.M. and G.M.; project administration, M.M. and G.M.; funding acquisition, M.M., G.M. and M.R.R.; All authors have read and agreed to the published version of the manuscript.

**Funding:** This research received no external funding.

**Conflicts of Interest:** The authors declare no conflict of interest.

## References

1. Sullivan, G.J.; Ohm, J.; Han, W.; Wiegand, T. Overview of the High Efficiency Video Coding (HEVC) Standard. *IEEE Trans. Circuits Syst. Video Technol.* **2012**, *22*, 1649–1668. [CrossRef]
2. Naccari, M.; Gabriellini, A.; Mrak, M.; Blasi, S.G.; Zupancic, I.; Izquierdo, E. HEVC Coding Optimisation for Ultra High Definition Television Services. In Proceedings of the Picture Coding Symposium, Cairns, Australia, 31 May–3 June 2015; pp. 20–24. [CrossRef]
3. Masera, M.; Fiorentin, L.R.; Masala, E.; Masera, G.; Martina, M. Analysis of HEVC Transform Throughput Requirements for Hardware Implementations. *Elsevier Signal Process. Image Commun.* **2017**, *57*, 173–182. [CrossRef]
4. Chen, Z.; Han, Q.; Cham, W. Low-Complexity Order-64 Integer Cosine Transform Design and its Application in HEVC. *IEEE Trans. Circuits Syst. Video Technol.* **2018**, *28*, 2407–2412. [CrossRef]
5. Sun, H.; Cheng, Z.; Gharehbaghi, A.M.; Kimura, S.; Fujita, M. Approximate DCT Design for Video Encoding Based on Novel Truncation Scheme. *IEEE Trans. Circuits Syst. I Regul. Pap.* **2019**, *66*, 1517–1530. [CrossRef]
6. Oliveira, R.S.; Cintra, R.J.; Bayer, F.M.; da Silveira, T.L.T.; Madanayake, A.; Leite, A. Low-complexity 8-point DCT approximation based on angle similarity for image and video coding. *Multidimens. Syst. Signal Process.* **2019**, *30*, 1363–1394. [CrossRef]
7. Fracastoro, G.; Fosson, S.M.; Magli, E. Steerable Discrete Cosine Transform. *IEEE Trans. Image Process.* **2017**, *26*, 303–314. [CrossRef] [PubMed]
8. Masera, M.; Fracastoro, G.; Martina, M.; Magli, E. A Novel Framework for Designing Directional Linear Transforms with Application to Video Compression. In Proceedings of the ICASSP 2019—2019 IEEE International Conference on Acoustics, Speech and Signal Processing (ICASSP), Brighton, UK, 12–17 May 2019; pp. 1812–1816. [CrossRef]

*Sensors* **2020**, *20*, 1405

9.      Sole, L.; Peloso, R.; Capra, M.; Roch, M.R.; Masera, G.; Martina, M.  VLSI Architectures for the Steerable-Discrete-Cosine-Transform (SDCT). Applications in Electronics Pervading Industry, Environment and Society (ApplePies) 2019, to Appear.  Available online: https://applepies.eu/ (accessed on 13 September 2019).

10.     Sole, L. VLSI architectures for the Steerable Discrete Cosine Transform.  Master's Thesis, Politecnico di Torino, Turin, Italy, 2018.

11.     Zeng, B.; Fu, J. Directional Discrete Cosine Transforms—A New Framework for Image Coding. *IEEE Trans. Circuits Syst. Video Technol.* **2008**, *18*, 305–313. [CrossRef]

12.     Shuman, D.I.; Narang, S.K.; Frossard, P.; Ortega, A.; Vandergheynst, P.  The emerging field of signal processing on graphs: Extending high-dimensional data analysis to networks and other irregular domains. *IEEE Signal Process. Mag.* **2013**, *30*, 83–98. [CrossRef]

13.     Meher, P.K.; Park, S.Y.; Mohanty, B.K.; Lim, K.S.; Yeo, C. Efficient Integer DCT Architectures for HEVC. *IEEE Trans. Circuits Syst. Video Technol.* **2014**, *24*, 168–178. [CrossRef]

14.     Daubechies, I.; Sweldens, W. Factoring wavelet transforms into lifting steps. *J. Fourier Anal. Appl.* **1998**, *4*, 247–269. [CrossRef]

15.     Bjontegard, G.  Calculation of Average PSNR Differences between RD-curves.  In Proceedings of the ITU—Telecommunications Standardization Sector STUDY GROUP 16 Video Coding Experts Group (VCEG), 13th Meeting, Austin, TX, USA, 2–4 April 2001.

16.     Zhao, W.; Onoye, T.; Song, T. High-performance multiplierless transform architecture for HEVC.  In Proceedings of the 2013 IEEE International Symposium on Circuits and Systems (ISCAS), Beijing, China, 19–23 May 2013, pp. 1668–1671. [CrossRef]

17.     Ahmed, A.; Shahid, U.; Rehman, A.  N point DCT VLSI architecture for emerging HEVC standard. *VLSI Design* **2012**, *2012*. [CrossRef]

18.     Masera, M.; Martina, M.; Masera, G. Adaptive Approximated DCT Architectures for HEVC. *IEEE Trans. Circuits Syst. Video Technol.* **2017**, *27*, 2714–2725. [CrossRef]

*Article*

# Distillation of an End-to-End Oracle for Face Verification and Recognition Sensors [†]

**Francesco Guzzi** [1,2,*], **Luca De Bortoli** [1], **Romina Soledad Molina** [1,3], **Stefano Marsi** [1], **Sergio Carrato** [1] **and Giovanni Ramponi** [1]

[1] Engineering and Architecture department, Image Processing Laboratory (IPL), University of Trieste, 34127 Trieste, Italy; luca.debortoli@dia.units.it (L.D.B.); ROMINASOLEDAD.MOLINA@phd.units.it (R.S.M.); marsi@units.it (S.M.); carrato@units.it (S.C.); ramponi@units.it (G.R.)

[2] Elettra Sincrotrone Trieste, Scientific Computing, 34149 Basovizza, Italy

[3] The Abdus Salam International Centre for Theoretical Physics (ICTP), Multidisciplinary Laboratory, 34151 Trieste, Italy

[*] Correspondence: francesco.guzzi@phd.units.it

[†] This paper is an extended version of our paper published in Applications in Electronics Pervading Industry, Environment and Society, Lecture Notes in Electrical Engineering 627, S. Saponara and A. De Gloria (eds.).

Received: 4 February 2020; Accepted: 28 February 2020; Published: 2 March 2020

**Abstract:** Face recognition functions are today exploited through biometric sensors in many applications, from extended security systems to inclusion devices; deep neural network methods are reaching in this field stunning performances. The main limitation of the deep learning approach is an inconvenient relation between the accuracy of the results and the needed computing power. When a personal device is employed, in particular, many algorithms require a cloud computing approach to achieve the expected performances; other algorithms adopt models that are simple by design. A third viable option consists of model (oracle) distillation. This is the most intriguing among the compression techniques since it permits to devise of the minimal structure that will enforce the same I/O relation as the original model. In this paper, a distillation technique is applied to a complex model, enabling the introduction of fast state-of-the-art recognition capabilities on a low-end hardware face recognition sensor module. Two distilled models are presented in this contribution: the former can be directly used in place of the original oracle, while the latter incarnates better the end-to-end approach, removing the need for a separate alignment procedure. The presented biometric systems are examined on the two problems of face verification and face recognition in an open set by using well-agreed training/testing methodologies and datasets.

**Keywords:** face recognition; face verification; biometric sensors; deep learning; distillation; convolutional neural networks; spatial transformer network

---

## 1. Introduction

### 1.1. Face Recognition Sensors

Face recognition systems represent now a pervasive reality. Smartphones, computers and social networks provide verification, similarity and recognition functions for both security and entertainment purposes. The basic hardware setup exploits a webcam and a single-board PC, that are already used in the device; a separate 'face sensor module' may be included to perform face recognition functions. Especially due to the latter class of sensors, many small devices emerged whose output is an identity or a biometric signature. Indeed, they are sold as biometric sensors on distributors' websites.

Face recognition has pushed more than any other research topic on Convolutional Neural Networks (CNNs) because the impact of human-like performances in this type of Artificial Intelligence is huge [1]. Making a self-learning system understand where and what to observe for reliable identification and the use of biomimicry for the design of the trainable structures provided the biggest performance boost with respect to the results of the previous hand-crafted algorithm.

Neural networks for face recognition are typically trained by the use of a supervised training procedure, in which the right answer to the problem is provided for each input; during the training, the model learns to give the correct answer analyzing the distribution of each input (the pixels) and then, through error back-propagation, enforcing a set of parameters (the weights of convolutional and fully-connected layers). Even if the aforementioned biomimicry has inspired a particular structure and hierarchy for signal processing, presently it is impossible to determine detailed reasons why a certain weight takes a certain value. The same happens for the meaning of each axis in the multi-dimensional output space in which a particular identity is defined. The main point of the current deep learning methods is to avoid defining handcrafted methods that are fully understood, but provide inevitably lower performances. That is why the term "oracle" is used in the literature. In this paper, we use oracle as a synonym of "trained model".

The common thinking is that the problem of face recognition is completely solved, but in fact, it is not. Present-day systems can be used in consumer products and for a small database of users, but they are unable to provide the high accuracy desired e.g., for a banking system. Until recently, 'the complex-the better' paradigm has been the only viable solution, leading to complex and very deep oracles that can operate on a High-Performance Computing (HPC) platform only. For example in [2] a multi-feature fusion algorithm is proposed, but its applicability is constrained to CUDA capable devices. The trend of the latest research [3] follows a complete end-to-end approach (fully neural detection and classification in the same structure), including accurate but complex object detectors (RCNN [4], FCN [5], MMod [6], MTCNN [7]), but unfortunately leading to slow models, even on high-end GPU [8]. For personal devices, instead, two options are available: to accept an inevitable performance loss (in terms of speed or accuracy) or to use a cloud computing infrastructure with an online API. The situation is a bit different when active-sensor designs are exploited: in that case, the identity of an individual is evaluated using not only the 2D image grabbed by a webcam but also analyzing a 3D map of the face. Such a system is more robust but way more expensive, and its distance range is limited by the projector power. In this paper, only monocular passive face sensor systems will be taken into account.

### 1.2. Framework

In a previous work [9] an open-source framework (Dlib [10]) for face-recognition has been identified and exhaustively tested. The framework, presented in Figure 1, consists of a selection of a face detector, an alignment procedure, an embedding oracle (feature extractor) and an identity classifier. The latter component can be implemented from scratch in the form of a shallow multi-layer perceptron (MLP) neural network (highest accuracy, short mandatory training phase) or with a simple distance metric (lower accuracy, insertion of identities in the database at runtime). Each classification algorithm is run on the features provided by a features extractor CNN (dlib-resnet-v1) [11] that is released as a part of the Dlib library in the form of a pre-trained model; in conjunction with the subsequent classifiers, this embedding network proved to be sufficiently discriminative.

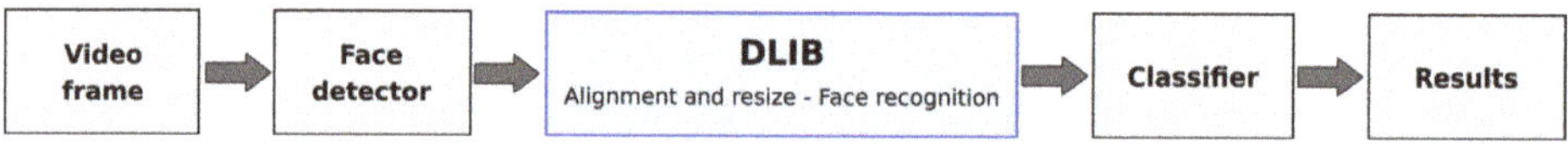


**Figure 1.** View of the former Dlib face recognition framework signal chain.

Besides this, while on a PC the presence of a CUDA-compatible GPU permits a reasonable processing rate of 5–10 fps, on mobile hardware with an ARM CPU the average speed is in the order of roughly 0.5 fps (with Dlib compiled using ARM NEON [12] instructions), making mobile use impractical. Another macroscopic problem of this pre-trained model is that it has been created within the Dlib framework. As a consequence, further modifications, fine-tuning and research, as well as a simple conversion of the model represent an unnecessarily difficult burden [1].

### 1.3. Paper Outcome

In a previous work, we shared our findings on the distillation of dlib-resnet-v1 into a smaller model that could be implemented on mobile hardware for the recognition of tens of identities. Previous results were encouraging but were trained using a private dataset, composed of a mixture of the ones available online. The crucial point lies in the fact that we wanted to make a dataset as much as possible similar to the one used for the training of the teacher network. The present contribution, on the contrary, describes two feature extractors (the distilled models) obtained by distilling with a standard dataset only (CASIA): the first one can be directly inserted in the place of dlib-resnet-v1, while the second one provides a novel distilled network obtained including a spatial transformer component [13] in the structure; this not only removes the landmark detector and the face aligner (moving towards an end-to-end neural approach), but also allows to obtain a higher accuracy in the case of non-perfectly-frontal faces. Commonly used face detectors (Viola–Jones [14] and HOG [15]) are sensitive to pose and fail to detect most non-frontal faces. The use of newer face detection algorithms, typically CNN-based (SSD [16] and MMOD [6]), solves the aforementioned problem; however, the subsequent phase of landmark detection may provide wrong results, reducing the accuracy of the entire system. The described the Spatial Transformer Network (STN) boosted model correctly gives attention to the face part in a wide ROI and produces an aligned and tight crop of the face for the subsequent feature extractor, also for difficult poses. As far as we know, the distillation of an STN model is a novel procedure, fighting the idea that STN are difficult to train.

Differently from the previous work, this paper includes the LFW face verification test, which is a standard test procedure composed of a dataset and a testing protocol. This test is crucial to estimate the embedding skill of a neural model. In most research, the final model is trained on face recognition tasks that involve a closed set of individuals, just like any classification, neglecting the fact that in order to actually use this model it is necessary to expand the set. If a classifier is completely embedded into the model, the entire procedure is cumbersome. In this work, we emphasize the need for a modular procedure with a clear distinction of a feature extractor and a classifier, where the latter is a simple customizable structure. Furthermore, as far as we know, in this paper, a face recognition test procedure on an open set is formalized and described for the first time. We hope this will fill an empty spot in the field of the evaluation of the face recognition ability of the model.

This paper is organized as follows: the first section describes the framework, the distillation technique and the network design; the second section describes the testing methodologies, while the last one provides a discussion of the results.

## 2. Materials and Methods

### 2.1. Transfer Learning and Model Compression

When a model is trained to accomplish a task, it is convenient not to start from scratch (e.g., Gaussian or Xavier initialization) but to apply 'transfer learning', that is to copy as much as possible the weights of a previous well-trained network to the one that has to be trained. It has been demonstrated that starting in this way is generally more effective (not in any case, e.g., [17]) than starting with no description of knowledge at all (random initialization), even if the tasks of the two networks are different. One of the best starting points for computer vision tasks is the set of pre-trained weights obtained for the classification of the Imagenet dataset. It is important to note that in order to take advance of this sweet spot in the Loss-Parameters space the network configuration and structure have to be kept equal, possibly also wasting resources (e.g., replicating the number of channels of a grayscale image, or scaling the input image to fit the size of the first layer).

As reported in [18], applying concepts from the vast topic of model compression is the first step for reducing model complexity, and this result is obtained just by reducing the computational cost of each operation, without changing the structure of the model. The reduction of the time and memory complexity is instead a process that involves both structure simplification and a reduction in the number of parameters; the sweet spot is given by a reduced set of parameters and a smart choice for the data processing flow that maintains the same level of accuracy as the original network [1].

*2.2. Model Distillation and Teacher–Student Approach*

The first section highlighted the need for a complex structure to achieve the complex goal of face recognition. In fact, what requires complexity is the extraction of general characteristics from the provided samples (during the supervised learning process), rather than their actual representation. This means that when this knowledge has been inferred, it can be eventually represented by a simpler structure that can, in turn, be deployed to mobile hardware [18].

A recent and detailed survey on the general principles of distillation and model compression is presented in [19].

The form of compression [18,20–22] used in this work decorrelates the accuracy that a model achieves when performing a task from its learned weights: what is important to transfer (to distill) into a new model is the I/O relationship of the model itself, or the capacity to reveal the latent conditional distribution $p(T|X)$ that relates the inputs $X$ and the outputs $T$. This capacity is called 'dark knowledge' [21] and the act of transferring it from a slow but well-trained model (the teacher) to a student model is called 'knowledge distillation' [22].

The training set for the distillation process carried out as supervised learning is composed of the tuple $(X, T)$, i.e., the input and the corresponding target. The distillation is carried out as a regression process, forcing the student network to provide the same descriptor generated by the teacher; in the case of an embedding network, this can be directly described in a distance metric framework, where a distance larger than the hypersphere radius of each cluster automatically flags bad learning. This motivates to choose as a loss metric the Euclidean Distance $L_d$ [18] calculated between the target feature vector $T$ and the corresponding predicted descriptor $Y$.

The recent paper [23] proposes a peculiar knowledge distillation method composed of two different phases that explicitly takes into account smaller size and low-quality faces. In a complex training procedure, firstly the teacher network is frozen and a trainable structure of fully connected layers is attached to it. This model is then trained using a classification loss. The student model is distilled in a similar fashion of the previous "annealing based distillation" of Hinton. Unfortunately, this complex procedure leads to poor results in terms of the LFW face verification test.

The paper [24] presents a model for person re-identification, distilled from an ensemble of teacher models. Again a complex framework is exploited, in which a log-Euclidean distance is used as a loss function over sample similarity matrices. The framework automatically decides the reliability of each teacher in an adaptive fashion.

The paper [25] explores different techniques for using pre-digested information or in the paper called "privileged information". In the paper, the term distillation is again used to denote the student training of output probability vectors, while the term "knowledge transfer" is used to denote a procedure that only slightly resembles our method: a mapping function is estimated that manipulates the features of the teacher adapting them for the student network.

Summarizing, the majority of the knowledge transfer methods based on distillation supervise the learning of intermediate features, or of output probability distribution (classification, soft-classification), eventually with the help of samples similarity-like matrixes. The only cases in which an output descriptor is somehow distilled [20,25] take into account just the adapted version of these features. In our work we designed a simple procedure for distillation in the metric framework that results in the training of a model completely different (and smaller) from the teacher, exploiting a different and smaller dataset composed of samples of low-quality images (image are reduced to a fourth). The testing of the distilled models is carried out on completely different identities (not only different images of the same id) unseen during the distillation, so the real generalization power of the model is tested. Making this entire training framework straightforward allows us to use distillation as an effective technique also for the initialization of a newer model, where training from scratch would require weeks of training.

### 2.3. Alignment Procedure and Spatial Transformer Network

Conventional network models, in general, do not have a high degree of spatial invariance. This makes the ROI cropped by a face detector not usable directly without a huge classification accuracy drop. If correctly realized, a face alignment procedure solves this problem by applying a spatial transform that brings face parts (eyes, mouth, nose, chin) on fixed points in the frame; the aforementioned procedure relies on a landmark detector (LD) in charge of searching for those landmarks within the frame. In the Dlib framework, the LD used is an implementation of the Kazemi-Sullivan algorithm [26] based on regression trees. Other approaches use local binary patterns [27] or a joint face detector/aligner structure based on SVM [28]. MTCNN [7] is one of the most effective CNN-based face-detector/landmark detectors and its recent implementation in Keras [29] increased its popularity. Research on multi-pose LD opened the way to 3D alignment: however, even if the most powerful methods (GAN [30] and symmetrization [31]) are optimal for restoration or entertainment purposes, 3D alignment did not show to provide significant advantages in terms of recognition accuracy over its 2D version [32].

A Spatial Transform layer [13] is a clever solution that has been introduced to provide spatial invariance to feature maps by applying a predefined spatial transformation on it; while stride and pool are fixed hyperparameters, STN transformation has parameters that are learned during the training of the entire model. The component that is responsible for the generation of suitable parameters is the so-called localizer, a shallow CNN which is responsible for the efficiency of the entire structure. A sampling grid is generated on-the-fly starting from the inferred transformation parameters and the gradients are calculated for the sampled points. When an STN is used as the input layer, an interesting effect happens: the network focuses on the portion of the input frame that it deems relevant for the task at hand. This is recognizable by observing the output image generated after the STN sampler. This fact can be used to localize a single object or a ROI within the frame or, like in this work, to localize a face in wide a crop (e.g., as provided by an uncertain face detector). In [33] an STN is used as well as the base of a neural face-detector STN, but with an important difference: the first stage is composed of a multi-task Region Proposal Network, which produces candidate ROI within the frame. Only in second place, the STN is used for the alignment of this candidate regions onto a canvas of predefined landmarks, whose positions represent some of the parameters to be learned. If the exploited transformation in an STN has at least four degrees of freedom (DoF) (e.g., it is a similarity transform), the byproduct of this method is a simple yet effective alignment of the face. In the influential [34] a shallow input STN (exploiting affine transform) is used as the input structure, and the following recognition model is simultaneously trained from scratch using a combination of loss functions.

In our work, we cascade an STN similar to the one above, with a different recognition model (the topic of the next section). The entire structure is then trained using distillation, following the teacher-student approach.

### 2.4. Contribution

#### 2.4.1. Network Architecture

The teacher Dlib network 'dlib-resnet-v1' is based on a ResNet-34 structure [35] with few layers removed and the number of filters per layer reduced by half [10]: It has a 150 x 150-pixel input size,

29 convolutional layers and one fully-connected output layer for a total of roughly 6 M parameters. The network is provided pre-trained (for two weeks) on a dataset composed of roughly 3 M images. Due to its training procedure design, it is referred to as an embedding network because, for any given input image (an aligned face), the model provides a 128-dimensional features vector which virtually belongs to the embedding of that particular identity. During the training, a fixed distance margin is imposed between different identities meaning that all the possible images of a defined person would lie in a hyper-sphere of radius lower than the margin (0.6).

The student network design is crucial because, in principle, the computationally lightest model that allows us to obtain the performances of the teacher has to be defined. We can state the problem similar to the search for an ad-hoc optimal lossy compression for an average input distribution, evaluating a similarity metric.

Different CNNs based on the Densenet121 model [36] were designed searching for a structure with fewer weights than the original dlib model. This network design uses a combination of dense blocks, where features at different convolutional layers are concatenated, and transition blocks, where the features are processed and reduced to limit the pyramidal growth. Compared to Resnet [35] or Unet [37], this structure produces a stronger gradient flow and is computationally more efficient.

After training and testing four different variants [1,18], obtained cutting the Densenet at a different number of dense-transition blocks (also in the middle), we decided to choose the second biggest network (Net 2.0) as our base for the evolution of the network with the STN. Net 2.0 yields a reduction by a factor of 3.7 in size and by one order of magnitude in processing time (with HW accelerator), which is considered acceptable. The performance gain This can be seen in detail in Figures 2 and 3.

| Layers | DenseNet121 | Net 2.5 | Net 2.0 | Net 1.0 | Net 0.5 |
|---|---|---|---|---|---|
| **Input** | Shape (80x80x3) | | | | |
| **Convolution** | Conv 7x7, stride 2, 64 kernels | | | | |
| Output shape | (40x40x64) | | | | |
| **Pooling** | Max pool 3x3, stride 2 | | | | |
| Output shape | (20x20x64) | | | | |
| **Dense Block 1** | 6x basic block | | | | **2x** basic block |
| Output shape | (20x20x256) | | | | (20x20x128) |
| **Transition** | Conv 1x1, 128 kernel + Avg pool 2x2, stride 2 | | | | |
| Output shape | (10x10x128) | | | | |
| **Dense Block 2** | 12x basic block | | | | |
| Output shape | (10x10x512) | | | | |
| **Transition** | Conv 1x1, 256 kernels + Avg pool 2x2, stride 2 | | **Conv 1x1, 256 kernels** | | |
| Output shape | (5x5x256) | | (10x10x256) | | |
| **Dense Block 3** | 24x basic block | **20x** basic block | | | |
| Output shape | (5x5x1024) | (5x5x896) | | | |
| **Transition** | Conv 1x1, 512 kernels + Avg pool 2x2, stride 2 | **Conv 1x1, 256 kernels** | | | |
| Output shape | (2x2x512) | (5x5x256) | | | |
| **Dense Block 4** | 16x basic block | | | | |
| Output shape | (2x2x1024) | | | | |
| **Pooling** | Avg pool 2x2 | Avg pool 5x5 | Avg pool 10x10 | Avg pool 20x20 | Avg pool 20x20 |
| Output shape | (1x1024) | (1x256) | (1x256) | (1x256) | (1x128) |
| **Fully Connected** | Output Shape (1x128) | | | | |
| **N° of Parameter** | 7,17 M | 3,94 M | 1,48 M | 381 K | 123 K |

| Layers | Basic Block |
|---|---|
| Input | Shape (NxNxK) |
| **Bottleneck** | Conv 1x1, 128 kernel |
| Output shape | (NxNx128) |
| **Compression** | Conv 3x3, 128 kernel |
| Output shape | (NxNx32) |
| **Concatenate** | ConCat(Input, Compression) |
| Output shape | (NxNxK+32) |

**Figure 2.** Schematic representation of the original Densenet121 model (first column) and our four different variants. In this work, Net 2.0 (3rd column) is chosen as our base recognition network.

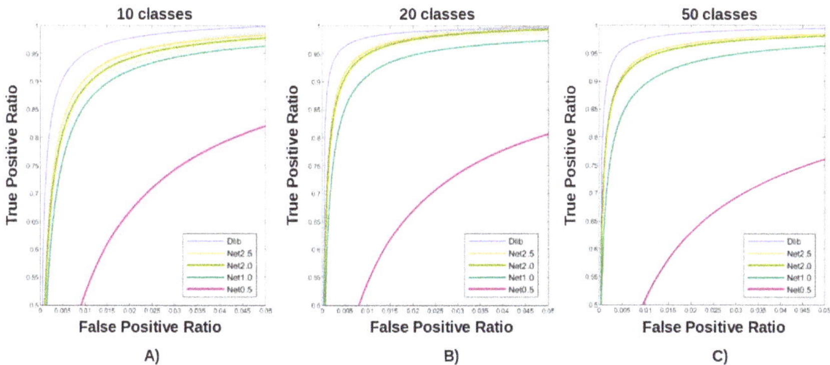

**Figure 3.** Performance evaluation of the different variants of the base distilled network [Net 0.5, Net 1.0, Net 2.0, Net 2.5] for the face recognition task in the case of 10 (**A**), 20 (**B**), 50 (**C**) classes. The procedure used was similar to the one described in section 2.4.2. The performance over # parameters ratio is extremely competitive for Net 2.0 (1.48 Mparameters), while Net 2.5 (3.94 Mparameters) provides only a limited amount of performance gain for its number of parameters with respect to Net 2.0.

A strong reduction in computational complexity is achieved also by limiting the image input size at 80 × 80 pixels, thus forcing smaller faces (trough distillation) to be described by the same point computed with a frame four times larger (Figure 4).

**Figure 4.** Signal chain of the hybrid framework composed by keeping the former face detector and alignment process; our distilled Convolutional Neural Networks (CNNs) block substitutes 'dlib-resnet-v1'.

In order to cope with difficult poses and to enforce a better distillation, we modified the previously described network, adding an STN structure that acts as a neural face aligner, as shown in Figure 5.

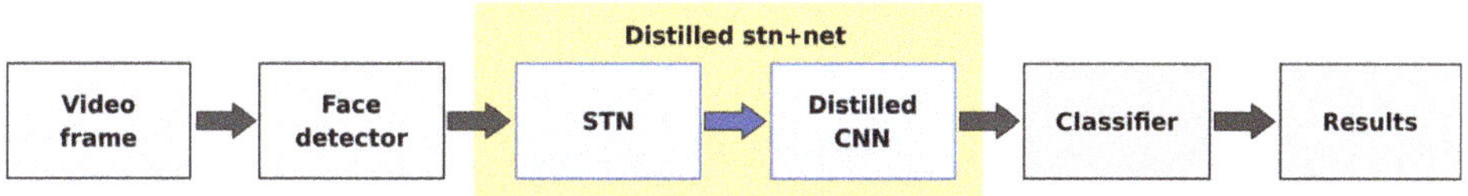

**Figure 5.** Signal chain of the novel hybrid framework composed by removing from the former the alignment procedure, which is substituted by the Spatial Transformer Network (STN) component in our 'distilled stn+net'; the end-to-end structure encloses also our feature extractor network, in place of 'dlib-resnet-v1'.

Like in [13], for the localization network we experimented with a shallow cascade of convolutional layers followed by a sequence of two fully connected layers whose output provides the six parameters of an affine transform. Differently from the previous case, the input size of the STN component is set to 120 × 120 pixels to help the localization network and provide a stronger free-data augmentation, but the final size of the transformed image is still 80 × 80; no modifications are needed in the recognition network. The structure detail is presented in Figure 6.

As far as we know, distillation on an STN based structure has never been attempted in the literature.

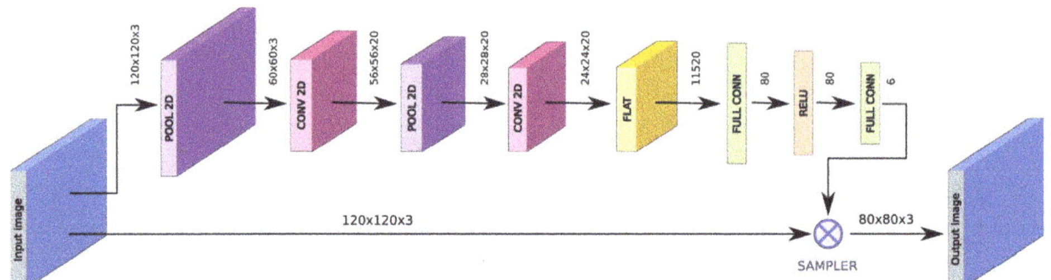

**Figure 6.** Structure of the STN component: an input image is processed by a shallow CNN-based localization network. The convolutional feature maps are then processed trough two fully connected layers that generate the six parameters of the affine transform. A grid generator (not represented) generates the corresponding sampling grid, that will be actually sampled by the sampler, producing an automatically aligned and cropped version of the input.

### 2.4.2. Multiclass Open Set Problem

Intuitively, the problem that a face recognition network will solve is to correctly classify identities. In an example access control system, subjects belonging to the group of "friends" have to be recognized not only as members of that group, but in their specific identity too, in order to avoid authentication errors. Concurrently, for "unknowns" the access must not be granted.

To emulate this problem (and evaluate our models), a multi-class classifier has been designed starting from the features generated from each image: the training process consumes the features of friends only, resulting in an n-classifier for "n-friends", with n-outputs. During a test procedure, the classifier decides and we keep track of its decision, counting how many times a correct or a wrong choice has been made. In a closed set, the procedure is limited since all the possible cases, as well as all the possible individuals, can be evaluated. An example of this kind of classification is object recognition, in which a trainable oracle has to decide among a limited number of objects.

In the case of an open set, on the contrary, the cases to be considered are non-numerable. A common-sense way to tackle this problem is to estimate a confidence index related to the classifier decision. By adopting such an index it is possible to discriminate unknown subjects (for whom a classifier has not been trained), basing on the probably lower confidence of their identification. Since in most applications, a false-positive error is more dangerous than a false negative, the identification accuracy of known subjects can be increased granting access only to those with a high confidence index.

Defining the performance of a multiclass classifier depends on the scenario in which the classifier operates. In fact, some indexes or parameters which are usually adopted for a binary classifier can hardly be fostered in the case of a multi-class classifier.

More formally, we define a set of positive examples ($Ni$) belonging to the group of known identity ("friends") that must be correctly classified (n-classes) and a set of negative samples ($F$), belonging to unknown individuals, that are used only to test the classifier, as no sample of this set have been seen during the classifier training. These latter samples, if correctly classified, represent the true-negative ($TN$), in respect of each class of known people and therefore they should not contribute in the evaluation of the global true-positive rate ($TPR$) like in the binary case. On the other hand, if they were incorrectly classified, they would represent false-positives ($FP$) for our system. Moreover, if a sample of a "friend" is erroneously classified as an 'unknown', this does not lead to an increase in the $FP$, but rather represents a false-negative ($FN$), whose impact on the evaluation of the classifier performance acts in a different way. Thus, we propose the following formulas for the calculation of the $TPR$ and $FPR$ in the case of multi-classification in an open-set scenario. A demonstration of these formulas is provided in Appendix A.

$$TPR = \frac{\sum_{i=1}^{K} TP_i}{N} \tag{1}$$

$$FPR = \frac{\sum_{i=1}^{K} FP_i}{K * F + (K-1)N} \tag{2}$$

where $K$ is the number of classes used, $F$ represents the total number of the negative samples ('unknown' or 'others' ID) and $N$ is the total number of the positive sample (known ID or 'friends').

## 3. Distillation Experiments

### 3.1. Distillation Process

In this section, it will be described how the distillation takes place. Compared to the Dlib network, the two design choices that allowed for an extensive parameter reduction (5.58 M vs. 1.48 M) in the distilled network are the use of modern network design and the reduction of the image input size from $150 \times 150 \times 3$ to $80 \times 80 \times 3$ pixels. The computational complexity has been reduced maintaining a comparable recognition accuracy. In the following text, we will refer to this realization as 'distilled net'. Besides the differences, this first distilled oracle can be used as a direct substitution in the former framework (Figure 7).

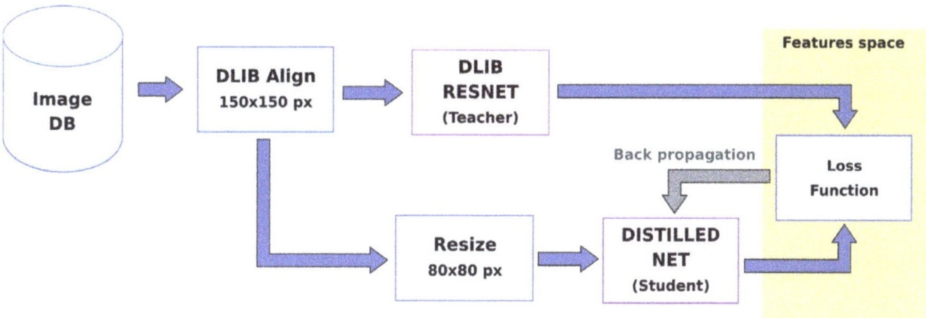


**Figure 7.** Proposed system with Dlib Resnet as teacher and Distilled network as student.

Furthermore, in our second realization, which can be seen in Figure 8, an input STN structure is added to the model: the net effect of this change is in an improved recognition accuracy especially in a less constrained scenario. An STN with $120 \times 120 \times 3$ pixel input and 0.93 M parameters is proposed; the overall network, which we will call 'distilled stn+net', uses 2.41 M parameters.

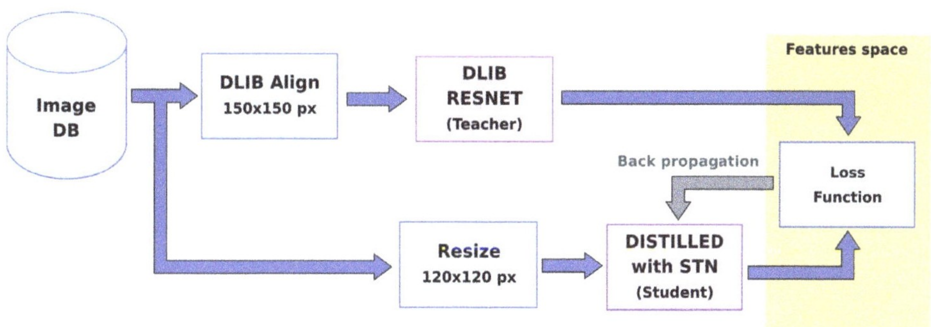


**Figure 8.** Proposed system with Dlib Resnet as teacher and Distilled network with STN as student.

To train the two distilled networks we adopt the CASIA Web Face [38] dataset, composed of approximately 500 k images for 10.6 k identities, while the LFW [39] dataset is used for the subsequent tests; these two datasets have an overlap of 16 identities, which have been removed from the training dataset, in order to test the generalization capacity of the distilled model. Each dataset was "filtered" with Dlib's HoG face detector: in this way, images with multiple faces were discarded. In the end, for each image, we have collected the corresponding features vector generated by the 'dlib-resnet-v1' model.

A set of ($image, target$) tuples is consumed during the training procedure, carried out forcing the student network to regress the target features vector for each image. For each RGB sample, the preprocessing step involves just a [0, 1] normalization and a per-channel shifting, while for each target vector the dataset average feature vector is subtracted. This procedure will simply change the origin of the 128-dimensional feature space.

In the case of the 'distilled net', all the color images have been aligned following the Dlib framework procedure and have been resized to 80 × 80 pixels. In contrast, for the 'distilled stn+net' no landmark detection and alignment was needed, and the dataset images have been only resized to 120 × 120 pixels.

For each distillation, we decided not to use any data-augmentation procedure because, in the described regression teacher-student approach, for each augmented sample we would have to generate the corresponding descriptor, inflating enormously the dataset dimension. In the case of 'distilled stn+net', again no data augmentation is enforced. During the training, the small fluctuations in the STN parameters (due to infinitesimal but nonzero gradient components) lead to a different image at the input of the recognition network, providing an effective data augmentation. At the same time, we have seen no signs of overfitting for the STN (that undergoes no augmentation).

The training of the nets continued for 100 epochs on batches of 128 images using Adam as the optimizer of choice. The supervised learning procedure evaluates the target error in terms of the Euclidean distance. The validation set consists of 1% of the train tuples, isolated at the start of each training.

*3.2. Model Testing*

The comparison and evaluation of the two distilled network, with respect to the former 'dlib-resnet-v1' is carried out using the LFW [39] dataset on two computer vision problems: face verification and multi-class face recognition in an open set. In order to make decisions, a form of classification has to be inevitably introduced. In this section, we are not only testing the models, but also the entire procedure that a potential user of the network has to fulfill (train of the ad-hoc classifier) in order to actually use the network itself. In fact, the features are just a mere representation of the identity, made invariant to lighting, pose and system conditions (within the input image). Training with the CASIA dataset and testing with LFW is a pretty well-standardized procedure and permits a robust and immediate comparison among methods; results on other datasets (e.g., Megaface [40]) are less widespread. In this work, we tested our solutions against 30 IDs, because the number of images available for each subject (in the testing dataset) was limited. In a previous work [1] however, we successfully tested our teacher network with a larger number of individuals, observing a limited performance drop.

### 3.2.1. LFW Face Verification Test

The first problem is tackled by the use of the standard LFW test, consisting of a binary verification between pairs of images. The test represents a standard because in [39] the entire procedure to follow is described in [41] and then it is widely used in the Computer Vision community. A face verification procedure is the one used for automated airport check-in, where the same identity in the image grabbed by a camera has to appear also in the passport picture. In order to pass the face verification test, the algorithm under test has to correctly provide the answer to the question: "does the same identity appear in the two images provided?". To do so, the LFW test provides 10 lists of 600 pairs of images (300 same ID, 300 different IDs). This is a binary test (2 classes: same ID, different ID), and a binary classifier has thus to be designed, since the output of our models is a feature vector, not a class. A classifier will produce a class response starting from the features. In this work, for the verification test, we opted for a Rocchio classifier, that exploits a simple distance metric; in such a classifier, only one trainable parameter is present, in the form of a distance threshold calculated ad-hoc on the validation dataset.

The procedure is split into two phases, called "View1" and "View2": in the first phase the classification algorithm has to be designed (the design phase includes a testing of the classifier too) using a provided list of 3200 pairs, while in the second ($10 \times 600$ pairs), the system is tested; the output of this second test, the real test, is processed to produce the accuracy value that can be communicated and compared to other solutions within the computer vision community. The purpose of the ten lists is to average these results. It has to be noted that the accuracy value estimated from the results does not depend only on the face recognition oracle itself, but on the entire framework used to process the images (e.g., the alignment procedure): in the case of 'dlib-resnet-v1' and 'distilled net', the aforementioned preprocessing procedure is the same and model-only performance differences emerge; for 'distilled stn+net', instead, changes in the figures involve also the alignment protocol.

Another proposed indicator consists of setting the maximum acceptable value of $FPR$ and then evaluating the resulting $TPR$ over the ROC curve, obtained by varying the threshold. Running the same test utilizing the ten lists provided by the protocol, it is possible to calculate the average and the standard deviation for each point.

### 3.2.2. Multi Class Face Recognition in an Open Set

The objective of the second test is to evaluate the clustering ability of the embedding models, crucial for reliable recognition. During this test, the system has to recognize people that it knows (labeling the correct name) against images of not only the known ID (the so-called "friends") but also taken from random identities (the "unknowns"). In order to simulate the scenario of open-set in the standard LFW dataset an amount of identity is taken to form the group of "friends" and the remaining IDs compose the unknown set. Note that LFW has no overlap with the CASIA dataset that is used for the distillation of our features extractor model. As described in Section 2.4.2, a multiclass classifier is needed for face recognition on an open set. Following the work presented in [9,18], we adopted a shallow Multi-Layer Perceptron (MLP) formed by three fully connected layers: the first two consists in 100 neurons, while the number of outputs in the last one is the number of classes to recognize. The intermediate nonlinearity used is a ReLU, while for the final nonlinearity we opted for the Softmax activation function. In order to distinguish a subject that does not belong to known classes ('unknown'), we used the normalized distance as confidence index, for which the logit values are compared, according to Equation (3).

$$C = \frac{d_1 - d_2}{d_1 - d_n},$$

(3)

where $d_1$, $d_2$ and $d_n$ are respectively the largest, the second-largest and the smallest value of the output layers.

The final decision is taken not only by observing the class of highest probability, but also the confidence value, calculated with Equation (3). For each classifier, we studied the effect of both a different number of classes and a variable number of samples provided to the model during the training. Table 1 summarizes the testing conditions. We set the number of classes (the output of the classifier) and we trained the model using 2, 5, or 15 samples for each identity to recognize. Only these samples are seen during the learning. During the testing, the number of images for each known ID is kept constant to 10. To test how well the classifier rejects unknown subjects, other samples have to be added (open-set problem).

To do so, a number of images of unknown identities equal to the number of image friends are used, randomly choosing from all LFW IDs who are not used as a friends. Working in this way no bias is triggered during the procedure. Details for all the explored cases are given in Table 1.

**Table 1.** Performances in face recognition are evaluated for different classifiers trained with a different amount of outputs (classes). This table summarize the number of samples used in each case. A limited number of individuals (5, 15 or 30) is extracted from the dataset: these are the identities we want to recognize. During the training (and the validation), only known IDs samples are provided. Note that the desired number of training samples (2, 5 or 15) is kept constant for each class. This choice allows for a perfect balance between each class, giving no a priori information through sample distribution. During the testing, all the other identities outside of the closed set of known IDs can potentially provide unknown samples. The classifier is tested against the unknown IDs rejection providing a fixed number of images that is equal between the known and unknown individuals. Again, no predilection on a particular class or on the "unknown" is inferred.

| CLASSES | # TRAIN Imgs (Known IDs Only) | # VALIDATION Imgs (Known IDs Only) | # TESTING Imgs (Known IDs) | # TESTING Imgs Unknown IDs |
|---|---|---|---|---|
| 5 | $5 \times (2$ or 5 or 15) | $5 \times 5$ | $5 \times 10 = 50$ | 50 |
| 15 | $15 \times (2$ or 5 or 15) | $15 \times 5$ | $15 \times 10 = 150$ | 150 |
| 30 | $30 \times (2$ or 5 or 15) | $30 \times 5$ | $30 \times 10 = 300$ | 300 |

Following the procedure described in Section 2.4.2, we calculated the $TPR$ and the $FPR$ as a function of the estimated confidence $C$ for a multi-class problem and we plotted the ROC curve of the classifier using Equations (4) and (5):

$$TPR = \frac{\sum_{i=1}^{K} TP_i}{N} \tag{4}$$

$$FPR = \frac{\sum_{i=1}^{K} FP_i}{K * F + (K-1)N}, \tag{5}$$

where $TP$ is the number of correctly classified samples (with $C$ above the selected threshold of confIDence) and $N$ is the number of known samples provIDed during the test; $FP$ is the number of misclassified samples (the number of known people whose identity has been misclassified plus the number of the unknown people which are classified with a confidence index above the threshold, i.e., faces that have been erroneously classified as a known person) and $F$ is the number of all the unknown samples.

Using the LFW dataset, only 30 identities have at least 30 images each: according to this limit, the training of MLP was carried out using only 2, 5 or 15 images for each subject, reserving five images to the verification (early stopping in training) and 10 for the test. The remaining 10 samples of each known face are used for the test, while $10 \times Nc$ images of other identities are enrolled to form the unknown people corpus. The number of unknown samples is chosen in order to balance the testing set: the entire procedure is repeated ten times for different individuals, in a cross-validation approach. The results of the various tests, at different thresholds of confidence, were represented in the ROC plane highlighting the area that contains 99% of the results and tracing the average ROC curve described by these values.

*3.3. Hardware Implementation*

The distilled network has been tested on a Single Board Computer (Odroid XU-4); the inference time of 'dlib-resnet-v1' (using the CPU, compiling Dlib with the Arm-Neon [12] flag) was compared with the distilled network using TensorFlow Lite [42] (CPU approach) and a hardware accelerator such as the Intel Movidius Neural Compute Stick (NCS) [43]. The mean inference time for Dlib is 816 ms, while for the 'distilled net' 195 ms are needed for its TensorFlowLite porting and only 67 ms are needed if the hardware accelerator is used, providing a speed gain of one order of magnitude, keeping the same accuracy.

TensorFlowLite and the Intel embedded converter are able to synthesize standard layers only, such as dense, convolutional, activation and so on. Unfortunately, the conversion of the STN boosted network ('distilled stn+net') is currently impossible due to the presence of the unconventional sampling layer. We hope that in a future version of the tools this conversion can be done.

## 4. Results and Discussion

### 4.1. LFW Verification Test

The first test was conducted on the original Dlib network and on the two proposed distilled networks using LFW dataset in order to evaluate their verification ability.

Table 2 summarizes the average results obtained from the 10 tests proposed by the LFW test: the accuracy was calculated following the protocol defined by LFW, while $TPR$ value with desired $FPR$ constrain was calculated as explained in Section 3.2.

**Table 2.** Comparative results of the LFW identity verification test. Each row provides the figure for each model, the former 'dlib resnet-v1' and our two distilled model, 'distilled net' and 'distilled net+stn'. The first column shows the resulting accuracy, while the last two columns provide the $TPR$ value on the ROC curve (Figure 9) corresponding to an imposed $FPR$ value of choice.

| Network | Accuracy | TPR @ FPR=1% | TPR @ FPR=0.1% |
|---|---|---|---|
| dlib-resnet-v1 | $0.9918 \pm 0.0033$ | $0.9923 \pm 0.0049$ | $0.9344 \pm 0.1365$ |
| Distilled net | $0.9852 \pm 0.0050$ | $0.9819 \pm 0.0106$ | $0.8931 \pm 0.1051$ |
| Distilled stn+net | $0.9852 \pm 0.0058$ | $0.9908 \pm 0.0137$ | $0.9067 \pm 0.1241$ |

The table shows that the distillation of the dark knowledge was successful: The accuracy of the two distilled models is comparable to the one of the teacher. Another interesting view on the verification test result is obtained choosing a threshold on the maximum acceptable $FPR$ and reading on the ROC curve the corresponding value of $TPR$. The solution 'distilled stn+net' provides a $TPR$ value even higher than the one of 'distilled net'.

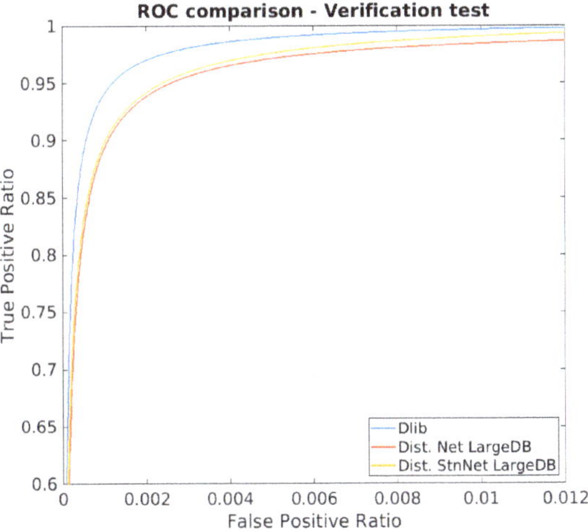


**Figure 9.** ROC curve for the LFW face verification test. Note that the graphs are highly zoomed portions of the entire curve.

### 4.2. Recognition Test

The second test aims to compare the performance of the networks considering the problem of face recognition in an open set. In the following Figures 10–12 the shadowed regions represent the areas that cover 99% of the results of the 10 tests, while the bold line represents the mean ROC curve for the former and the two 'distilled net' and 'distilled stn+net'.

In Figure 10 many different ROC curves (the result of different classifiers), are produced imposing a varying limit on the number of samples used during the training. As described in the previous section, this test has been performed fixing the number of classes and then using the 30-class classifier only. It should be noted that even with the training of only two images per ID, it is possible to recognize a person in the wild with acceptable accuracy. This setup is particularly interesting e.g., for the automatic checking of suspect subjects of whom only a few photos are available.

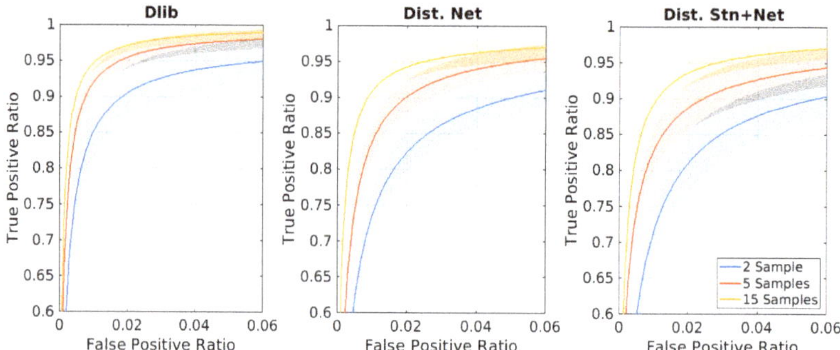

**Figure 10.** Average ROC curves estimated from a ten-fold cross-validation procedure on multiclass classifier. Each classifier is trained using a fixed amount of classes (30) and a varying number of training sample, using the features generated by 'dlib-resnet-v1', 'distilled-net', 'distilled stn+net'. Note that the graphs are highly zoomed portions of the entire curve.

In Figure 11, similar tests were proposed by fixing instead of the number of training samples to 15 and varying the number of classes (of known subjects) among 5, 15 and 30. Up to a certain limit, the entire framework is invariant to the class number, allowing for the best performances when the 'known person' database is composed of a few dozen identities.

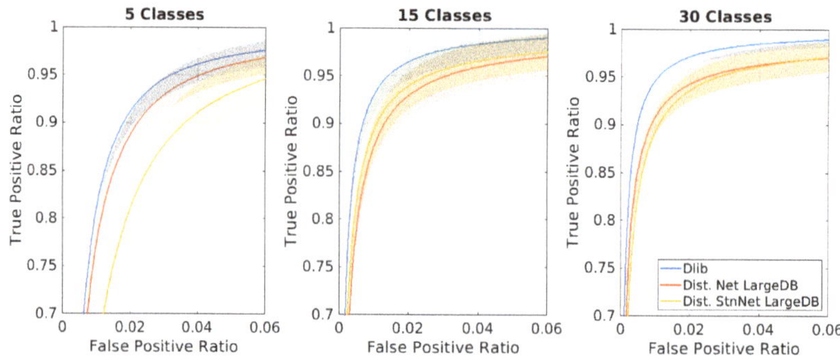

**Figure 11.** Average ROC curves estimated from a ten-fold cross-validation procedure on multiclass classifier. Each classifier is trained using a fixed amount of samples (15) and a varying number of classes, using the features generated by 'dlib-resnet-v1', 'distilled-net', 'distilled stn+net'. Note that the graphs are highly zoomed portions of the entire curve.

For clarity, we have summarized the two results in Figure 12, comparing the ROC of the networks under test with the teacher network in the case of optimal parameters (30-class classifier trained with 15 samples per ID).

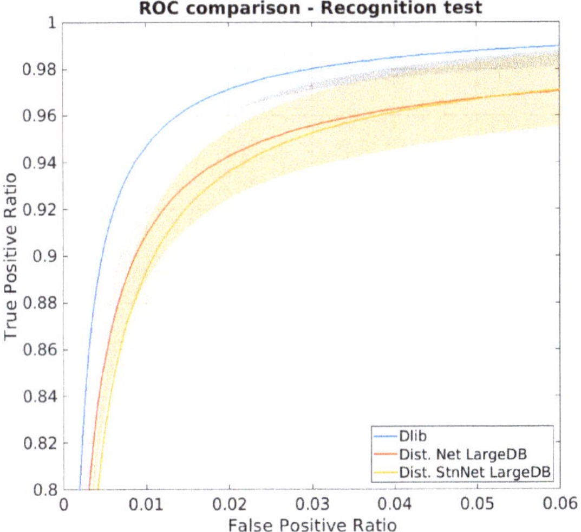


**Figure 12.** ROC curve comparison for the best case (30-class classifier trained with 15 samples per ID). Note that the graphs are highly zoomed portions of the entire curve.

## 4.3. STN Analysis: Co-Adaptation and Difficult Poses

The STN and the face recognition network are used in tandem after a common training phase. The only feature maps shared between the two is the STN output image which lies in the standard RGB image space. Analyzing this output image provides insight into the training of the entire model. At the end of each epoch, a callback launches the test, and for nine test samples, the output of the STN component is extracted and saved. Since this image is a mere feature map, we can analyze the co-adaptation between the STN and the recognition components and evaluate how the alignment skills are learned after each epoch. In Figure 13 the output of the STN aligner is reported for nine people after 4, 16, 64 and 128 training epochs.

We can observe that typically in ten epochs the STN component learned to isolate a face within the frame and found the best way to minimize the Euclidean distance loss function. Interestingly, the network automatically decided that the best possible alignment procedure (which minimize at most the loss function) consists of rotating the face by a few tens of degrees, in order to occupy the largest possible area, thus removing part of the background remaining around the hair and chin. It is reasonable that this behavior is forced also by the downsampling of the input image operated by the sampler in the STN.

In Figure 14 a similar experiment has been carried out using a pre-trained and frozen distilled net, in which the STN was the only trainable component: in the processed face the eyes are aligned to the horizon; the STN learned in just one epoch to emulate the Dlib alignment procedure, localizing and aligning the face.

**Figure 13.** This figure shows the evolution of the STN output during the training of 'distilled net+stn'. In less than 10 epochs, the STN correctly localizes the faces, while in 30 epochs the STN correctly learns to localize and align images for the recognition network.

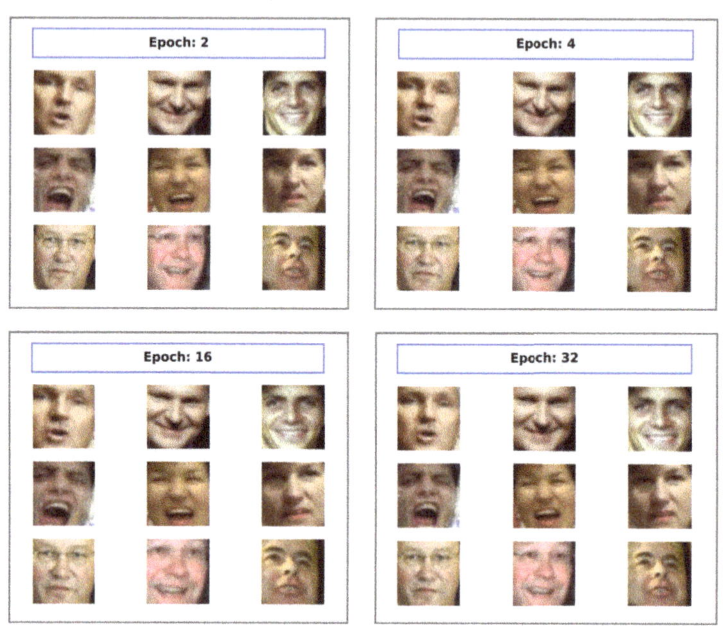

**Figure 14.** Differently from the previous case, this figure shows the evolution of the STN output for the training of the STN component only, providing a pre-trained 'distilled net' as the embedding model. In just one epoch, the STN learns to align each face putting the eyes horizontally, and emulating the crop factor of the former Dlib face aligner.

The results presented till now take into account only the samples that can be actually analyzed by the Dlib framework, e.g., the ones that have been selected as faces by the face detector. The real advantage of using the STN distilled network emerges when difficult poses are recorded in the frame. In order to evaluate this aspect, we selected the samples from the LFW dataset where no faces are found (for a deficiency of the face detector). If a landmark detection is carried out on these frames, the subsequent alignment will produce images for which dlib-resnet-v1 cannot produce meaningful features. In Figure 15 we compare the alignment of the Dlib algorithm with the alignment obtained with our proposed model: the STN is, in any case, able to give attention to the face and to align it in a manner that makes the subsequent model able to verify the identity (the points in the hyperspace are closer than the threshold used for the binary 'same–different' verification test). We point out that the shown results are carried out on test samples, which the network had never seen during the training phase.

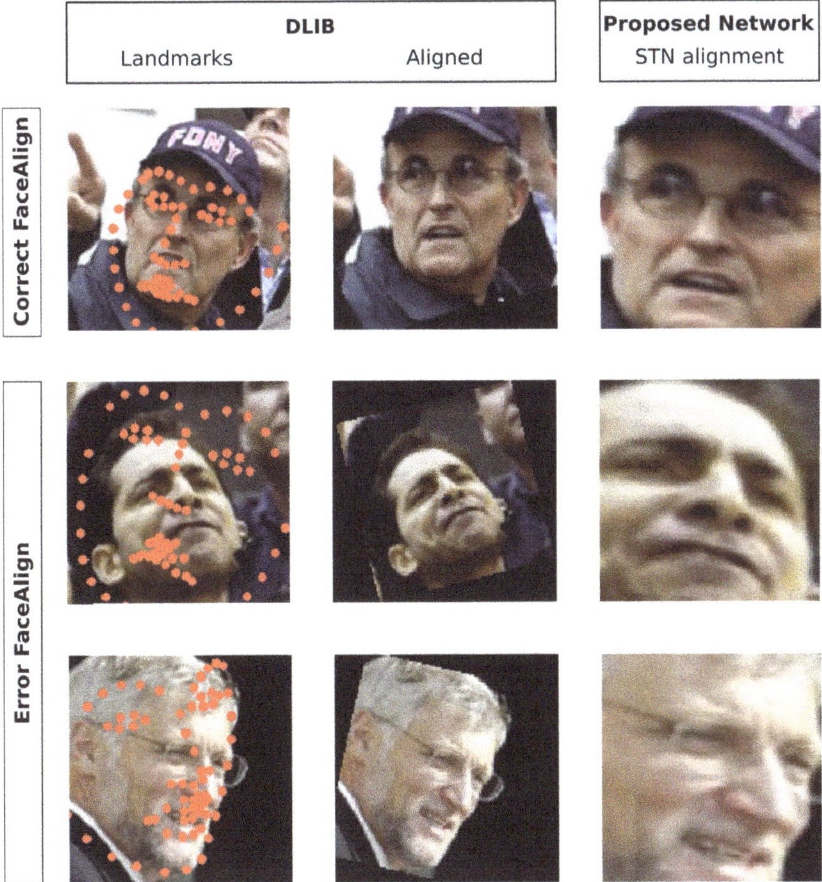


**Figure 15.** In this figure different alignments are compared for difficult face poses. In the first column, the landmark found with the dlib shape predictor are shown; in all the samples the error is heavily present. Only the face alignment procedure carried out on the first sample generates a correct recognition (a descriptor in the Euclidean space close to the centroid of its identity). The 'distilled stn+net' model is capable of correctly localizing and aligning the face, just like in any other pose.

*4.4. Distillation Strategy as a "Transfer Learning" for the New Model*

Two different distillation training strategies have been followed: the first one enrolls the entire Casia dataset blindly, while the second exploits a predefined sample presentation structure in each training batch; in the second case, we fixed the number of different IDs for each batch to 64 with two samples for each ID. In this manner, even if the number of samples per batch remained constant in the two cases (128), each epoch lasted more than 10 times less, allowing to train in half the time, for 1000 epochs. In Figure 16 the two-loss evolutions are compared: The resulting accuracy is highly comparable, highlighting that for a correct distillation it is crucial to have different cluster centroids in the sample space.

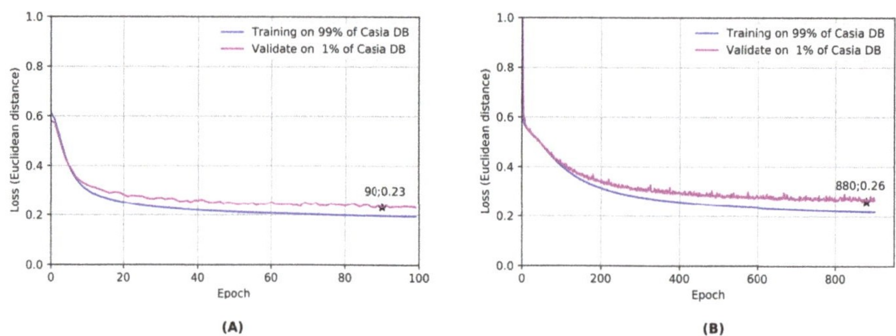

**Figure 16.** Two distillation trainings are compared: in panel (**A**), the 'full' learning is exploited, providing the best results. In panel (**B**), a different strategy is used: for each training batch a different identity is extracted and only two random images are given to the network. This procedure is way faster than the previous one (1000 epochs in 6–8 hours vs 100 epochs in 15 hours) but obviously a bit less performing. During the distillation, what counts is the number of identities (cluster centroid), rather than the number of samples for each centroid.

From the graphs a second observation can be drawn: distillation can be enforced as a fast initial training technique for the training of the new network, as a "transfer learning methodology" for newer networks, if the newer model under investigation has to answer to the same type of question.

## 5. Conclusions

In this work, we described two face recognition models that can be implemented on low-cost hardware, in the form of a face recognition sensor module. The key procedure exploited in this work is knowledge distillation, used to extract the dark knowledge of a dlib-resnet-v1 network in a teacher-student framework. Each distillation has been obtained in a simple metric framework, essential if distillation is used as an initialization technique. In this sense, a relatively fast distillation can be used as a "transfer learning" phase between different models. One model is a direct substitute of the original network, that can be then used without adaption layers; our second realization embraces instead of an end-to-end approach that permits to remove the separate alignment procedure. The second model proved to be definitely more robust in the case of difficult poses. To the best of our knowledge, a distillation of such a structure for face recognition has never been attempted. A well-acknowledged training and testing protocol has been exploited to evaluate the performances of each realization, in the form of the LFW face verification test and a novel face recognition in an open scenario test description. The outcome of this problem description is a procedure for unknown ID rejection that exploits a confidence measure and thus minimizes the false-positive error rate.

**Author Contributions:** F.G., S.M., S.C. and G.R. conceived the research, F.G., L.D.B. and S.M. conceived the algorithms, F.G. and L.D.B. designed the experiments; F.G., L.D.B., R.S.M., and S.M. performed and evaluated the experiments and analyzed the data; S.C., S.M. and G.R. reviewed the paper; F.G., L.D.B., R.S.M., G.R., S.C. and S.M. wrote the paper. All authors have read and agreed to the published version of the manuscript.

**Acknowledgments:** The support of the University of Trieste - FRA projects and of a fund in memory of Angelo Soranzo (1939-2012) are gratefully acknowledged.

**Conflicts of Interest:** The authors declare no conflict of interest. The funders had no role in the design of the study; in the collection, analyses, or interpretation of data; in the writing of the manuscript, or in the decision to publish the results.

## Appendix A

### Appendix A.1

Defining the performance of a multiclass classifier is not trivial and essentially depends on the scenario in which the classifier operates. In fact, some indexes or parameters which are usually adopted to define the performances of a binary classifier are not adequately suited to the case of a multiclass classifier.

One performance representation commonly used in binary classifiers is the ROC (receiver operating characteristic) curve. This curve represents the True Positive Rate ($TPR$) vs. the False Positive Rate ($FPR$), as achieved by the classifier at various threshold settings.

Such parameters in a binary classifier are defined as follows in Equations (A1) and (A2):

$$TPR = \frac{TP}{N} \tag{A1}$$

$$FPR = \frac{FP}{F} \tag{A2}$$

where $N$ is the number of positive samples and $F$ is the number of negative samples used to test the classifier. $TP$ (True positive) is the number of positive samples correctly classified and $FP$ (False Positive) is the number of negative samples erroneously classified as true. For completeness we usually define False Negative ($FN = N - TP$) the number of positive cases erroneously classified as negative and True Negative ($TN = F - FP$) the number of negative cases correctly classified.

The application of similar parameters in a multiclass classifier depends on the correct identification of the scenario. Suppose that the task is to classify an individual among a finite number $K$ of classes knowing that the individual belongs to one and just one of these classes. It is possible to define the true positive rate for each class $i$, $TPR_i$, as:

$$TPR_i = \frac{TP_i}{N_i} \tag{A3}$$

where $N_i$ is the number of samples, belonging to class $i$, submitted to the classifier, during the test, and $TP_i$ is the number of these samples correctly classified.

Now it is possible to extend the concept of True Positive Rate to the entire classifier, simply by averaging all the values of $TPR_i$ on all the classes, better if using a weighted average based on the number of samples submitted to the classifier for each class:

$$TPR = \sum_{i=1}^{K} \alpha_i * TPR_i \tag{A4}$$

where $\alpha_i$ must be constrained: it must be proportional to the number of samples $N_i$ and $\sum \alpha_i = 1$, thus:

$$\alpha_i = \frac{N_i}{\sum_{i=1}^{K} N_i} \tag{A5}$$

Substituting Equations (A3) and (A5) in Equation (A4) and simplifying we obtain:

$$TPR = \frac{\sum_{i=1}^{K} TP_i}{\sum_{i=1}^{K} N_i} = \frac{\sum_{i=1}^{K} TP_i}{N} \tag{A6}$$

where $N = \sum N_i$ is the number of samples provided to the classifier during the test.

Operating in the same way it is possible to evaluate the $FPR$:

$$FPR_i = \frac{FP_i}{N - N_i} \tag{A7}$$

where $N - N_i$ is the number of samples belonging to classes other than $i$ and $FP_i$ is the number of these samples erroneously classified. To define a global $FPR$ value it is possible to proceed, like before adopting a weighted average:

$$FPR = \sum_{i=1}^{K} \beta_i * FPR_i \tag{A8}$$

The constraints to the weights $\beta_i$ are $\beta_i \propto (N - N_i)$ and $\sum \beta_i = 1$ thus:

$$\beta_i = \frac{N - N_i}{\sum_{i=1}^{K}(N - N_i)} \tag{A9}$$

Substituting Equations (A7) and (A9) in Equation (A8):

$$FPR = \frac{\sum_{i=1}^{K} FP_i}{\sum_{i=1}^{K}(N - N_i)} = \frac{\sum_{i=1}^{K} FP_i}{(K - 1)N} \tag{A10}$$

However, the scenario we want to take into consideration in this article is slightly different from the one just mentioned. In our case there are both a number of positive examples ($N_i$) relating to the known subjects that must be correctly classified into a proper class, and a set of negative samples ($F$), belonging to unknown individuals, that have never been seen before by the classifier. These latter samples, if correctly classified, are $TN$ (True Negative) with respect to each class of known people and therefore they should not contribute to the computation of the global $TPR$ as in the previous case. On the other hand, if they are incorrectly classified, they represent false positives for our system.

On the other hand, an individual belonging to the positive samples who is erroneously classified as an unknown subject does not lead to an increase in the $FP$; rather, it represents a $FN$, whose impact on the evaluation of the classifier performance is very different.

Thus, while Equation (A6) for the calculation of the $TPR$ can remain unchanged, the calculation of the $FPR$ must be applied to all the cases in which an individual is attributed to the wrong class, considering as target all possible class but the one composed by unknown individuals. Thus Equation (A7) should be reviewed considering that for each class the negative samples come both from samples of the other classes ($N - Ni$) and from the ones which belong to unknown subjects ($F$).

$$FPR_i = \frac{FP_i}{F + N - N_i} \tag{A11}$$

Thus the weights adopted in Equation A8 to evaluate the weighted average should be revised as follows:

$$\beta_i = \frac{F + N - N_i}{\sum_{i=1}^{K}(F + N - N_i)} = \frac{F + N - N_i}{K(F + N) - N} \tag{A12}$$

Thus:

$$FPR = \frac{\sum_{i=1}^{K} FP_i}{K * F + (K - 1)N} \tag{A13}$$

## References


1.  DeBortoli, L.; Guzzi, F.; Marsi, S.; Carrato, S.; Ramponi, G. A fast face recognition CNN obtained by distillation. In Proceedings of the 7th International Workshop Applications in Electronics Pervading Industry, Environment and Society, ApplePies, Pisa, Italy, 11–13 September 2019.
2.  Yang, Y.; Wen, C.; Xie, K.; Wen, F.; Sheng, G.; Tang, X. Face Recognition Using the SR-CNN Model. *Sensors* **2018**, *18*, 4237. [CrossRef] [PubMed]
3.  Guo, G.; Zhang, N. A survey on deep learning based face recognition. *Comput. Vis. Image Underst.* **2019**, *189*. [CrossRef]
4.  Yang, S.; Luo, P.; Loy, C.C.; Tang, X. From Facial Parts Responses to Face Detection: A Deep Learning Approach. In Proceedings of the 2015 IEEE International Conference on Computer Vision, ICCV, Santiago, Chile, 7–13 December 2015; pp. 3676–3684.
5.  Huang, L.; Yang, Y.; Deng, Y.; Yu, Y. DenseBox: Unifying Landmark Localization with End to End Object Detection. *CoRR* **2015**. Available online: http://arxiv.org/abs/1509.04874 (accessed on 31 January 2020).
6.  King, D.E. Max-Margin Object Detection. *CoRR* **2015**. Available online: http://arxiv.org/abs/1502.00046 (accessed on 31 January 2020).
7.  Zhang, K.; Zhang, Z.; Li, Z.; Qiao, Y. Joint Face Detection and Alignment Using Multitask Cascaded Convolutional Networks. *IEEE Signal Process. Lett.* **2016**, *23*, 1499–1503. [CrossRef]
8.  Chen, D.; Hua, G.; Wen, F.; Sun, J. Supervised Transformer Network for Efficient Face Detection. In Proceedings of the Computer Vision—ECCV 2016—14th European Conference, Amsterdam, The Netherlands, 11–14 October 2016; pp. 122–138. [CrossRef]
9.  Marsi, S.; Bortoli, L.D.; Guzzi, F.; Bhattacharya, J.; Cicala, F.; Carrato, S.; Canziani, A.; Ramponi, G. A Face Recognition System Using Off-the-Shelf Feature Extractors and an Ad-Hoc Classifier. In *Applications in Electronics Pervading Industry, Environment and Society, APPLEPIES 2018*; Saponara, S., Gloria, A.D., Eds.; Springer: Berlin, Germany, 2018; pp. 145–151. [CrossRef]
10. Dlib API. Available online: http://blog.dlib.net/2017/02/high-quality-face-recognition-with-deep.html (accessed on 31 January 2020).
11. Dlib Models. Available online: https://github.com/davisking/dlib-models (accessed on 31 January 2020).
12. ARM Neon. Available online: https://developer.arm.com/architectures/instruction-sets/simd-isas/neon (accessed on 31 January 2020).
13. Jaderberg, M.; Simonyan, K.; Zisserman, A.; Kavukcuoglu, K. Spatial Transformer Networks. In Proceedings of the Advances in Neural Information Processing Systems 28: Annual Conference on Neural Information Processing Systems 2015, Montreal, QC, Canada, 7–12 December 2015; pp. 2017–2025.
14. Viola, P.A.; Jones, M.J. Robust Real-Time Face Detection. *Int. J. Comput. Vis.* **2004**, *57*, 137–154. [CrossRef]
15. Dalal, N.; Triggs, B. Histograms of Oriented Gradients for Human Detection. In Proceedings of the 2005 IEEE Computer Society Conference on Computer Vision and Pattern Recognition (CVPR 2005), San Diego, CA, USA, 20–26 June 2005; pp. 886–893. [CrossRef]
16. Liu, W.; Anguelov, D.; Erhan, D.; Szegedy, C.; Reed, S.E.; Fu, C.; Berg, A.C. SSD: Single Shot MultiBox Detector. In Proceedings of the Computer Vision—ECCV 2016—14th European Conference, Amsterdam, The Netherlands, 11–14 October 2016; pp. 21–37. [CrossRef]
17. Song, X.; Herranz, L.; Jiang, S. Depth CNNs for RGB-D Scene Recognition: Learning from Scratch Better than Transferring from RGB-CNNs. In Proceedings of the Thirty-First AAAI Conference on Artificial Intelligence, San Francisco, CA, USA, 4–9 February 2017; pp. 4271–4277
18. Guzzi, F.; DeBortoli, L.; Marsi, S.; Carrato, S.; Ramponi, G. Distillation of a CNN for a high accuracy mobile face recognition system. In Proceedings of the 42nd International Convention on Information and Communication Technology, Electronics and Microelectronics, MIPRO 2019, Opatija, Croatia, 20–24 May 2019; pp. 989–994.
19. Cheng, Y.; Wang, D.; Zhou, P.; Zhang, T. Model Compression and Acceleration for Deep Neural Networks: The principles, progress, and challenges. *IEEE Signal Process Mag.* **2018**, *35*, 126–136. [CrossRef]
20. Luo, P.; Zhu, Z.; Liu, Z.; Wang, X.; Tang, X. Face Model Compression by Distilling Knowledge from Neurons. 2016. pp. 3560–3566. Available online: https://www.researchgate.net/publication/319770258_Face_Model_Compression_by_Distilling_Knowledge_from_Neurons (accessed on 28 February 2020).

21. Hinton, G.E.; Vinyals, O.; Dean, J. Distilling the Knowledge in a Neural Network. *CoRR* **2015**. Available online: http://arxiv.org/abs/1503.02531 (accessed on 28 February 2020).
22. Ba, J.; Caruana, R. Do Deep Nets Really Need to be Deep? In Proceedings of the Advances in Neural Information Processing Systems 27: Annual Conference on Neural Information Processing Systems 2014, Montreal, QC, Canada, 8–13 December 2014; pp. 2654–2662.
23. Zhao, S.; Gao, X.; Li, S.; Ge, S. Low-Resolution Face Recognition in the Wild with Mixed-Domain Distillation. In Proceedings of the Fifth IEEE International Conference on Multimedia Big Data, BigMM 2019, Singapore, 11–13 September 2019; pp. 148–154. [CrossRef]
24. Wu, A.; Zheng, W.S.; Guo, X.; Lai, J.H. Distilled Person Re-Identification: Towards a More Scalable System. In Proceedings of the IEEE Conference on Computer Vision and Pattern Recognition, CVPR 2019, Long Beach, CA, USA, 15–20 June 2019. [CrossRef]
25. Kim, J.; Ahn, G.; Kim, S.; Kim, Y.; López, J.; Kim, T., Eds. Detection under Privileged Information. In Proceedings of the 2018 on Asia Conference on Computer and Communications Security, AsiaCCS 2018, Incheon, Korea, 4–8 June 2018. [CrossRef]
26. Kazemi, V.; Sullivan, J. One Millisecond Face Alignment with an Ensemble of Regression Trees. In Proceedings of the 2014 IEEE Conference on Computer Vision and Pattern Recognition, CVPR 2014, Columbus, OH, USA, 23–28 June 2014; pp. 1867–1874.
27. Ren, S.; Cao, X.; Wei, Y.; Sun, J. Face Alignment at 3000 FPS via Regressing Local Binary Features. In Proceedings of the 2014 IEEE Conference on Computer Vision and Pattern Recognition, CVPR 2014, Columbus, OH, USA, 23–28 June 2014; pp. 1685–1692.
28. Chen, D.; Ren, S.; Wei, Y.; Cao, X.; Sun, J. Joint Cascade Face Detection and Alignment. In Proceedings of the Computer Vision—ECCV 2014—13th European Conference, Zurich, Switzerland, 6–12 September 2014; pp. 109–122.
29. Keras MTCNN Implementation. Available online: https://github.com/xiangrufan/keras-mtcnn (accessed on 31 January 2020).
30. Zhuang, W.; Chen, L.; Hong, C.; Liang, Y.; Wu, K. FT-GAN: Face Transformation with Key Points Alignment for Pose-Invariant Face Recognition. *Electronics* **2019**, *8*, 807. [CrossRef]
31. Hassner, T.; Harel, S.; Paz, E.; Enbar, R. Effective face frontalization in unconstrained images. In Proceedings of the IEEE Conference on Computer Vision and Pattern Recognition, CVPR 2015, Boston, MA, USA, 7–12 June 2015; pp. 4295–4304.
32. Zhou, E.; Cao, Z.; Sun, J. GridFace: Face Rectification via Learning Local Homography Transformations. In Proceedings of the Computer Vision—ECCV 2018—15th European Conference, Munich, Germany, 8–14 September 2018; pp. 3–20.
33. Chen, T.; Goodfellow, I.J.; Shlens, J. Net2Net: Accelerating Learning via Knowledge Transfer. In Proceedings of the 4th International Conference on Learning Representations, ICLR 2016, San Juan, Puerto Rico, 2–4 May 2016.
34. Zhong, Y.; Chen, J.; Huang, B. Toward End-to-End Face Recognition Through Alignment Learning. *IEEE Signal Process. Lett.* **2017**, *24*, 1213–1217. [CrossRef]
35. He, K.; Zhang, X.; Ren, S.; Sun, J. Deep Residual Learning for Image Recognition. In Proceedings of the 2016 IEEE Conference on Computer Vision and Pattern Recognition, CVPR 2016, Las Vegas, NV, USA, 27–30 June 2016; pp. 770–778. [CrossRef]
36. Huang, G.; Liu, Z.; van der Maaten, L.; Weinberger, K.Q. Densely Connected Convolutional Networks. In Proceedings of the 2017 IEEE Conference on Computer Vision and Pattern Recognition, CVPR 2017, Honolulu, HI, USA, 21–26 July 2017; pp. 2261–2269. [CrossRef]
37. Ronneberger, O.; Fischer, P.; Brox, T. U-Net: Convolutional Networks for Biomedical Image Segmentation. *CoRR* **2015**. Available online: http://arxiv.org/abs/1505.04597 (accessed on 31 January 2020).
38. Yi, D.; Lei, Z.; Liao, S.; Li, S.Z. Learning Face Representation from Scratch. *CoRR* **2014**. Available online: http://arxiv.org/abs/1411.7923 (accessed on 31 January 2020).
39. Huang, G.B.; Mattar, M.; Berg, T.; Learned-Miller, E. Labeled Faces in the Wild: A Database forStudying Face Recognition in Unconstrained Environments. In Proceedings of the Workshop on Faces in 'Real-Life' Images: Detection, Alignment, and Recognition, Erik Learned-Miller and Andras Ferencz and Frédéric Jurie, Marseille, France, October 2008. Available online: https://hal.inria.fr/inria-00321923 (accessed on 28 February 2020).

40. Nech, A.; Kemelmacher-Shlizerman, I. Level Playing Field for Million Scale Face Recognition. In Proceedings of the 2017 IEEE Conference on Computer Vision and Pattern Recognition, CVPR 2017, Honolulu, HI, USA, 21–26 July 2017; pp. 3406–3415. [CrossRef]
41. LFW Protocol. Available online: http://vis-www.cs.umass.edu/lfw/#views (accessed on 31 January 2020).
42. TensorFlow Lite. Available online: https://www.tensorflow.org/lite (accessed on 31 January 2020).
43. Intel NCS. Available online: https://software.intel.com/en-us/neural-compute-stick (accessed on 31 January 2020).

*Article*

# A Model-Based Design Floating-Point Accumulator. Case of Study: FPGA Implementation of a Support Vector Machine Kernel Function [†]

**Marco Bassoli, Valentina Bianchi * and Ilaria De Munari**

Department of Engineering and Architecture, University of Parma, Parco Area delle Scienze, 181/A, 43124 Parma, Italy; marco.bassoli@unipr.it (M.B.); ilaria.demunari@unipr.it (I.D.M.)

* Correspondence: valentina.bianchi@unipr.it; Tel.: +39-0521-906284

† This paper is an extended version of the conference paper published in: Bassoli, M.; Bianchi, V.; De Munari, I. A Simulink Model-based Design of a Floating-point Pipelined Accumulator with HDL Coder Compatibility for FPGA Implementation. Appl. Electron. Pervading Ind. Environ. Soc. ApplePies 2019. Lect. Notes Electr. Eng. 2019, 1–9, in press.

Received: 31 January 2020; Accepted: 28 February 2020; Published: 2 March 2020

**Abstract:** Recent research in wearable sensors have led to the development of an advanced platform capable of embedding complex algorithms such as machine learning algorithms, which are known to usually be resource-demanding. To address the need for high computational power, one solution is to design custom hardware platforms dedicated to the specific application by exploiting, for example, Field Programmable Gate Array (FPGA). Recently, model-based techniques and automatic code generation have been introduced in FPGA design. In this paper, a new model-based floating-point accumulation circuit is presented. The architecture is based on the state-of-the-art delayed buffering algorithm. This circuit was conceived to be exploited in order to compute the kernel function of a support vector machine. The implementation of the proposed model was carried out in Simulink, and simulation results showed that it had better performance in terms of speed and occupied area when compared to other solutions. To better evaluate its figure, a practical case of a polynomial kernel function was considered. Simulink and VHDL post-implementation timing simulations and measurements on FPGA confirmed the good results of the stand-alone accumulator.

**Keywords:** model-based design; FPGA; HDL code generation; wearable sensors; embedded devices

## 1. Introduction

In recent years, the concept of a smart home has been extended from the simple automation and automatic control of the home appliances to a more complex management of the user interaction with several sensors and actuators deployed in the home environment in order to pursue the users' wellbeing and energy sustainability [1–9]. The development of wearable sensors has expanded the possibilities available in this context, pushing research towards new solutions based on behavioral monitoring [10–13]. The role of wearable sensors in this framework is very wide, but recent research has focused on human activity recognition (HAR) as a new service to monitor the amount of activity for health purposes; this can be assessed and considered in order to early detect anomalies possibly relevant to users' wellbeing [14–17].

The most advanced HAR algorithms are based on machine learning (ML) techniques, which are usually very computationally demanding [14]. The development of wearable devices leads to implementation of ML algorithms directly on board [18,19], allowing for the reduction of the amount of data to be transmitted, and with consistent advantages in terms of power consumption and system usability [14]. To address the issues related to the need for platforms with good computing capacity, instead of general-purpose processors, dedicated hardware architectures such as field programmable gate arrays (FPGAs) can be selected for the implementation of the algorithms [20–23]. This allows for the control of the resources needed for the task and to optimize the system for performance or physical size, depending on the use case. Recent advantages in FPGA technologies allow these platforms to also be used in applications with low cost [24] and/or low power consumption requirements [23].

The design of dedicated hardware architectures is traditionally done by using hardware description languages (HDL). However, as proven by different works [25–27], higher abstraction level frameworks can support the designer in helping to focus the attention on system functionalities and in reducing time-to-market. This is possible, for example, by using MATLAB/Simulink software [28]. A high-level approach can be developed, and the benefits of a model-based design can be exploited [29,30]. Moreover, with the dedicated HDL Coder tool, a HDL code can be automatically generated from the system block diagram and hence used to program the selected platform.

In this paper, we present the development of a model-based floating-point accumulator. To better study the performance of the designed model with respect to available solutions, it was applied to a practical case—a Simulink model-based kernel function conceived as the core of a support vector machine (SVM) classifier to be embedded in an FPGA-based wearable device. The SVM is a widely used algorithm for solving classification problems, also utilized in the field of HAR. The classification work consists in finding a line or a hyperplane that allows data to be divided into different regions. In the case of non-separable sample sets, the kernel function has to be introduced into the SVM algorithm [31]. Different approaches can be found in the literature—the linear, the polynomial, the Gauss radial basis, and others. A combination of them is often also proposed [32]. In the present work, the polynomial solution was adopted to evaluate the proposed accumulator because it has been recognized as the one with strong generalization capability [32]. The polynomial kernel function, as well as other kernels, involves the dot product of the input vectors, resulting in a sum of products to be implemented in arithmetic blocks. Among these, the accumulator has an important role.

The simplest accumulator architecture can be designed by using an adder in which the first input receives the operand element and the second input is the feedback of the output [33]. It is worth noting that general FPGA-based SVM architectures deal with data with high dynamic range; thus, they are based on floating-point arithmetic, as this is the best solution with data with this requirement [34]. A typical approach when dealing with hardware floating-point arithmetic is to introduce pipelined architectures to reduce the critical path timing, potentially increasing the system clock frequency [35]. When used in the simple accumulator architecture described before, pipelined floating-point adders become critical. In fact, new input should be presented only when the output of the last addition can be fed back to ensure correct operation and avoid data hazards independently on the number and the length of the input vectors [33]. This would limit the applicability of the system, and different accumulator architectures should be individuated. However, when the boundary conditions allow this solution to be exploited, the latency of the whole accumulator for an input vector of $n$ elements is $T = np$, where $p$ is the length of the adder pipeline.

Considering model-based designs and, in particular, the Simulink environment, several accumulator blocks are already available. However, some of them are not suitable for the specific application due to incompatibility with the HDL Coder workflow (such as the *Cumulative Sum* block) or with the Floating-Point HDL library (as for the *Multiply-Accumulate* block). Some others (*Sum of Elements* and *Matrix Sum*) implement the HDL code as a binary tree or a linear chain of floating-point adders presenting all the input elements in parallel—since the complexity of these solutions grows with the number of inputs to accumulate, the amount of resources used when these blocks are exploited in this field should be evaluated [36]. Hence, in this paper, a possible alternative Simulink model for an accumulation circuit based on a floating-point pipelined adder and fully compatible with the HDL Coder workflow is presented. The paper is organized as follows: in Section 2 a review of the state-of-the-art accumulation circuits is presented. The architecture of the developed accumulator and kernel are introduced in Section 3, whereas in Section 4, tests are described and results are discussed. In Section 5, conclusions are drawn.

## 2. Related Works

To select the architecture to be implemented, a review of the state-of-the-art accumulation circuits was carried out. In reference [37], Luo and Martonosi present an architecture of an accumulator in which a floating-point pipelined adder is broken down into its mathematical operations (i.e., the ones involving the sign, mantissa, and exponent) and an internal feedback loop is introduced to embed the accumulation feature. This solution is reported as providing a minimal accumulation latency of $T = p + (n - 1) + t_{norm}$, where $t_{norm}$ is the combinational logic delay of the last accumulation part of the architecture. However, the exact latency value cannot be determined a priori because it is strongly dependent on the target hardware architecture [37].

A similar approach is used by Nagar and Bakos in reference [38], in which the accumulation latency is independent from the hardware implementation, and $t_{norm}$ can be considered as a one clock cycle. However, as for reference [37], the model-based implementation of this solution implies the additional development of a new adder architecture in order to consider the required modification.

In reference [36], Zhuo et al. present two main architectures based on standard floating-point adders: the fully compacted binary tree (FCBT) and the single strided adder (SSA). FCBT is an accumulator derived from a binary adder tree in which the first level is replaced by one buffer and a single adder, and the rest of the levels are replaced by an additional adder shared by $\lceil \log n \rceil - 1$ buffers. With the proper control logic, the system can perform the accumulation in $T \leq 3n + (p - 1)\lceil \log n \rceil - 3$ for $n < n_{max}$, where $n_{max}$ is the maximum input vector length that the system has been designed to work with. Because two different floating-point adders were deployed, this solution turned out to be undesirable, as it requires large area resources [36].

To overcome this issue and to remove the $n_{max}$ limitation, the SSA architecture has been introduced. This architecture is based on a single adder, two buffers, and a control logic. With this system, the latency has proven to be $T \leq n + 2p^2$.

A different set of architectures are based on the work presented in [39]. Here, an implementation of an accumulator based on a standard pipelined floating-point adder is described. The input data vector is split in two different buffers and, at each clock cycle, one element from each buffer is given to the adder operands. Then, after $p$ cycles, the vector of adder results are split in two halves again, which serve as the new input elements. This procedure is repeated until no other couples of operands are present in the buffers, meaning the accumulation has ended and the result is ready. This architecture was found to produce the accumulation result in $T = (p - 1)\lceil \log n \rceil + 3(n - 1)$. The main limitation of this system is that only one input vector can be accumulated at a time, resulting in the fact that the subsequent vectors must wait for the current result to be produced before they can be processed.

The time and the resources needed to perform the accumulation are reduced in [40]. Compared to the work presented in [39], the input buffers are substituted by two multiplexers at the input of the adder. One multiplexer can switch between the input vector and a register holding the adder output, and the other can switch between a constant value and the direct adder output. With the proper control of the multiplexers and the register, the time needed to compute the accumulation is improved for $n > p$. Then, the resulted latency is

$$T = \begin{cases} (p - 1)\lceil \log n \rceil + 3(n - 1) & , \ n \leq p \\ n + (p - 1)\lceil \log p \rceil + 4(p - 1) & , \ n > p \end{cases} . \tag{1}$$

From this work, the total accumulation time of this circuits is found to be

$$T = T_f + T_m + T_d , \tag{2}$$

where $T_f$ is the time needed for all the input elements to get inside the accumulator (feed phase), $T_m$ is the time needed to process all the partial results given by the couples of adder input operands (merging phase), and $T_d$ is the time needed for the last result to exit the adder pipeline (drain phase). It can be shown that this formula is applicable to every accumulator based on the architecture presented in [40]. Moreover, it can be easily observed that $T_f = n$ and $T_d = p - 1$.

In reference [41], an improved control algorithm (i.e., asymmetric method (AM)) for the merging time is presented. In this case, the merging time was found as

$$T_m^{AM} = \begin{cases} n\lceil \log n\rceil - 2^{\lceil \log n\rceil} + n + (k-n)\lceil \log n\rceil \;, & n < p \\ p\lceil \log p\rceil - 2^{\lceil \log p\rceil} + p & , \; n \geq p \end{cases} \tag{3}$$

which shortened the total accumulation time by $3(p-1)$.

In reference [42], a modified AM is proposed, with an improvement for every $n < 2p$:

$$T_m^{AM}(n) = \begin{cases} T_m^{AM}(n) - p & , \; n \leq p \\ T_m^{AM}(n) - p + 1 & , \; n = p + 1 \\ T_m^{AM}(\lfloor n/2\rfloor) - D(\lfloor n/2\rfloor) & , \; p + 1 < n < 2p \\ T_m^{AM}(n) & , \; n \geq 2p \end{cases} \tag{4}$$

where $D$ is a displacement function that compensates the irregular merging pattern that characterizes the control logic.

In reference [43], Tai et al. propose a modified version of [42], introducing the delayed buffering (DB) algorithm, in which the control logic can further reduce the merging time $T_m^{DB}$ in respect to $T_m^{MA}$ for certain input set lengths:

$$T_m^{DB} = \begin{cases} p\lceil \log n\rceil + 2^{\lceil \log n\rceil} + n - p \;, & n \leq p \\ p\lceil \log p\rceil + 2^{\lceil \log p\rceil} + n - p \;, & n = p + 1 \\ pL - 2^L + \lfloor G\rfloor - D + 1 \;, & n > p + 1 \end{cases} \tag{5}$$

where $L$ and $G$ are functions of $n$ and $p$, which, as the $D$ function, compensate the irregular merging pattern.

In reference [44], a solution requiring variable number of adders is presented—this increased the reuse and portability of the accumulator, but with higher occupied area. For example, for the area-efficient modular fully pipelined architecture (AeMFPA), two adders are required. More recently, reference [45] presents an accumulator circuit that can simultaneously add multiple independent vectors; however, the input buffer size is dependent on the number of the inputs, limiting the portability over different applications. Finally, in reference [46], a more flexible solution is reported—the core of the idea is a new state-based method (SBM) algorithm, a scheduling strategy for buffer management aiming at a lower latency and smaller area.

In Table 1, a summary of the performance of the mentioned architectures is reported, along with some practical examples that were computed considering an adder latency of $p = 11$, as the latency of Simulink floating point adder intellectual property (IP), and an input length of $n = 15$, as well as an adder latency of $p = 14$ and $n = 16$ for comparison with the solution reported in [46].

**Table 1.** State-of-the-art hardware accumulator architectures.

| Method | Accumulator Latency | | |
|---|---|---|---|
| | Generic | $p=11, n=15$ | $p = 14, n = 16$ |
| SSA [36] | $\leq n + 2p^2$ | 257 | 408 |
| FCBT [36] | $\leq 3n + (p-1)\lceil \log n \rceil$ | 85 | 100 |
| AM [41] | $n + p - 1 + T_m^{AM}$ | 64 | 83 |
| MA [42] | $n + p - 1 + T_m^{MA}$ | 58 | 71 |
| A2eMFPA [44] | $n + p\lceil \log p + 2 \rceil$ | 81 | 100 |
| [45] | $n + T_m^{AM} + \lceil p/2 \rceil$ | 60 | 104 |
| SBM [46] | not available | - | 75 |
| DB [43] | $n + p - 1 + T_m^{DB}$ | 57 | 71 |

As can be seen the system presented in [43], it offers the lowest latency for the accumulation of an input set of data. In this case, the total accumulation time depends on the input vector length with respect to the pipelined adder latency as expressed in Equation (6).

$$T = T_f + T_m^{DB} + T_d = \begin{cases} n + p - 1 + p\lceil \log n \rceil + 2^{\lceil \log n \rceil} + n - p\ , & n \leq p \\ n + p - 1 + p\lceil \log p \rceil + 2^{\lceil \log p \rceil} + n - p\ , & n = p + 1 \\ n + p - 1 + pL - 2^L + \lfloor G \rfloor - D + 1\quad , & n > p + 1 \end{cases} \tag{6}$$

This model was exploited in the proposed model-based implementation and in the SVM kernel. It was fully tested in an FPGA implementation and, to validate the results, it was compared with the simple iterative accumulator solution [33], SBM [46], and the built-it *Sum of Elements* Simulink block. Then, the proposed accumulator was used in a model-based implementation of the SVM kernel function.

## 3. Materials and Methods

### 3.1. Accumulator Architecture

In reference [43], two versions of the DB algorithm with different input processing properties are described. The first one, the single-set DB, is able to process one input vector at a time. If more than one vector has to be accumulated, each vector has to wait for the result of the previous one to be processed. The second algorithm is the multi-set DB, which is able to process a continuous stream of input vectors without the need to wait for the output results to be produced. An implementation of the single-set DB is presented in [28]. Because the data is processed in a streaming fashion in the SVM context, in this paper we focused on the multi-set DB version, although it required a more complex design with respect to the single-set design. In Figure 1, the proposed Simulink model-based accumulator is shown.

To design the proposed model, basic Simulink blocks were used. The core of the architecture is the adder that must handle floating point inputs. When dealing with floating point arithmetic, pipelined structures were introduced to ease the timing closure and achieve the desired operating frequency. In fact, a reduction in the total propagation delay, and then a higher clock frequency, can be obtained at the expense of an increase in the occupied area and in data path latency, due to the introduction of registers to segment the combinational logic. Pipelined adders can be modeled in Simulink as a cascade of an adder and a delay block; for this purpose, an *Adder With Latency* block was introduced. This configuration also allows for the configuration of the latency $p$ of the adder with a customizable value. Moreover, in this implementation, the adder was set to manage input data with 32 bit length compatible with the IEEE 754 single format; however, if higher precision is requested, the adder can be set accordingly and the whole architecture automatically scales consequently. The remainder of the architecture features two multiplexers (modeled with Simulink Switch blocks *A_Switch* and *B_Switch*); two multiple-word registers (*Input Buffer* (IBUF) and *Result Buffer* (RBUF)); and a control logic that is composed of three blocks: *Set IDentification (SID) Generator, Adder Supervisor Logic,* and

*Main Control Logic.* The *SID Generator* takes care of the tagging of the input elements to track each element of different sets. Each time the *data_last* flag is asserted together with the *data_valid* flag, the SID value is increased by one and it is merged into the internal bus together with the data value. Thus, all the data in the operands path are a pair of input data and a SID. The size of the counter (i.e., the maximum value of the SID) can be precisely set by knowing that, as found in [43], there cannot be more than $\lceil 5p/3 \rceil$ sets at the same time inside the architecture. The *Adder Supervisor Logic,* instead, tracks all the SIDs to notify whether the current adder output is of the same set of any other set inside the adder pipeline (*sum_internal_compare*) or the current adder output is of the same set of the input (*sum_input_compare*) or, finally, a new adder output is produced (*sum_valid*). To achieve this result, the internal architecture exploits comparators and simple logic functions. The *Main Control Logic* is the core control unit of the system. The model-based implementation relies on the pseudocode presented in [43] and exploits full combinational logic. The inputs of the logic function are flags indicating relevant events (if new data are available, if a new sum is ready, etc.) and, at the output, produce the configuration setups for all the involved elements (i.e., IBUF, RBUF, A and B working modes, and when the output accumulation is ready). No sequential logic was used for this block, resulting in an output update rate independent from the system clock.

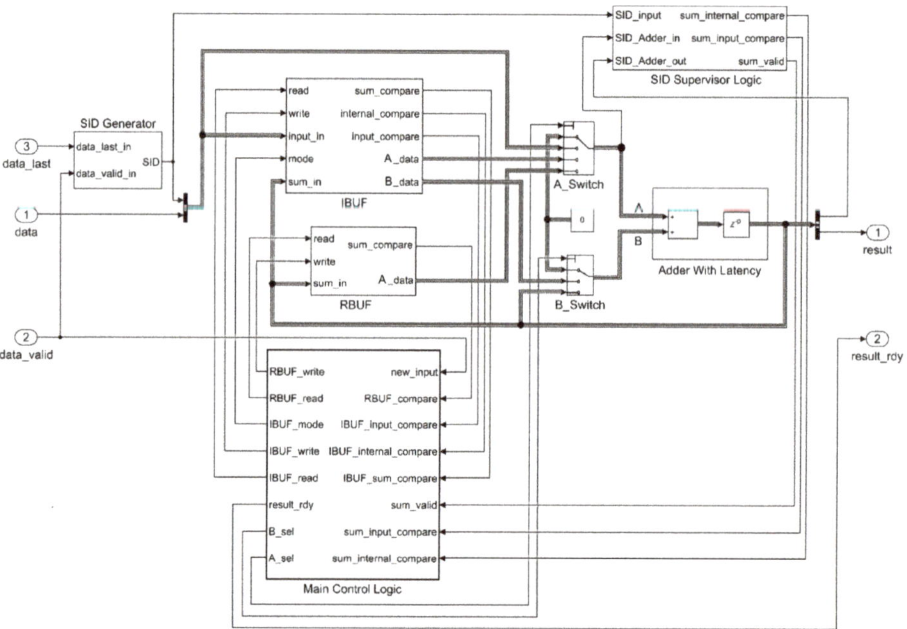

**Figure 1.** Simulink multi-set delayed buffering (DB) accumulator implementation.

As mentioned, in a multi-set DB version, two different buffers are needed. The *IBUF* buffer stores all the input elements that cannot enter the adder immediately because a couple of the same set (i.e., with the same SID) is not yet available. It is composed of an array of memory cells and two controllers, one for read and one for write operations. The model-based architecture is shown in Figure 2.

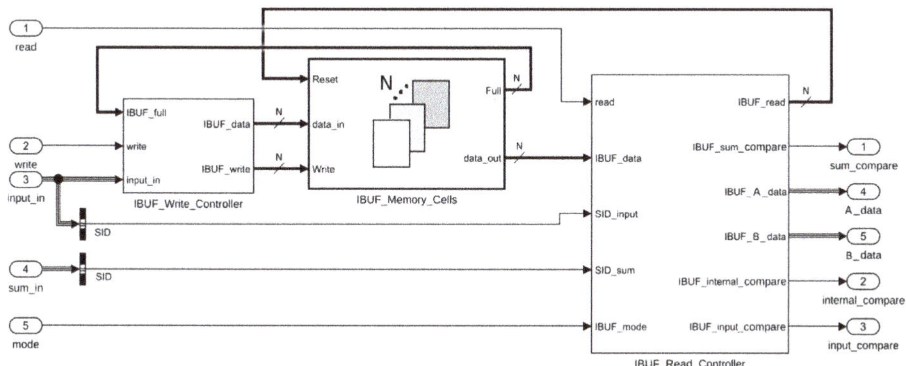

**Figure 2.** Architecture of the *Input Buffer* (*IBUF*) block.

The memory cells array can store the data values along with their SIDs. The model-based implementation of this part exploits the *For Each* Simulink subsystem, which can scale and replicate its internal architecture (i.e., the single memory cell, in this case) based on a parameter N. According to the work presented in [43], N was set to $\lceil p/2 \rceil$ to guarantee no storage overflow. When a write operation is issued, the write controller drives the input data to the first empty cell of the array by setting the proper *IBUF_write* array value to 1.

The read controller takes as input the content of all the cells, the SIDs of the input and of the sum, the mode of operation, and the *read* flag. When a read operation is requested by setting the *read* bit, the content of one or two cells are presented at the *A_data* and *B_data* ports depending on the *mode* input signal. When *mode* is equal to 1, one value is read and is assigned to *A_data*, which is managed by the A-Switch (Figure 1) accordingly to the main control logic combinational function, and eventually set as input of the adder. In this case *B_data* value is kept un-set, leaving the other input of the internal accumulator adder decided by the main control logic through the B_Switch (Figure 1). If *mode* is equal to 2, a pair of data of the same set has to be read. The controller logic automatically selects the pair having the same and oldest SID, giving priority to the oldest accumulation result production. The read data are assigned to *A_data* and *B_data*. When *mode* is equal to 3, the behavior is specular to mode 1: the single read value is assigned to *B_data* and *A_data* is not used.

The remainder of the output signals serve as inputs for the *Main Control Logic* and are produced by combinational logic. In particular, the *sum_compare* and *input_compare* signals are produced by looking for the cells with the same *SID_sum* and the *SID_input*. The *internal_compare* signal is computed by comparing the internal content of the memory cells and it is used to notify if two or more memory cells hold data of the same set.

The purpose of the *RBUF* is quite similar, except that it holds all the adder outputs that cannot be re-introduced yet as inputs, as there are not a couple of operands with the same tag to be processed already. The implementation results in a subset of the architecture of *IBUF*. In fact, its input signals are only the read, write, and *data_in* values and its output signals are the *sum_compare* and *A_data* values. For this architecture, the value of N was set to $\lceil 2p/3 \rceil$.

## 3.2. Kernel Architecture

In order evaluate the accumulation circuits described thus far, they were exploited in the development of a model-based design of a polynomial kernel. As defined in [47], the kernel equation is

$$k(x,x') = (\langle x,x'\rangle + c)^n \; , \tag{7}$$

where $\langle x, x' \rangle$ is the dot product between the input vector $x'$ (also called the features vector) and the training input matrix (also called support vectors) $x$. As can be seen by Equation (7), the process involves a dot product, and hence an accumulation stage that directly affects the performance of the system.

To evaluate the optimal values for the parameters $c$ and $n$ in Equation (7), an offline training process was performed on the training set employed in [14], which contains the acquisitions of an inertial measurement unit (IMU) sampled at 50 Hz. From this training phase, the $n$ and $c$ parameters were found to be equal to 3 and 1, respectively, according to works dealing with similar problems [48,49]. Then, the resulting function can be expressed as

$$k_j(x, x') = (\langle x, x' \rangle + 1)^3 = \left( \sum_{i=1}^{N} \left( x_i \cdot x'_{ij} \right) + 1 \right)^3,$$

(8)

where $N$ is the length of the features vector $x$.

In Figure 3, a model-based implementation of the cubic kernel is shown. It embeds a multiplier, an accumulator, and a cubic power block. It was tested with different architectures for the accumulator, as explained in the following sections.

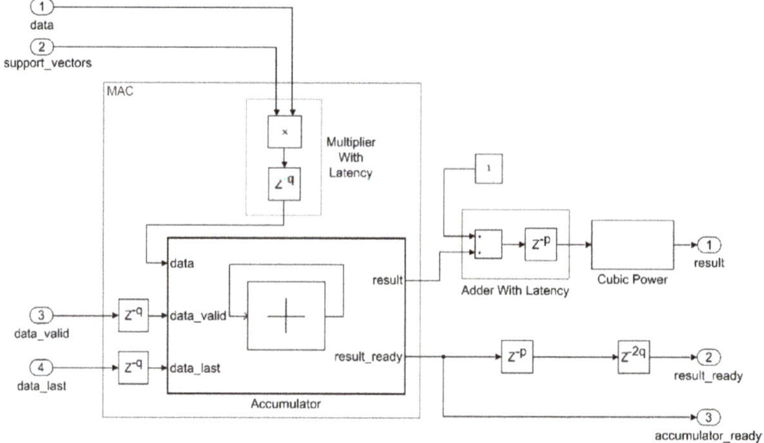


**Figure 3.** Simulink cubic kernel implementation.

All the data flowing inside the kernel are in floating-point 32 bit format. As for the adder block described earlier, the multiplier block was also modeled with a Simulink floating point IP cascaded with a delay block in order to take into account the introduced latency ($q$). To synchronize all the data paths, several delay lines were introduced to compensate the latency of the mathematical operations. For example, as the cubic power block is implemented as a cascade of two multipliers, the required synchronizing delay on the *result_rdy* signal is two times the delay of a single multiplier ($2q$). The input scheduling should be tailored according to the selected accumulator architecture.

## 4. Results and Discussion

### 4.1. Stand-Alone Model-Based Accumulator

To assess the performance of the proposed accumulator, Simulink simulations were carried out. The standard Simulink IP adder was exploited in the accumulator architecture. This block featured an internal latency of $p = 11$ cycles. Two mathematical series were exploited as input and the latency, and the correct accumulation results were evaluated. In particular, the inputs were the Euler's number $e$ and the Leibniz $\pi$ mathematical approximation series, defined as

$$e = \sum_{k=0}^{\infty} \frac{1}{k!} \; , \tag{9}$$

$$\pi = \sum_{k=0}^{\infty} \frac{4 \cdot (-1)^k}{2k+1} \; . \tag{10}$$

The series was generated in MATLAB environment as a 50 element vector for the Euler's number series and as a 200 element vector for the Leibniz $\pi$ series, in to obtain an approximation error lower than 0.5%. Then, the two vectors were imported in Simulink with the *From Workspace* block and presented as input to the accumulator.

In Figure 4, the results of the accumulation of two input vectors are shown.

The two vector series were presented to the input of the accumulator as a data stream, Euler's series first (at t = 0μs), and then the Leibniz series (at t = 0.5 μs) (Figure 4a). The last element of a single vector was highlighted by the *data_last* signal (Figure 4b). The *data_valid* signal was high until a valid data is given to the accumulator input (Figure 4c). The accumulation outputs (Figure 4d) were evaluated when the *result_ready* signal was asserted (Figure 4e). From this simulation, the correctness of the results can be assessed. From Equation (6), considering a latency of $p = 11$ cycles and a simulation step of 10 ns, the first result was produced 99 cycles after the first input element. Then, the second result was produced 200 cycles after the first result. These time intervals can be verified from Figure 4.

Once the functionality of the accumulator architecture was verified, VHDL code was automatically generated from the Simulink model and used to synthetize the circuits in Xilinx Vivado software. Here, results were evaluated in terms of FPGA look-up tables (LUTs), flip-flops (FFs), and (Digital Signal Processor) DSP usage, and maximum achievable clock frequency. For this purpose and to show portability, two different platforms were considered: a Xilinx Artix-7 XC7A100T FPGA device along with Xilinx Vivado 2019.1 software and an Altera Cyclone 10 LP 10CL010 with Quartus 19.1. All the simulations and timing results were carried out considering a clock frequency of 100 MHz. In these experiments, a stream of 200 vectors, each one of 100 elements, was considered to highlight the capability of the models to process subsequent vectors in a short timeframe, without the need of complex input synchronization logic.

The performance of the proposed accumulator model was compared to that of the available Simulink solution. The Simulink IP block takes as input a set of data in parallel to perform the sum. If the input values are fed serially, an input buffer is needed to host all the elements. The time needed for this buffering stage is equal to the length of the input stream, and the length of the buffer represents the maximum vector length the system can accumulate. This, in a VHDL implementation, limits the input streaming vector length. During the VHDL generation process, the accumulator architecture is designed as a binary tree adder or a linear adder chain. For this comparison, the input buffer was set to 100 samples and the architecture to the one offering the lowest implementation resource usage, i.e., the linear adder chain. In Table 2, the post-implementation results for Xilinx are reported, whereas data for Altera are shown in Table 3.

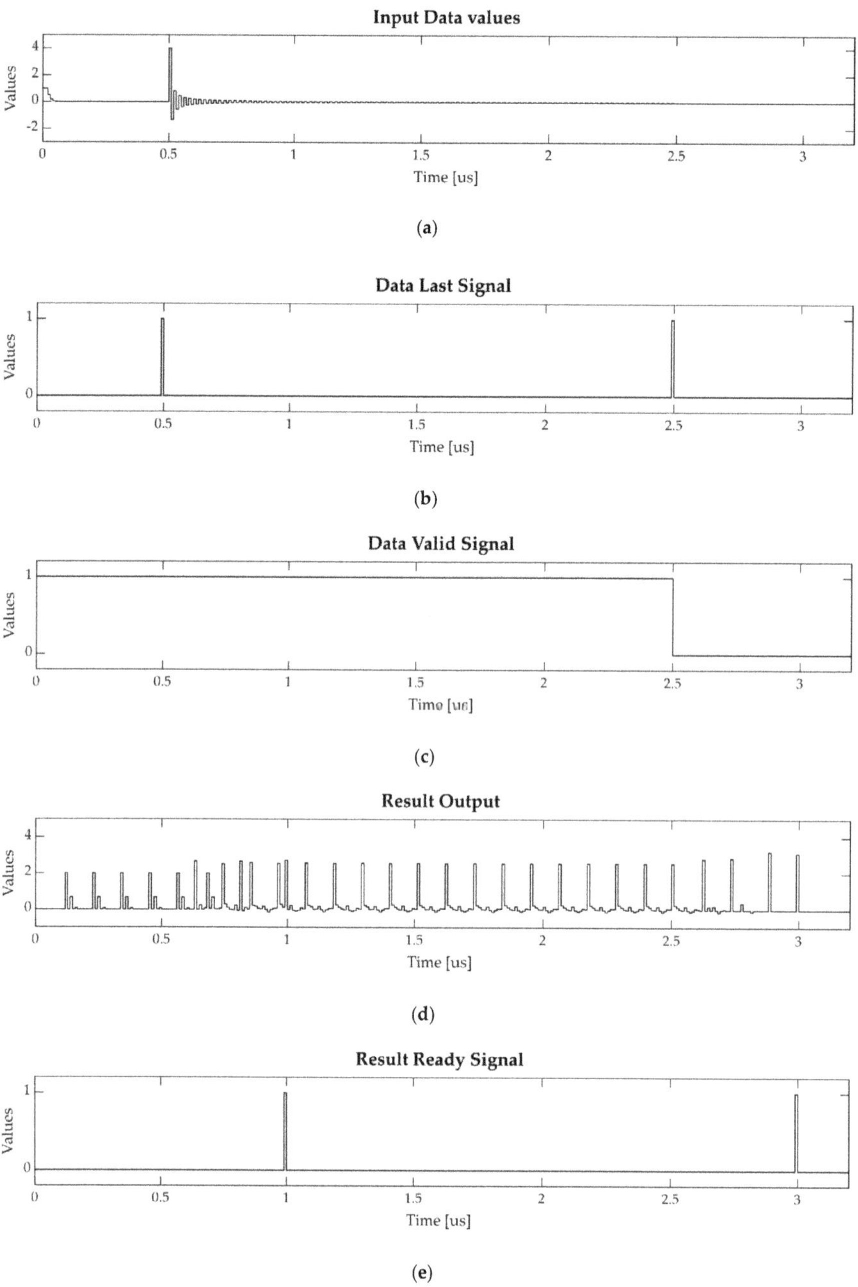

**Figure 4.** Input and output signals of the multi-set DB accumulator in Simulink simulation: (**a**) Input vector values, (**b**) *Data_last* signal, (**c**) *Data_valid* signal, (**d**) Accumulator output value, and (**e**) *Result_ready* signal.

**Table 2.** Simulink accumulator resource usage, maximum frequency, and latency on Xilinx Artix 7.

|  | **Proposed Model** | **Simulink IP** |
|---|---|---|
| Slice LUTs | 1643 | 40198 |
| Slice registers | 1239 | 33450 |
| DSPs | 0 | 0 |
| BRAM | 0 | 0 |
| Fmax (MHz) | 105 | 109 |
| Latency (cycles) | 49 | 989 |

**Table 3.** Simulink accumulator resource usage, maximum frequency, and latency on Altera Cyclone 10 LP.

|  | **Proposed Model** | **Simulink IP** |
|---|---|---|
| Logic Elements | 2483 | 47430 |
| DSPs | 0 | 0 |
| Memory (bits) | 154 | 436648 |
| Fmax (MHz) | 108 | N.A. |
| Latency (cycles) | 49 | 989 |

As can be seen from Tables 2 and 3, the proposed accumulator outperformed Simulink IP in both area and time. In particular, from the area point of view, in Xilinx implementation, the proposed model used 2.6% of available LUTs and 0.97% of slice registers, whereas Simulink IP used 63.4% and 26.4%, respectively. Moreover, in the Altera implementation, although the quantity of logic elements (LEs) of the proposed accumulator corresponded to 24%, the Simulink IP could not be implemented; in fact, the occupied area saturated the resources, resulting in a 460% quantity of logic elements. For this reason, the achievable maximum frequency was not reported in this case. The advantage of the new model over the available IP appears evident. It is worth noting that, despite the low area occupied, the proposed solution does not require DSP slices, resulting in independence from the presence of these blocks in the selected platform, enhancing portability.

To evaluate the performance of the proposed model-based accumulator, once generated, in respect to other solutions, some comparisons were made with other accumulator architectures and available IPs: iterative accumulator, single-set DB, SBM, and Vivado floating-point accumulator IP. These architectures, Vivado IP in particular, are not suitable for automatic code generation; however, these data can give some information about the applicability of the whole process and can confirm the choice of the architecture. In Table 4, the Xilinx Artix 7 post-implementation results of the compared accumulator architectures are reported.

**Table 4.** Post implementation accumulator resource usage, maximum frequency, and latency on Xilinx Artix 7 FPGA.

|  | **Proposed Accumulator** | **Single-Set DB [28]** | **Iterative [33]** | **SBM [46]** | **Vivado IP** |
|---|---|---|---|---|---|
| Slice LUTs | 1643 | 749 | 658 | 1411 | 3245 |
| Slice registers | 1239 | 811 | 534 | 1027 | 3120 |
| DSPs | 0 | 0 | 0 | 0 | 0 |
| BRAM | 0 | 8.5 | 8.5 | 0 | 0 |
| Fmax (MHz) | 105 | 112 | 126 | 102 | 134 |
| Latency (cycles) | 49 | 9800 | 200,200 | 54 | 23 |

The Xilinx Floating-Point IP is made of a fixed-point accumulator wrapped by floating-point conversions at the input and the output stages. To support the full precision and range of the 32 bits floating-point format, the internal fixed-point accumulator register must be correctly configured. The DSP slice usage was disabled to make a fair comparison with the proposed model. Moreover, the architecture optimization was set to produce the lowest latency—with this configuration, the internal fixed-point adder latency value resulted in 23 cycles. As can be seen from Table 4, the proposed model occupied less than a half in the area, without a significant difference in maximum frequency. Furthermore, the achievable frequency of the proposed accumulator is compatible with the maximum frequency allowed by the target FPGA.

Regarding the comparison with other architectures presented in the literature, the proposed accumulator outperforms the iterative and the single-set DB architectures in terms of latency needed to produce the result. It is important to note that this difference arises from the fact that the selected architecture is designed to process a stream of consecutive vectors, whereas both the iterative and the single-set DB solutions do not have this capability. The greater the number of vectors to be processed, the greater the latency associated with the latter two architectures. Furthermore, the buffers used for the management of the inputs synchronization must be carefully designed by considering the size of the vectors stream. SBM architecture performs well in terms of the occupied area. As a percentage, it occupies the 2.3% of the available LUTs and the 0.8% of the available slice registers. However, these numbers are close enough to that observed for the proposed model. The same goes for the maximum frequency, with a slight advantage for the selected accumulator. The new model also presents good results in latency, confirming the correct choice of the architecture also compared to the newer solutions presented in the literature.

### 4.2. Evaluation of the Proposed Model in a Practical Context: The Case of SVM Kernel Function

To frame the accumulator performance in a practical context, we evaluated it in the design of a cubic kernel function architecture conceived for an SVM applied to HAR. The inputs for the kernel were computed from datasets described in [14], where nine different daily activities have to be recognized. Data from a 9 degree of freedom (DoF) inertial measurement unit (IMU) were collected and processed, resulting in a dataset of 15,616 instances. This dataset was divided into a training set, used to train the SVM algorithm, and a test set, used in the inference phase. The support vectors $x'$, described in Equation (8), were computed during the training phase. In particular, each instance was labelled as belonging to an activity and was processed to extract a vector of nine features representing statistical values (mean, standard deviation, etc.) of the nine DoF data, resulting in a vector of 81 elements. The support vector in Equation (8) refers to a binary problem—as this dataset refers to a multidimensional problem, 36 support vectors are need to resolve the whole classification. In this experiment, a single support vector of $207 \times 81$ elements related to a single binary problem was selected and used as *support_vectors* input of the presented kernel architecture. The same statistical elaboration was applied to data in the test set—one vector of 81 elements, representing one instance of the test set, was exploited as the *data* input of the kernel architecture.

The kernel function was designed as a model-based block in Simulink. For the accumulation process, we compared our solution with the Simulink IP. An HDL code was generated and implemented on the same Xilinx Artix-7 FPGA exploited for the stand-alone accumulators, with a clock frequency of 100 MHz.

In this practical evaluation, other than Simulink and VHDL post-implementation simulations, measurements on hardware implementation were performed.

Simulink simulations were performed to compare the proposed model latency with the ones of the kernel implementation with Simulink IP. In both the implementations, standard Simulink floating-point adder and multiplier IP were exploited. Similar to the adder IP, which had an already mentioned latency of $p = 11$, the internal latency of the multiplier IP was found as $q = 6$.

In Figure 5, Simulink simulations are shown, in which the *result ready* signals are plotted. The dashed line refers to the time taken to complete the processing of the dot product of the whole $207 \times 81$ support vectors and the $1 \times 81$ data vector. In the case of the proposed model (Figure 5a), the time needed to accumulate the first vector at the input was equal to 161 cycles. Then, $206 \times 81$ cycles were needed for the remaining vectors. Considering a clock frequency of 100 MHz, this corresponded to 168.5 µs.

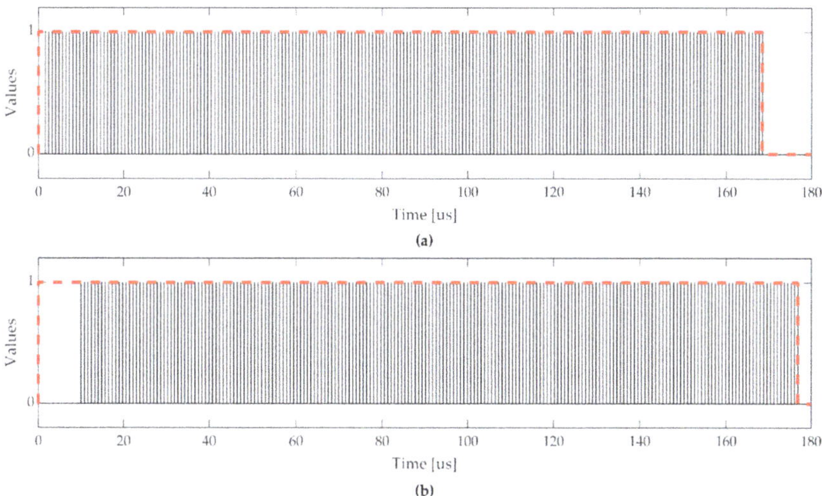

**Figure 5.** Kernel performance in Simulink simulations: (**a**) Kernel with proposed accumulator, (**b**) Kernel with Simulink IP accumulator.

In the case of Simulink IP (Figure 5b), with an input stream of 81 elements and $p = 11$, a total time of 891 cycles were required to obtain the correct accumulation, along with 111 cycles for the remainder of the kernel operations, starting from when the first element was available. Hence, the kernel processing for the first vector took 1002 cycles, and then $206 \times 81$ cycles were needed to complete the processing, corresponding to 176.9 µs.

The kernel models' VHDL codes were automatically generated, and performance was evaluated in Vivado environment in terms of resources usage and maximum achievable frequency. Moreover, the latencies resulting from Simulink were verified in the Vivado post-implementation timing simulations. A busy signal was configured in order to be high from the first element presented at the input to the last kernel output produced. Examples are shown in Figure 6.

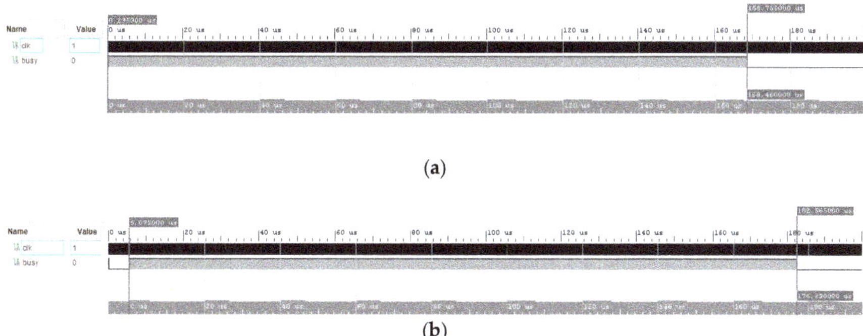

**Figure 6.** Xilinx Vivado post-implementation results of the kernel with (**a**) Kernel with proposed accumulator, (**b**) Kernel with Simulink IP accumulator.

Performance in terms of resources usage, maximum achievable frequency, and latency are summarized in Table 5.

**Table 5.** Post implementation kernel resource usage and maximum frequency on Xilinx Artix 7 FPGA.

|  | Proposed Accumulator | Simulink IP |
| --- | --- | --- |
| Slice LUTs | 3266 | 41,354 |
| Slice registers | 2791 | 34,700 |
| DSPs | 3 | 3 |
| BRAM | 0 | 0 |
| Fmax (MHz) | 106 | 106 |
| Latency (clock cycles) | 161 | 1002 |

The resulted latencies confirmed the Simulink simulations and the results on the stand-alone accumulators. The proposed model definitely performed better in terms of occupied area—it used only 5.2% of the available LUTs and 2.2% of the available registers for the whole kernel. Contrarily, the Simulink IP appeared critical in this context, with 65% and 27% of the LUTs and registers, respectively. Considering that many other logic blocks need to be instantiated together with the kernel in a complete SVM implementation, our solution appears a possible valid approach in this context. Moreover, it is worth noting that in wearable sensors, low power consumption has particular relevance. With the technology advancement in the FPGA field, as already mentioned, many low power models have been made available and can be exploited in this context, even considering floating point arithmetic [23]. The lowest power platforms have generally a low number of resources available; for this reason, the occupied area aspect is of utmost importance in these kinds of applications.

Although the maximum operating frequency was the same for both solutions, the resulting latency for our model was definitely lower.

To further confirm the simulation values, the FPGA was configured with the generated code and the performance was measured directly on hardware. The busy signals were measured using a Tektronix MSO 2024 oscilloscope. In Figure 7 the experimental setup is shown.

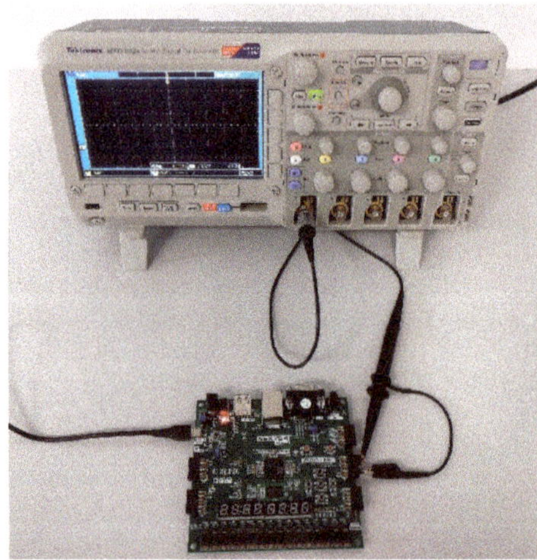

**Figure 7.** Experimental setup for the hardware measurement.

Results are reported in Figures 8 and 9, which are related to the proposed architecture and the Simulink IP, respectively.

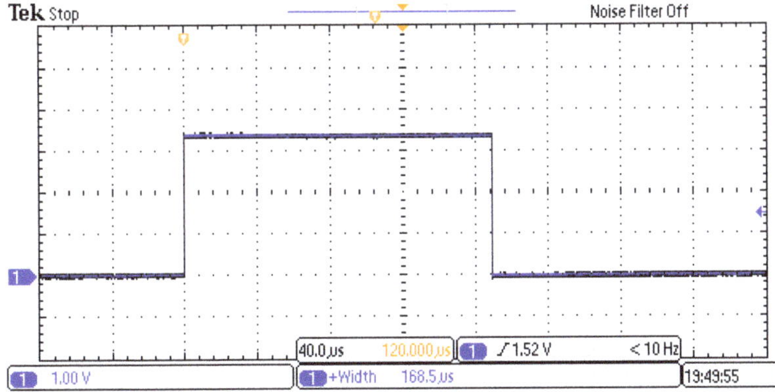

**Figure 8.** Measurement of the processing time of the kernel with the proposed accumulator implemented on the FPGA.

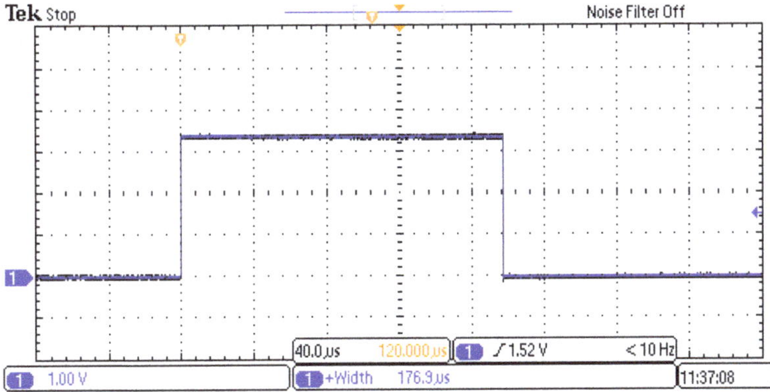

**Figure 9.** Measurement of the processing time of the kernel with Simulink IP implemented on the FPGA.

Measurements confirm the latencies of the simulations and the correctness of the result—the difference between the two processing times was of about 8.4 µs, which corresponds to 840 clock cycles. As can be seen from Table 5, this result corresponds with the difference in latency of the two solutions.

## 5. Conclusions

In this paper, a floating-point Simulink model-based accumulator architecture was presented. The functionality of the proposed accumulator was first tested with behavioral simulation in the Simulink environment. The tests were carried out using two mathematical series vectors as inputs. Results show the correct output accumulation values for both the series. Then, VDHL code was automatically generated and performance was assessed with post-implementation timing simulations on two different target FPGAs, a Xilinx Artix 7 and an Altera Cyclone 10 LP, in order to demonstrate portability. Results were compared with available Simulink IP supporting HDL code generation, demonstrating a significant reduction of about 95% in both area and time. Other solutions presented in the literature [28,33,46] and Vivado IP were compared, as well as demonstrating the applicability of the HDL code generation process and to confirm the choice of architecture. To frame the accumulator performance in a practical context, we evaluated it in the design of a polynomial cubic kernel function architecture conceived for an SVM applied to HAR. Additionally in this context, better performance was confirmed, greatly reducing the occupied area and making the solution particularly attractive for implementation in the context of wearable sensors, in which low resource platforms are usually exploited. The simulation results were also validated with hardware measurements on the target FPGA.

**Author Contributions:** Conceptualization, V.B., M.B., and I.D.M.; methodology, V.B., M.B., and I.D.M.; validation, V.B., M.B., and I.D.M.; formal analysis, V.B. and M.B.; investigation, V.B., M.B., and I.D.M.; data curation, V.B. and M.B.; writing—original draft preparation, V.B., M.B., and I.D.M.; writing—review and editing, V.B., M.B., and I.D.M.; visualization, V.B., M.B., and I.D.M.; supervision, I.D.M.; project administration, I.D.M.; funding acquisition, I.D.M. All authors have read and agreed to the published version of the manuscript.

**Funding:** This research received no external funding.

**Conflicts of Interest:** The authors declare no conflict of interest.

## References

1. Bassoli, M.; Bianchi, V.; De Munari, I. A plug and play IoT wi-fi smart home system for human monitoring. *Electronics* **2018**, *7*, 200. [CrossRef]
2. Petrosanu, D.M.; Carutasu, G.; Carutasu, N.L.; Pîrjan, A. A review of the recent developments in integrating machine learning models with sensor devices in the smart buildings sector with a view to attaining enhanced sensing, energy efficiency, and optimal building management. *Energies* **2019**, *12*, 4745. [CrossRef]
3. Marin, I.; Vasilateanu, A.; Molnar, A.J.; Bocicor, M.I.; Cuesta-Frau, D.; Molina-Picó, A.; Goga, N. I-light—intelligent luminaire based platform for home monitoring and assisted living. *Electronics* **2018**, *7*, 220. [CrossRef]
4. Rana, S.P.; Dey, M.; Ghavami, M.; Dudley, S. Signature inspired home environments monitoring system using IR-UWB technology. *Sensors* **2019**, *19*, 385. [CrossRef]
5. Ghayvat, H.; Awais, M.; Pandya, S.; Ren, H.; Akbarzadeh, S.; Mukhopadhyay, S.C.; Chen, C.; Gope, P.; Chouhan, A.; Chen, W. Smart aging system: Uncovering the hidden wellness parameter for well-being monitoring and anomaly detection. *Sensors* **2019**, *19*, 766. [CrossRef]
6. Guerra, C.; Bianchi, V.; De Munari, I.; Ciampolini, P. CARDEAGate: Low-cost, ZigBee-based localization and identification for AAL purposes. In Proceedings of the 2015 IEEE International Instrumentation and Measurement Technology Conference (I2MTC), Pisa, Italy, 11–14 May 2015; pp. 245–249.
7. Grossi, F.; Matrella, G.; De Munari, I.; Ciampolini, P. A Flexible Home Automation System Applied to Elderly Care. In Proceedings of the 2007 Digest of Technical Papers IEEE International Conference on Consumer Electronics, Las Vegas, NV, USA, 10–14 January 2007.
8. Leitao, J.; Gil, P.; Ribeiro, B.; Cardoso, A. A survey on home energy management. *IEEE Access* **2020**, *8*, 5699–5722. [CrossRef]
9. Son, H.; Kim, H. A pilot study to test the feasibility of a home mobility monitoring system in community-dwelling older adults. *Int. J. Environ. Res. Public Health* **2019**, *16*, 1512. [CrossRef]
10. Haratian, R. Assistive Wearable Technology for Mental Wellbeing: Sensors and Signal Processing Approaches. In Proceedings of the 2019 5th International Conference on Frontiers of Signal Processing (ICFSP), Marseille, France, 18–20 September 2019; pp. 7–11.
11. Montalto, F.; Guerra, C.; Bianchi, V.; De Munari, I.; Ciampolini, P. MuSA: Wearable multi sensor assistant for human activity recognition and indoor localization. *Biosyst. Biorobotics* **2015**, *11*, 81–92.
12. Moufawad El Achkar, C.; Lenoble-Hoskovec, C.; Paraschiv-Ionescu, A.; Major, K.; Büla, C.; Aminian, K. Physical behavior in older persons during daily life: Insights from instrumented shoes. *Sensors* **2016**, *16*, 1225. [CrossRef]
13. Yang, S.; Gao, B.; Jiang, L.; Jin, J.; Gao, Z.; Ma, X.; Woo, W.L. IoT structured long-term wearable social sensing for mental wellbeing. *IEEE Internet Things J.* **2019**, *6*, 3652–3662. [CrossRef]
14. Bianchi, V.; Bassoli, M.; Lombardo, G.; Fornacciari, P.; Mordonini, M.; De Munari, I. IoT wearable sensor and deep learning: An integrated approach for personalized human activity recognition in a smart home environment. *IEEE Internet Things J.* **2019**, *6*, 8553–8562. [CrossRef]
15. Ahmed, N.; Rafiq, J.I.; Islam, M.R. Enhanced human activity recognition based on smartphone sensor data using hybrid feature selection model. *Sensors* **2020**, *20*, 317. [CrossRef] [PubMed]
16. Irvine, N.; Nugent, C.; Zhang, S.; Wang, H.; Ng, W.W.Y. Neural network ensembles for sensor-based human activity recognition within smart environments. *Sensors* **2020**, *20*, 216. [CrossRef] [PubMed]

17. Manjarres, J.; Narvaez, P.; Gasser, K.; Percybrooks, W.; Pardo, M. Physical workload tracking using human activity recognition with wearable devices. *Sensors* **2019**, *20*, 39. [CrossRef] [PubMed]

18. Janidarmian, M.; Fekr, A.R.; Radecka, K.; Zilic, Z. A comprehensive analysis on wearable acceleration sensors in human activity recognition. *Sensors* **2017**, *17*, 529. [CrossRef]

19. Ravi, D.; Wong, C.; Lo, B.; Yang, G.Z. Deep learning for human activity recognition: A resource efficient implementation on low-power devices. In *BSN 2016-13th Annual Body Sensor Networks Conference*; Institute of Electrical and Electronics Engineers Inc.: Piscataway, NJ, USA, 2016; pp. 71–76.

20. Mikos, V.; Heng, C.H.; Tay, A.; Yen, S.C.; Chia, N.S.Y.; Koh, K.M.L.; Tan, D.M.L.; Au, W.L. A wearable, patient-adaptive freezing of gait detection system for biofeedback cueing in Parkinson's disease. *IEEE Trans. Biomed. Circuits Syst.* **2019**, *13*, 503–515. [CrossRef] [PubMed]

21. De Venuto, D.; Annese, V.F.; Mezzina, G.; Defazio, G. FPGA-based embedded cyber-physical platform to assess gait and postural stability in parkinson's disease. *IEEE Trans. Compon. Packag. Manuf. Technol.* **2018**, *8*, 1167–1179. [CrossRef]

22. Wang, H.; Shi, W.; Choy, C.S. Hardware design of real time epileptic seizure detection based on STFT and SVM. *IEEE Access* **2018**, *6*, 67277–67290. [CrossRef]

23. Gaikwad, N.B.; Tiwari, V.; Keskar, A.; Shivaprakash, N.C. Efficient FPGA implementation of multilayer perceptron for real-time human activity classification. *IEEE Access* **2019**, *7*, 26696–26706. [CrossRef]

24. Wisniewski, R.; Bazydlo, G.; Szczesniak, P. Low-cost FPGA hardware implementation of matrix converter switch control. *IEEE Trans. Circuits Syst. II Express Briefs* **2019**, *66*, 1177–1181. [CrossRef]

25. Giardino, D.; Matta, M.; Re, M.; Silvestri, F.; Spanò, S. IP generator tool for efficient hardware acceleration of self-organizing maps. *Lect. Notes Electr. Eng.* **2019**, *550*, 493–499.

26. Hai, J.C.T.; Pun, O.C.; Haw, T.W. Accelerating Video and Image Processing Design for FPGA Using HDL Coder and Simulink. In Proceedings of the 2015 IEEE Conference on Sustainable Utilization and Development in Engineering and Technology (CSUDET 2015), Selangor, Malaysia, 15–17 October 2015; pp. 28–32.

27. Michael, T.; Reynolds, S.; Woolford, T. Designing a generic, software-defined multimode radar simulator for FPGAs using simulink®HDL coder and speedgoat real-time hardware. In *2018 International Conference on Radar, RADAR 2018*; Institute of Electrical and Electronics Engineers Inc.: Piscataway, NJ, USA, 2018.

28. Bassoli, M.; Bianchi, V.; De Munari, I. A Simulink Model-based Design of a Floating-point Pipelined Accumulator with HDL Coder Compatibility for FPGA Implementation. *Appl. Electron. Pervading Ind. Environ. Soc. ApplePies 2019. Lect. Notes Electr. Eng.* **2019**, 1–9, in press.

29. Perry, S. Model Based Design Needs High Level Synthesis: A Collection of High Level Synthesis Techniques to Improve Productivity and Quality of Results for Model Based Electronic Design. In Proceedings of the 2009 Design, Automation and Test in Europe (DATE; 2009), Nice, France, 20–24 April 2009; pp. 1202–1207.

30. Choe, J.M.; Arnedo, L.; Lee, Y.; Sorchini, Z.; Mignogna, A.; Agirman, I.; Kim, H. Model-Based Design and DSP Code Generation using Simulink®for Power Electronics Applications. In Proceedings of the 10th International Conference on Power Electronics and ECCE Asia (ICPE 2019-ECCE Asia), Busan, Korea, 27–30 May 2019; pp. 923–926.

31. Ghosh, S.; Dasgupta, A.; Swetapadma, A. A study on support vector machine based linear and non-linear pattern classification. In *Proceedings of the International Conference on Intelligent Sustainable Systems, ICISS 2019*; Institute of Electrical and Electronics Engineers Inc.: Piscataway, NJ, USA, 2019; pp. 24–28.

32. Wang, A.N.; Zhao, Y.; Hou, Y.T.; Li, Y.L. A Novel Construction of SVM Compound Kernel Function. In Proceedings of the 2010 International Conference on Logistics Systems and Intelligent Management (ICLSIM 2010), Harbin, China, 9–10 January 2010; Volume 3, pp. 1462–1465.

33. De Dinechin, F.; Pasca, B.; Creţ, O.; Tudoran, R. An FPGA-Specific Approach to Floating-Point Accumulation and Sum-of-Products. In Proceedings of the 2008 International Conference on Field-Programmable Technology (ICFPT 2008), Taipei, Taiwan, 8–10 December 2008; pp. 33–40.

34. Flynn, M.J.; Oberman, S.F. *Advanced Computer Arithmetic Design*; Wiley: New York, NY, USA, 2001.

35. Hallin, T.G.; Flynn, M.J. Pipelining of arithmetic functions. *IEEE Trans. Comput.* **1972**, *100*, 880–886.

36. Zhou, L.; Morris, G.R.; Prasanna, V.K. High-performance reduction circuits using deeply pipelined operators on FPGAs. *IEEE Trans. Parallel Distrib. Syst.* **2007**, *18*, 1377–1392. [CrossRef]

37. Luo, Z.; Martonosi, M. Accelerating pipelined integer and floating-point accumulations in configurable hardware with delayed addition techniques. *IEEE Trans. Comput.* **2000**, *49*, 208–218. [CrossRef]

38. Nagar, K.K.; Bakos, J.D. A High-Performance Double Precision Accumulator. In Proceedings of the 2009 International Conference on Field-Programmable Technology (FPT'09), Sydney, NSW, Australia, 9–11 December 2009; pp. 500–503.
39. Kuck, D.J. *The Structure of Computers and Computations*; Wiley: New York, NY, USA, 1978; ISBN 9780471027164.
40. Kogge, P.M. *The Architecture of Pipelined Computers*; Hemisphere Pub. Corp.: Washington, DC, USA, 1981; ISBN 9780891164944.
41. Ni, L.M.; Hwang, K. Vector-reduction techniques for arithmetic pipelines. *IEEE Trans. Comput.* **1985**, *100*, 404–411. [CrossRef]
42. Sips, H.J.; Lin, H. An improved vector-reduction method. *IEEE Trans. Comput.* **1991**, *40*, 214–217. [CrossRef]
43. Tai, Y.G.; Lo, C.T.D.; Psarris, K. Accelerating matrix operations with improved deeply pipelined vector reduction. *IEEE Trans. Parallel Distrib. Syst.* **2012**, *23*, 202–210. [CrossRef]
44. Huang, M.; Andrews, D. Modular design of fully pipelined reduction circuits on FPGAs. *IEEE Trans. Parallel Distrib. Syst.* **2013**, *24*, 1818–1826. [CrossRef]
45. Wei, M.; Huang, Y.H. A Tag Based Vector Reduction Circuit. In Proceedings of the 2015 IEEE High Performance Extreme Computing Conference (HPEC 2015), Waltham, MA, USA, 15–17 September 2015; pp. 1–6.
46. Tang, L.; Cai, G.; Yin, T.; Zheng, Y.; Chen, J. A Resource Consumption and Performance Overhead Optimized Reduction Circuit on FPGAs. In Proceedings of the 2019 International Conference on Field-Programmable Technology (ICFPT), Tianjin, China, 9–13 December 2019; pp. 287–290.
47. Hofmann, T.; Schölkopf, B.; Smola, A.J. Kernel methods in machine learning. *Ann. Stat.* **2008**, *36*, 1171–1220. [CrossRef]
48. Wu, D.; Wang, Z.; Chen, Y.; Zhao, H. Mixed-kernel based weighted extreme learning machine for inertial sensor based human activity recognition with imbalanced dataset. *Neurocomputing* **2016**, *190*, 35–49. [CrossRef]
49. Althloothi, S.; Mahoor, M.H.; Zhang, X.; Voyles, R.M. Human activity recognition using multi-features and multiple kernel learning. *Pattern Recognit.* **2014**, *47*, 1800–1812. [CrossRef]

MDPI

St. Alban-Anlage 66

4052 Basel

Switzerland

Tel. +41 61 683 77 34

Fax +41 61 302 89 18

www.mdpi.com

*Sensors* Editorial Office

E-mail: sensors@mdpi.com

www.mdpi.com/journal/sensors

www.ingramcontent.com/pod-product-compliance
Lightning Source LLC
LaVergne TN
LVHW070125170726
843515LV00007B/1754